Applied Cognitive Sciences

Applied Cognitive Sciences

Editors

Attila Kovari
Cristina Costescu

MDPI • Basel • Beijing • Wuhan • Barcelona • Belgrade • Manchester • Tokyo • Cluj • Tianjin

Editors
Attila Kovari
University of Dunaujvaros
Hungary

Cristina Costescu
Babes-Bolyai University
Romania

Editorial Office
MDPI
St. Alban-Anlage 66
4052 Basel, Switzerland

This is a reprint of articles from the Special Issue published online in the open access journal *Applied Sciences* (ISSN 2076-3417) (available at: https://www.mdpi.com/journal/applsci/special_issues/Cognitive_Sciences).

For citation purposes, cite each article independently as indicated on the article page online and as indicated below:

LastName, A.A.; LastName, B.B.; LastName, C.C. Article Title. *Journal Name* **Year**, *Volume Number*, Page Range.

ISBN 978-3-0365-4859-3 (Hbk)
ISBN 978-3-0365-4860-9 (PDF)

© 2022 by the authors. Articles in this book are Open Access and distributed under the Creative Commons Attribution (CC BY) license, which allows users to download, copy and build upon published articles, as long as the author and publisher are properly credited, which ensures maximum dissemination and a wider impact of our publications.

The book as a whole is distributed by MDPI under the terms and conditions of the Creative Commons license CC BY-NC-ND.

Contents

Preface to "Applied Cognitive Sciences" vii

Attila Kovari
Research Directions of Applied Cognitive Sciences
Reprinted from: *Appl. Sci.* 2022, 12, 5789, doi:10.3390/app12125789 1

Boris M. Velichkovsky, Artemiy Kotov, Nikita Arinkin, Liudmila Zaidelman, Anna Zinina and Kirill Kivva
From Social Gaze to Indirect Speech Constructions: How to Induce the Impression That Your Companion Robot Is a Conscious Creature
Reprinted from: *Appl. Sci.* 2021, 11, 10255, doi:10.3390/app112110255 7

Tor Finseth, Michael C. Dorneich, Nir Keren, Warren D. Franke and Stephen B. Vardeman
Manipulating Stress Responses during Spaceflight Training with Virtual Stressors
Reprinted from: *Appl. Sci.* 2022, 12, 2289, doi:10.3390/app12052289 25

Rahul Sharma, Bernardete Ribeiro, Alexandre Miguel Pinto and Amílcar Cardoso
Emulating Cued Recall of Abstract Concepts via Regulated Activation Networks
Reprinted from: *Appl. Sci.* 2021, 11, 2134, doi:10.3390/app11052134 47

Hiroomi Hikawa, Yuta Ichikawa, Hidetaka Ito and Yutaka Maeda
Dynamic Gesture Recognition System with Gesture Spotting Based on Self-Organizing Maps
Reprinted from: *Appl. Sci.* 2021, 11, 1933, doi:10.3390/app11041933 73

Jong-Gyu Shin, Ga-Young Choi, Han-Jeong Hwang and Sang-Ho Kim
Evaluation of Emotional Satisfaction Using Questionnaires in Voice-Based Human–AI Interaction
Reprinted from: *Appl. Sci.* 2021, 11, 1920, doi:10.3390/app11041920 87

Cristina Costescu, Iulia Chelba, Adrian Roșan, Attila Kovari and Jozsef Katona
Cognitive Patterns and Coping Mechanisms in the Context of Internet Use
Reprinted from: *Appl. Sci.* 2021, 11, 1302, doi:10.3390/app11031302 101

Adrian Rodriguez Aguiñaga, Luis Muñoz Delgado, Víctor R. López-López and Andrés Calvillo Téllez
EEG-Based Emotion Recognition Using Deep Learning and M3GP
Reprinted from: *Appl. Sci.* 2022, 12, 2527, doi:10.3390/app12052527 115

Andreea Bianca Popescu, Ioana Antonia Taca, Cosmin Ioan Nita, Anamaria Vizitiu, Robert Demeter, Constantin Suciu, Lucian Mihai Itu
Privacy Preserving Classification of EEG Data Using Machine Learning and Homomorphic Encryption
Reprinted from: *Appl. Sci.* 2021, 11, 7360, doi:10.3390/app11167360 131

Qasim Ali, Ilona Heldal, Carsten G. Helgesen, Gunta Krumina, Cristina Costescu, Attila Kovari, Jozsef Katona and Serge Thill
Current Challenges Supporting School-Aged Children with Vision Problems: A Rapid Review
Reprinted from: *Appl. Sci.* 2021, 11, 9673, doi:10.3390/app11209673 149

Jan Francisti, Zoltán Balogh, Jaroslav Reichel, Martin Magdin, Štefan Koprda and György Molnár
Application Experiences Using IoT Devices in Education
Reprinted from: *Appl. Sci.* 2020, 10, 7286, doi:10.3390/app10207286 173

Robert Pinter, Sanja Maravić Čisar, Attila Kovari, Lenke Major, Petar Čisar and Jozsef Katona
Case Study: Students' Code-Tracing Skills and Calibration of Questions for Computer Adaptive Tests
Reprinted from: *Appl. Sci.* **2020**, *10*, 7044, doi:10.3390/app10207044 . . . . **187**

María Consuelo Sáiz-Manzanares, Ismael Ramos Pérez, Adrián Arnaiz Rodríguez, Sandra Rodríguez Arribas, Leandro Almeida and Caroline Françoise Martin
Analysis of the Learning Process through Eye Tracking Technology and Feature Selection Techniques
Reprinted from: *Appl. Sci.* **2021**, *11*, 6157, doi:10.3390/app11136157 . . . . **209**

Cecilia Hammar Wijkmark, Maria Monika Metallinou and Ilona Heldal
Remote Virtual Simulation for Incident Commanders—Cognitive Aspects
Reprinted from: *Appl. Sci.* **2021**, *11*, 6434, doi:10.3390/app11146434 . . . . **233**

Žolt Namestovski and Attila Kovari
Framework for Preparation of Engaging Online Educational Materials—A Cognitive Approach
Reprinted from: *Appl. Sci.* **2022**, *12*, 1745, doi:10.3390/app12031745 . . . . **253**

Milan Gnjatović, Ivan Košanin, Nemanja Maček and Dušan Joksimović
Clustering of Road Traffic Accidents as a Gestalt Problem
Reprinted from: *Appl. Sci.* **2022**, *12*, 4543, doi:10.3390/app12094543 . . . . **269**

Preface to "Applied Cognitive Sciences"

Cognitive science is an interdisciplinary field of investigation of the mind and intelligence. The term cognition refers to different mental processes, including perception, problem solving, learning, decision-making, language use, and emotional state and experience. The contributions of philosophy and computer science to the investigation of cognition are the basis of cognitive sciences. Computer science is very important in the investigation of cognition, because computer-aided research, machine learning and decision-making methods helps to develop the mental processes, and computers are useful in testing scientific hypotheses about mental organization and functioning. In addition, the emergence of human-computer interfaces, such as eye movement tracking, allows the observation and examination of cognitive load in relation to a more complex cognitive process. Empirical theories are very important for guiding practice (including education, pedagogy, or psychology) and operational research and engineering, in particular, the design of human-computer interfaces that can be used efficiently without placing too much emphasis on human intellectual abilities. Studies using psychological experiments and computational models are also very important in mental health diagnosis and treatment. Cognitive science plays a significant role in the field of mental illnesses, such as depression, and neurodevelopmental disorders. More specifically the understanding of the possible mechanisms that underlie them and the way interventions work require an understanding of how the mind works. This book provides a platform for a review of these disciplines and the presentation of cognitive research as an independent field of study.

Attila Kovari and Cristina Costescu
Editors

Editorial

Research Directions of Applied Cognitive Sciences

Attila Kovari

Department of Computer Science, GAMF Faculty of Engineering and Computer Science, John von Neumann University, 6000 Kecskemét, Hungary; kovari.attila@o365.uni-neumann.hu

1. Introduction

Cognitive science is an interdisciplinary field of investigation of the mind and intelligence. The term cognition refers to different mental processes, including perception, problem solving, learning, decision-making, language use, and emotional state and experience [1]. The contributions of philosophy and computer science to the investigation of cognition are the basis of cognitive sciences. Computer science is very important in the investigation of cognition, because computer-aided research, machine learning [2] and decision-making [3] methods help to develop the mental processes, and computers are useful in testing scientific hypotheses about mental organization and functioning. In addition, the emergence of human-computer interfaces, such as eye movement tracking, allows the observation and examination of cognitive load in relation to a more complex cognitive process. [4,5] Empirical theories are very important for guiding practice (including education, pedagogy, or psychology) and operational research and engineering, and in particular, the design of human-computer interfaces that can be used efficiently without placing too much emphasis on human intellectual abilities. Studies using psychological experiments and computational models are also very important in mental health diagnosis and treatment. Cognitive science plays a significant role in the field of mental illnesses, such as depression, and neurodevelopmental disorders. More specifically the understanding of the possible mechanisms that underlie them and the way interventions work require an understanding of how the mind works. This special issue provides a platform for a review of these disciplines and the presentation of cognitive research as an independent field of study.

The main focus of this special issue includes the next topics:

1. cognition;
2. user experiences; user satisfaction;
3. human–AI interaction; interaction design;
4. human–robot interaction;
5. emotional interfaces;
6. voice-based intelligent system;
7. dynamic gesture recognition; gesture spotting;
8. internet addiction;
9. dysfunctional emotions; emotional problems; stress;
10. computational psychology; computational cognitive modeling; computational creativity;
11. privacy-preserving computations; homomorphic encryption;
12. problem solving and decision making;
13. learning and assessment; computer adaptive testing;
14. code tracing; basic programming skills;
15. functional vision; vision screening; vision training;
16. eye-tracking; eye–brain–computer interfaces;
17. machine learning; deep learning;
18. clustering; spatially prolonged risk;
19. instance selection; clustering; information processing

20. virtual reality; virtual environment; virtual simulation;
21. internet of things; heart rate.

2. Overview of Research Directions Based on the Spec Issue Papers

The main content and results of the articles published in the spec issue are summarized below.

Cecilia Hammar Wijkmark et al. [6] presents the necessary enablers for setting up Remote Virtual Simulation (RVS) and its influence on cognitive aspects of assessing practical competences. Data were gathered through observations, questionnaires, and interviews from students and instructors, using action-case research methodology. The results show the potential of RVS for supporting higher cognitive processes, such as recognition, comprehension, problem solving, decision making, and allowed students to demonstrate whether they had achieved the required learning objectives. Other reported benefits were the value of not gathering people (imposed by the pandemic), experiencing new, challenging incident scenarios, increased motivation for applying RVS based training both for students and instructors, and reduced traveling.

Qasim Ali et al. [7] showing a rapid review to a better understanding of the importance of vision screening for school-aged children, and to investigate the possibilities of how eye-tracking (ET) technologies can support this. The authors review interdisciplinary research on performing vision investigations and discuss current challenges for technology support. The focus of the paper is on exploring the possibilities of ET technologies to better support screening and handling of vision disorders, especially by non-vision experts. The data originate from a literature survey of peer-reviewed journals and conference articles complemented by secondary sources, following a rapid review methodology. The authors highlight current trends in supportive technologies for vision screening and identify the involved stakeholders and the research studies that discuss how to develop more supportive ET technologies for vision screening and training by non-experts.

Robert Pinter et al. [8] analyzes the application of Computer adaptive testing (CAT) which enables an individualization of tests and give better accuracy of knowledge level determination [9]. The authors compare the results of questions' difficulty determination given by experts (teachers) and students. Analyzing the correct answers shows that the basic programming knowledge, taught in the first year of study, evolves very slowly among senior students. The comparison of estimations on questions difficulty highlights that the senior students have a better understanding of basic programming tasks; thus, their estimation of difficulty approximates to that given by the experts.

Jan Francisti et al. [10] describes the possibilities of using commonly available devices such as smart wristbands (watches) and eye tracking technology, i.e., using existing technical solutions and methods that rely on the application of sensors while maintaining non-invasiveness. By comparing the data from these devices, the authors observed how the students' attention affects their results. The results show a correlation between eye tracking, heart rate, and student attention and how it all impacts their learning outcomes.

Cristina Costescu et al. [11] investigate the relationship between coping mechanisms, dysfunctional negative emotions, and Internet use. The authors measured participants' coping strategies, emotional distress, social and emotional loneliness, and their online behavior and Internet addiction using self-report questionnaires. The results showed that maladaptive coping strategies and Internet use were significant predictors of dysfunctional negative emotions. Moreover, passive wishful thinking, as a pattern of thinking, was associated with anxious and depressed feelings. The relation between Internet use and dysfunctional negative emotions was mediated by participants' coping mechanisms. The authors conclude that the level of negative feelings is associated with the coping strategies used while showing an increased level of Internet addiction.

Jong-Gyu Shin et al. [12] proposed a method to evaluate user satisfaction during interaction between a voice-based intelligent systems (VIS) and a user-centered intelligent system. As a user satisfaction evaluation method, a VIS comprising four types of design parameters was developed and user satisfaction was measured using Kansei words (KWs). The questionnaire scores collected through KWs were analyzed using exploratory factor analysis and ANOVA was used to analyze differences in emotion. On the "pleasurability" and "reliability" axes, it was confirmed that among the four design parameters, "sentence structure of the answer" and "number of trials to acquire the right answer for a question" affect the emotional satisfaction of users.

Hiroomi Hikawa et al. [13] proposed a real-time dynamic hand gesture recognition system with gesture spotting function. In the proposed system, input video frames are converted to feature vectors, and they are used to form a posture sequence vector that represents the input gesture. The introduced gesture recognition method was tested by simulation and real-time gesture recognition experiment. Results shows that the system could recognize nine types of gesture with an accuracy of 96.6%, and it successfully outputted the recognition result at the end of gesture using the spotting result.

Rahul Sharma et al. [14] used the Toy-data problem in their paper to illustrate the regulated activation network modeling and recall procedure to report the computational simulation of the cued recall of abstract concepts by exploiting their learned associations. The results show how regulation enables contextual awareness among abstract nodes during the recall process. The authors show that every recall process converges to an optimal image. With more cues, better images are recalled, and every intermediate image obtained during the recall iterations corresponds to the varying cognitive states of the recognition procedure.

María Consuelo Sáiz-Manzanares et al. [15] analyze the results obtained with the eye tracking methodology by applying statistical tests and supervised and unsupervised machine learning techniques, and to contrast the effectiveness of each one. The parameters of fixations, saccades, blinks and scan path, and the results in a puzzle task were found. The statistical study concluded that no significant differences were found between participants in solving the crossword puzzle task; significant differences were only detected in the parameters saccade amplitude minimum and saccade velocity minimum. On the other hand, this study, with supervised machine learning techniques, provided possible features for analysis, some of them different from those used in the statistical study. Regarding the clustering techniques, a good fit was found between the algorithms used (k-means ++, fuzzy k-means and DBSCAN).

Andrea Bianca Popescu et al. [16] propose an encoding method that enables typical homomorphic encryption schemes to operate on real-valued numbers of arbitrary precision and size. The approach is evaluated on two real-world scenarios relying on EEG signals: seizure detection and prediction of predisposition to alcoholism. The results show that the prediction performance of the models operating on encoded and encrypted data is comparable to that of standard models operating on plaintext data.

Boris M. Velichkovsky et al. [17] implemented different modes of social gaze behavior in a companion robot, F-2, to evaluate the impression of the gaze behaviors on humans in three communicative situations. The authors extended the computer model of the robot in order to simulate realistic gaze behavior in the robot and create the impression of the robot changing its internal cognitive states. They used an iterative approach, extending the applied cognitive architecture in order to simulate the balance between different behavioral reactions and to test it in the experiments.

Žolt Namestovski and Attila Kovari [18] examine the process of creating successful, engaging, interactive, and activity-based online educational materials, while taking the cognitive aspects of learners into account [19]. The quality of online educational materials has become increasingly important in the recent period, and it is crucial that content is created that allows our students to learn effectively and enjoyably. The authors present the milestones of curriculum creation and the resulting model, the criteria of selecting online learning environments and also introduce some principles of instructional design, as well as a self-developed model that can be used to create effective online learning materials and online courses.

Tor Finseth et al. [20] investigate and validate different virtual reality (VR) stressor levels from existing emergency spaceflight procedures. Experts in spaceflight procedures and the human stress response helped design a VR spaceflight environment and emergency fire task procedure. Since stress is a complex construct, physiological data (heart rate, heart rate variability, blood pressure, electrodermal activity) and self-assessment (workload, stress, anxiety) were collected for each stressor level. The results suggest that the environmental-based stressors can induce significantly different, distinguishable levels of stress in individuals.

Adrian Rodriguez Aguiñaga et al. [21] presents the proposal of a method to recognize emotional states through EEG analysis. The novelty of this work lies in its feature improvement strategy, based on multiclass genetic programming with multidimensional populations (M3GP), which builds features by implementing an evolutionary technique that selects, combines, deletes, and constructs the most suitable features to ease the classification process of the learning method. After implementing the M3GP, the results showed an increment of 14.76% in the recognition rate without changing any settings in the learning method. The proposed methodology achieves a mean classification rate of 92.1%, and simplifies the feature management process by increasing the separability of the spectral features.

Milan Gnjatović et al. [22] introduces and illustrates an approach to automatically detecting and selecting "critical" road segments, intended for application in circumstances of limited human or technical resources for traffic monitoring and management [23]. The presented approach is psychologically inspired to the extent that it introduces a clustering criterion based on the Gestalt principle of proximity.

3. Conclusions

This short review of the papers shows that the Research Directions of Applied Cognitive Sciences offer a very wide range of research opportunities, which multidisciplinary will remain an important field of research for the future.

Funding: This research received no external funding.

Conflicts of Interest: The authors declare no conflict of interest.

References

1. Magdin, M.; Balogh, Z.; Reichel, J.; Francisti, J.; Koprda, Š.; Molnar, G. Automatic detection and classification of emotional states in virtual reality and standard environments (LCD): Comparing valence and arousal of induced emotions. *Virtual Real.* **2021**, *25*, 1029–1041. [CrossRef]
2. Muhi, K.; Johanyak, Z.C. Dimensionality reduction methods used in Machine Learning. *Műsz. Tud. Közl.* **2020**, *13*, 148–151. [CrossRef]
3. Koncz, A.; Johanyak Zs, C.; Pokoradi, L. Multiple criteria decision-making method solutions based on failure mode and effect analysis. *Ann. Fac. Eng. Hunedoara-Int. J. Eng.* **2021**, *19*, 6.
4. Katona, J. Measuring Cognition Load Using Eye-Tracking Parameters Based on Algorithm Description Tools. *Sensors* **2022**, *22*, 912. [CrossRef] [PubMed]
5. Katona, J. Clean and Dirty Code Comprehension by Eye-tracking Based Evaluation using GP3 Eye Tracker. *Acta Polytech. Hung.* **2021**, *18*, 79–99. [CrossRef]
6. Wijkmark, C.H.; Metallinou, M.M.; Heldal, I. Remote Virtual Simulation for Incident Commanders—Cognitive Aspects. *Appl. Sci.* **2021**, *11*, 6434. [CrossRef]
7. Ali, Q.; Heldal, I.; Helgesen, C.G.; Krumina, G.; Costescu, C.; Kovari, A.; Katona, J.; Thill, S. Current Challenges Supporting School-Aged Children with Vision Problems: A Rapid Review. *Appl. Sci.* **2021**, *11*, 9673. [CrossRef]
8. Pinter, R.; Maravić Čisar, S.; Kovari, A.; Major, L.; Čisar, P.; Katona, J. Case Study: Students' Code-Tracing Skills and Calibration of Questions for Computer Adaptive Tests. *Appl. Sci.* **2020**, *10*, 7044. [CrossRef]
9. Szucs, R.S. Cognitive level of consumers' knowledge in the case of a few food products. *Eur. Sci. J.* **2014**, *10*, 58–65.
10. Francisti, J.; Balogh, Z.; Reichel, J.; Magdin, M.; Koprda, Š.; Molnár, G. Application Experiences Using IoT Devices in Education. *Appl. Sci.* **2020**, *10*, 7286. [CrossRef]
11. Costescu, C.; Chelba, I.; Roşan, A.; Kovari, A.; Katona, J. Cognitive Patterns and Coping Mechanisms in the Context of Internet Use. *Appl. Sci.* **2021**, *11*, 1302. [CrossRef]
12. Shin, J.-G.; Choi, G.-Y.; Hwang, H.-J.; Kim, S.-H. Evaluation of Emotional Satisfaction Using Questionnaires in Voice-Based Human–AI Interaction. *Appl. Sci.* **2021**, *11*, 1920. [CrossRef]

13. Hikawa, H.; Ichikawa, Y.; Ito, H.; Maeda, Y. Dynamic Gesture Recognition System with Gesture Spotting Based on Self-Organizing Maps. *Appl. Sci.* **2021**, *11*, 1933. [CrossRef]
14. Sharma, R.; Ribeiro, B.; Pinto, A.M.; Cardoso, A. Emulating Cued Recall of Abstract Concepts via Regulated Activation Networks. *Appl. Sci.* **2021**, *11*, 2134. [CrossRef]
15. Sáiz-Manzanares, M.C.; Pérez, I.R.; Rodríguez, A.A.; Arribas, S.R.; Almeida, L.; Martin, C.F. Analysis of the Learning Process through Eye Tracking Technology and Feature Selection Techniques. *Appl. Sci.* **2021**, *11*, 6157. [CrossRef]
16. Popescu, A.B.; Taca, I.A.; Nita, C.I.; Vizitiu, A.; Demeter, R.; Suciu, C.; Itu, L.M. Privacy Preserving Classification of EEG Data Using Machine Learning and Homomorphic Encryption. *Appl. Sci.* **2021**, *11*, 7360. [CrossRef]
17. Velichkovsky, B.M.; Kotov, A.; Arinkin, N.; Zaidelman, L.; Zinina, A.; Kivva, K. From Social Gaze to Indirect Speech Constructions: How to Induce the Impression That Your Companion Robot Is a Conscious Creature. *Appl. Sci.* **2021**, *11*, 10255. [CrossRef]
18. Namestovski, Ž.; Kovari, A. Framework for Preparation of Engaging Online Educational Materials—A Cognitive Approach. *Appl. Sci.* **2022**, *12*, 1745. [CrossRef]
19. Hardi, J. An overview of metacognitive strategies in young learners' vocabulary learning. *GRADUS* **2015**, *2*, 46–60.
20. Finseth, T.; Dorneich, M.C.; Keren, N.; Franke, W.D.; Vardeman, S.B. Manipulating Stress Responses during Spaceflight Training with Virtual Stressors. *Appl. Sci.* **2022**, *12*, 2289. [CrossRef]
21. Rodriguez Aguiñaga, A.; Muñoz Delgado, L.; López-López, V.R.; Calvillo Téllez, A. EEG-Based Emotion Recognition Using Deep Learning and M3GP. *Appl. Sci.* **2022**, *12*, 2527. [CrossRef]
22. Gnjatović, M.; Košanin, I.; Maček, N.; Joksimović, D. Clustering of Road Traffic Accidents as a Gestalt Problem. *Appl. Sci.* **2022**, *12*, 4543. [CrossRef]
23. Boldizsar, A. Statistical analysis of road freight transport in Catalonia. *Prod. Eng. Arch.* **2022**, *28*, 40–49. [CrossRef]

Article

From Social Gaze to Indirect Speech Constructions: How to Induce the Impression That Your Companion Robot Is a Conscious Creature

Boris M. Velichkovsky [1,2,3,*], Artemiy Kotov [1,3,4], Nikita Arinkin [1,3], Liudmila Zaidelman [1,3], Anna Zinina [1,3,4,*] and Kirill Kivva [3,5]

[1] National Research Center "Kurchatov Institute", 123182 Moscow, Russia; kotov_aa@nrcki.ru (A.K.); arinkin_na@nrcki.ru (N.A.); laburnum-yz@yandex.ru (L.Z.)
[2] Faculty of Psychology, Technische Universitaet Dresden, 01062 Dresden, Germany
[3] Scientific Department, Scientific Center of Cognitive Programs and Technologies, Russian State University for the Humanities, 125047 Moscow, Russia
[4] Laboratory of Cognitive Research of the Foundations of Communication, Moscow State Linguistic University, 1190034 Moscow, Russia
[5] Computer Science & Control Systems Faculty, Bauman Moscow State Technical University, 105005 Moscow, Russia; k.kivva@gmail.com
* Correspondence: velcih@applied-cognition.org (B.M.V.); anna.zinina.22@gmail.com (A.Z.)

Featured Application: To ensure our fluent and partner interaction with the new generation of interactive artificial agents, these agents must use intermodal communication channels, including eye movements and diverse behavioral manifestations of intellectual and emotional attitudes towards the referents, denoted in communication. We intended to implement an imitation of the so-called social gaze, as it is of the primary significance for human–robot interactions. The suggested model can be used in the next generation of companion robots to maintain a feeling of eye contact and natural personal interaction. Such robots may include personal companions, smart home interfaces, information assistants in public transport, offices, or shops, toys, and educational robots.

Abstract: We implemented different modes of social gaze behavior in our companion robot, F-2, to evaluate the impression of the gaze behaviors on humans in three symmetric communicative situations: (a) the robot telling a story, (b) the person telling a story to the robot, and (c) both parties communicating about objects in the real world while solving a Tangram puzzle. In all the situations the robot localized the human's eyes and directed its gaze between the human, the environment, and the object of interest in the problem space (if it existed). We examined the balance between different gaze directions as the novel key element to maintaining a feeling of social connection with the robot in humans. We extended the computer model of the robot in order to simulate realistic gaze behavior in the robot and create the impression of the robot changing its internal cognitive states. Other novel results include the implicit, rather than explicit, character of the robot gaze perception for many of our subjects and the role of individual differences, especially the level of emotional intelligence, in terms of human sensitivity to the robotic gaze. Therefore, in this study, we used an iterative approach, extending the applied cognitive architecture in order to simulate the balance between different behavioral reactions and to test it in the experiments. In such a way, we came to a description of the key behavioral cues that suggest to a person that the particular robot can be perceived as an emotional and even conscious creature.

Keywords: human–robot interaction; social gaze; eye-to-eye contact; emotional interfaces; eye–brain–computer interfaces; attention; reflection; usability; brain hemispheric lateralization

1. Introduction

Gaze and speech are the primary cues of attention and intellect of another person. Within personal communication, gaze is the first communicative sign that we encounter, and which conveys information about the mental state, attention, and attitude of our counterpart. The direction of the opponent's gaze is easily detected by another person (and measured by conventional eye trackers), so gaze not only has a perceptive function but a significant communicative effect as well, which is extensively described in the literature on human–human [1,2] and human–robot [3,4] interactions. Overall, behavioral patterns of the eyes, including the eyelids and brows, can be studied and modeled as a symptom of the internal cognitive operations of the subject, or as a communicative cue, exposing to the addressee the internal state of the subject, or intentionally conveying meanings in communication [5]. We examined the perception of gaze behavior for short periods of direct eye contact, as well as for longer durations of human–robot interactions, in which the robot had to change its gaze direction between several objects of interest in the environment. In this work, we used an iterative approach, trying (a) to extend the robot control architecture in order to simulate gaze dynamics, (b) to test the architecture in experiments within typical communicative situations, and (c) to observe the key features—communicative cues or strategies—that have to be included in applied and theoretical architecture in order to make robots' behavior more reflective from the point of view of a human. Additionally, we considered the interaction of eye movements with other activities that evolve within the framework of continuous meaningful interactions. Not immediately obvious responses, such as indirect speech acts, may be of particular importance for inducing the feeling that your partner is not an automat.

As studied from the point of view of a human subject, vision and, therefore, gaze are the most important channels of perception. Gaze is directed by the attention system and can be used in humans and in primates in the ambient mode (for orientation in the space) or in the focal mode (to examine a particular object of interest) [6,7]. During a simulation, it is important that ambient and focal systems compete to gain control over the direction of the eyes, as attention to different objects of interest may raise the internal competition to shift the gaze to each of the objects of attention. This creates a bottleneck in a situation of time pressure, and the gaze system becomes a limited resource that must meet the numerous requirements of the attention and communication systems—to direct the attention to each interesting object, to simulate communicative behavioral patterns, and to blink [1,3]. The units to control the gaze can experience not only positive activation—arousal—but also negative influence—suppression. Although the addressee is a natural object of interest and, thus, must attract the attention, a long and direct gaze constitutes a face threatening act [8] in many human cultures and should be limited in terms of the communicative theory of politeness. In this way, a separate unit should withdraw the gaze from the addressee after some critical time periods.

The eyelids and brows system, controlled by musculi or action units, according to the Facial Action Coding System [9], may also serve the attention system by squinting or widely opening the eyes, as well as expressing numerous cognitive and emotional states. On one hand, this behavior can be out of the subject's voluntary control, but the corresponding superficial cues can still convey information to the addressee, thus serving as normal signs in communication. On the other hand, the capacity of voluntary control over the system of the gaze/lids/brows allows the subject to express intended meanings; one may look at an object to designate it, or intentionally frown to express the concern. The variable degree of intentionality within the control of communicative cues is described as Kendon's continuum [10,11], where the unintentional and uncontrolled behavioral patterns stay at one end, while the controlled nonverbal signs, quite like the signs of natural language, are located at the other end. Departing from this point, the expressive possibilities of the gaze system are extensively studied within the theory of communication; a comprehensive review can be found in [12]. Gaze is also an important feature in the natural interfaces of companion robots, and thus, it is widely implemented and evaluated within applied robotic systems, for example, in the robots Cog [13], Kismet [14], and Infanoid [15], and others.

The attention of researchers and developers to artificial gaze systems is attracted by various features of the natural gaze.

(a) **Gaze direction** is the primary object of study, and in applied systems, the direction of gaze is critical for effective interaction. At the same time, the perception of gaze is quite complex and may include the evaluation of the position of the eyes relative to the rotation of the head. For example, the gaze direction of a back-projected robot is perceived more naturally and precisely if its eyes are moving relative to the head and not fixed in the central position [16].

(b) Saccades, pupil dilatation, and ocular torsion constitute the micromovements within the gaze system. They are the main objects of studies regarding the psychology of attention but are not widely implemented in artificial interfaces. An extensive study on gaze micro-features, including eye saccades, eyelid micro-movements, and blinking, and their relation to the level of trust that they arouse in users, is represented in [17].

(c) Movements of eyelids and brows and blinking. Most virtual agents and even some physical robots are designed to blink and move their eyelids and eyebrows. In [18], the behavioral modes of a physical robot were compared, including non-blinking and blinking according to statistical models, and blinking according to physiological data. According to this study, a robot with a physiological blink seemed more intelligent to the respondents. In [19], the researchers varied the number of blinks performed by the agent. It was determined that respondents have a feeling that the agent is looking at them when the number of blinks of the agent exceeds the number of blinks of the respondent. Thus, eyebrows and eyelids combined can express numerous expressive and emotional patterns, which are widely studied in the field of the psychology of emotion and are also used within the existing interfaces of Kismet, Mertz, iCat, and others.

(d) Responsive gaze is studied as the ability to reply to another's gaze by looking at the opponent and maintaining the balance of gazes during communication. This topic is extensively studied by Yoshikawa and his colleagues with regard to human–robot interactions. In experimental studies, the researchers showed that the robot that responded with its gaze to the gaze of the interlocutor provided the subjects with the feeling of gaze contact and was evaluated as a more positive counterpart [19,20].

(e) Joint attention is studied as the ability to concentrate on the object of interest of another person. A robot's ability to support joint attention with humans is considered an important feature for companion robots, as it makes them look more competent and socially interactive while solving spatial tasks with humans [21]. In a study from our lab, subjects learned to control a robotic device using the joint attention gaze patterns. The robot was prompted by a responsive gaze of a human and then followed the direction of the human gaze, simulating joint attention to select the location for the required action of the robot [22].

These studies on the superficial expressive patterns of gaze are combined with a major area, in which human gaze is used as a controller in interfaces to position a cursor, select objects, or navigate in the environment. Although one can hardly underestimate the importance of these studies, in this article, we want to address the communicative aspect of the gaze—the ability of the eyes to express the internal cognitive processes of an agent. This feature, if simulated by artificial architectures, can give us a better understanding of the dynamics of a human gaze in its connection to internal cognitive states, and when applied, can make companion robots look more intelligent and attractive. Such an interface can act not as an extension of a human body, but as a companion, utilizing its gaze to report to a human its internal or communicative states. In this respect, we extended an applied cognitive architecture to model gaze behavior for an experimental companion robot, F-2, and tested the model in two situations of storytelling as well as a collaborative game situation. We used gaze direction to balance looks at the addressee, the object of interest (game pieces), or side gazes. For all the experiments, we expected that the robot, controlled by the balance model, would look more attractive and/or intelligent than the robot controlled by a simpler direct gaze management system. Within the experiments, we also applied some movements of the eyelids and eyebrows to make the gazes more vivid. We tested the models of joint attention and responsive gazes in the collaborative game and storytelling. At the same time, we did not apply micromovements such as microsaccades, drifts, pupil dilatation, and ocular torsion because their influence on the impression of the robot was discovered to be insignificant during the preliminary tests.

2. Methods and Architecture of the Model

In our study, we relied on an iterative approach, including the basis of a multimodal corpus and previous experiments from which we selected communicative functions that might distinguish conscious behavior if simulated by a companion robot. We tried to extend the previously developed architecture of a companion robot to simulate the observed functions and attempted to combine them with other, previously developed behavioral cues and strategies. Furthermore, we tested the impact of the modeled cues within the experiments to evaluate their contribution to the perception of the robot as a conscious creature. In this series of experiments, we extended the model to simulate the gaze management patterns in three symmetrical communicative situations.

In previous experiments, we have noticed that people may follow some nonverbal cues of the robot and attribute internal states to some specific movements and to the changes of the robot's gaze direction. In the developed architecture, nonverbal cues are combined with verbal responses; a robot can perform gestures and/or utterances in response to different incoming stimuli, including users' utterances, tactile events, and users' movements. In order for the robot to react to the most essential stimuli, or to express the most essential internal states, we implemented a compound architecture in which the internal units compete and concurrently gain control over all or some of the robot's effectors to perform movements or phrases. This architecture is also applicable to the control of gaze direction; an applied gaze management system should be organized to hold and constantly solve the conflicts between numerous units of attention and expression, competing to gain control over the direction of eyes. M. Minsky [23] suggested the classic architecture of proto-specialists to handle such conflicts in a system, by which numerous cognitive or emotional units try to control the body of an artificial agent or robot. Following Minsky, each proto-specialist constitutes a simple cognitive unit, responsible for handling some simple stimulus (e.g., a threat) or goal (e.g., hunger). Proto-specialists change their activation over time; the leading unit gains control over the required effectors, and then discharge, being satisfied, while losing the initiative in favor of other proto-specialists. This architecture has been extended by A. Sloman within the Cognition and Affects Project (CogAff). He has distinguished reactive units into different cognitive levels; while emotions and drives stay at the primary level, deliberative mechanisms (second level) or reflective reasoning (third level) can also gain control over the body of the agent to suggest the execution of longer and more sophisticated behavioral programs. In this architecture, deliberative reasoning may suppress emotions, conciliating the agent via rational reasoning. On the other hand, emotions can return the reasoning process of the agent to the object of desire or anxiety, thus reactivating themselves via the mechanism of positive feedback [24].

Our companion robot, F-2, has quite simple hardware and is controlled by an extended software architecture, which processes natural language texts (oral speech and written sources), as well as the events from a computer vision system to provide the robot with reasonable reactions (see Figure 1). Its central component conceptually corresponds to the CogAff architecture and operates with a set of scripts that are activated by incoming events and sending their behavioral packages to be executed by the robot.

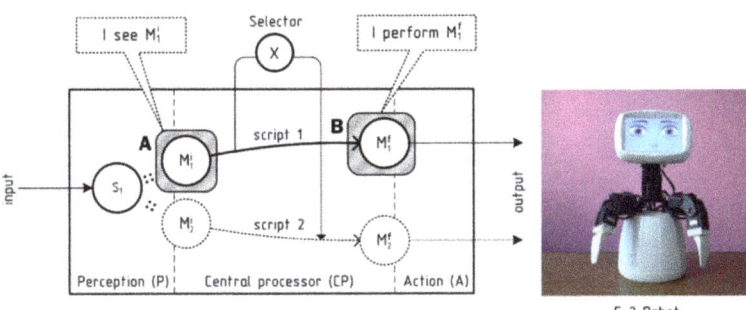

Figure 1. The general processing architecture of the F-2 robot.

Different incoming stimuli are processed within several components. Speech and text are processed by a semantic parser [25], which receives written text at its input, passes the stages of morphological and syntactic analysis, and constructs a semantic predication—a frame—for each sentence. For oral messages, an external speech-to-text service can be used to convert the signal to the written form (in our case, *Yandex speech API*). Several recognition variants can be processed in parallel in case they are generated by the recognition service. The ability to operate with semantic predications (frames) is the key feature of the suggested model and its main difference from the CogAff architecture. We used semantic predications in a form, which is usual for the representation of a sentence meaning in linguistics [26,27]. Each frame is divided by valencies, including the predicate, agent, patient, instrument, time, location, and so on. We used a list of 22 valencies, following the method by Fillmore [28]. Each valency within a frame is represented by a set of semantic markers—nuclear semantic units extracted from the meanings of words within this valency or assigned by the corresponding visual processing component. We used a list of 4835 semantic markers designed on the basis of (a) the list of semantic primes by A. Wierzbicka [29], (b) categories of the semantic dictionary [30], and (c) two-level clustering of word2vec semantic representations. Within the annotation (c), one marker is assigned to each word within a cluster, so two markers are assigned to each word in a two-level clustering tree [31]. The method was designed to handle automatic clustering with a tree of arbitrary depth. A two-level tree is presently used for the approbation of the approach.

Visual stimuli are processed by dedicated software components, and each of these also produces semantic predications to be processed by the central component of the robot—scripts. The robot keeps and updates the coordinates of recognized objects and events in 3D space, and generates semantic predications for the script component, such as "someone is looking at me". Face detection and orientation are processed with the help of the OpenCV library. The recognition of tangram puzzle elements is carried out with a specially developed tool using the *IGD* marker system. Each movement of a game piece is also converted to a semantic predication (such as "the user correctly moves game piece No 7"), and accounted for in the module, which monitors the progress in the solution of the puzzle. The extended processing architecture, with examples of semantic predications for different stimuli, is presented in Figure 2.

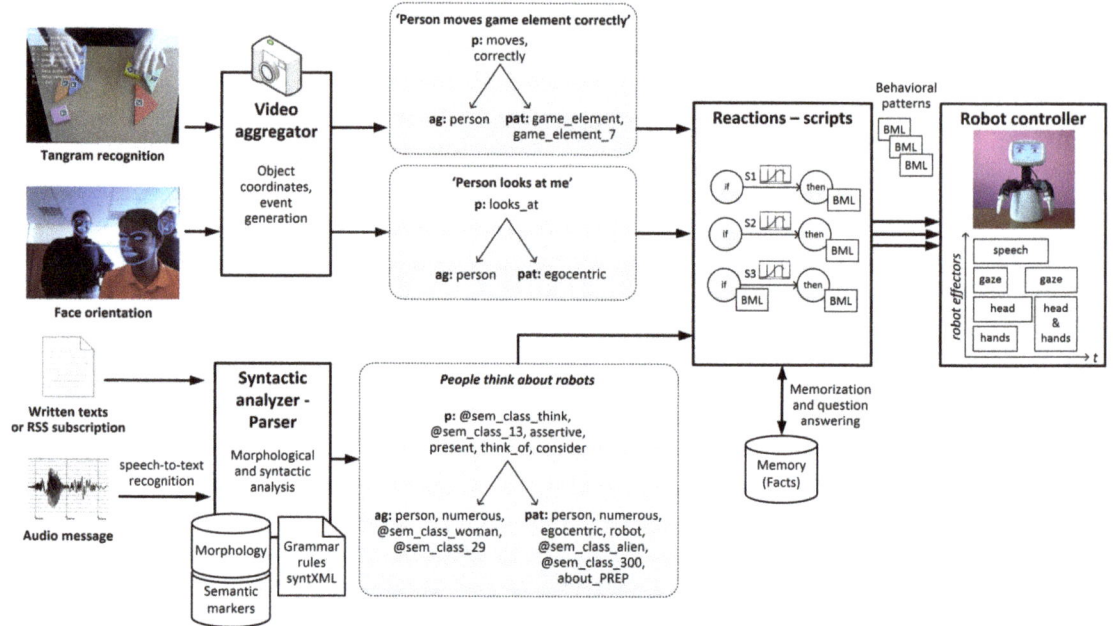

Figure 2. The extended processing architecture of the F-2 robot.

Within the suggested architecture, we used a list of scripts to process the stimuli and execute responsive reactions, as well as to support the competition between scripts in order to establish compound behavior (see further [25]). Each script is a type of production—an if-then operator. Its premise (if condition) and corollary (then condition) are also defined as semantic predications, so an input stimulus may invoke a script, which may, in turn, create a new semantic predication to invoke further scripts. A script may contain behavioral patterns to be executed once the script is activated so it can control the effectors of the robot, including the gaze direction. As the semantics of input sentences and real-world events are represented in a unified way, the system may (a) receive new knowledge from a visual observation or a text description, and (b) react to each incoming representation—such as an incoming user's gaze, a user's move within a game, or a user's utterance—and balance the reactions to each stimulus, regardless of its modality. It can also handle the competition between numerous scripts of the same type, for instance, when a user makes many moves within the game or when many people look at the robot simultaneously. For each incoming stimulus, we calculated its similarity with all the available premises of scripts via a modified Jaccard similarity coefficient to evaluate the number of semantic markers expected by a script premise and present in the stimulus. The best script is selected and activated proportionally to its similarity with the stimulus and its sensitivity, for example, its prior activation by preceding stimuli. Each activated script sends the desired behavioral pattern to the robot controller. The controller monitors the available effectors of the robot and the list of behavioral patterns suggested for executions by all the activated scripts; further, it executes a pattern from the most activated script as soon as the required effectors are available. A script loses its activation either immediately, when its behavioral patterns are executed on the robot, or gradually in time, while it waits for the expression and becomes irrelevant.

In this architecture, visual stimuli, as well as stimuli of other modalities, are naturally filtered by the agent depending on their subjective relevance. A relevant stimulus better matches scenarios and forces their activation, thus ensuring that the robot will respond to this stimulus with its behavior. Less relevant stimuli cause the moderate activation of scripts that can be expressed in the absence of a more relevant stimulus, or otherwise decelerate in time. Irrelevant stimuli can match no script and not get processed at all. This architecture allows the agent to respond to the stimuli, following their subjective relevance to the agent and depending on the available resources—most notably, time.

In our script component, we used 82 scripts for emotional processing and 3500 scripts for the rational processing of events. Rational scripts allow for the resolution of speech ambiguity and provide the foundation for the planned mechanism of rational inference. The most relevant script for an input event is selected based on the number of semantic markers of the script premise present within the input event, and it is calculated via the modified Jaccard similarity coefficient. For each stimulus, a separate instance of the script is created with the calculated activation. For example, if several people look at the robot, then a separate script instance with a responsive gaze behavioral pattern is created for each person. Each script is linked to a behavioral pattern, defined with the Behavior Markup Language (BML) [32,33]. Specific movements, such as gestures and head and eye movements, that are not attached to the coordinates of the surrounding objects, are selected from the REC emotional corpus [34], which was designed in Blender 3D rendering software and stored in the LiteDB database as the arrays of coordinates for each of the robot's effectors in time.

The materials, methods, human participants, and the results of three actual experiments with robots, which were developed on the basis of the architecture, are described in detail in Section 3.

3. Experimental Studies and Results

We consecutively applied the model to three major communicative situations, in which (a) the robot tells a human a story, so the denotatum is represented by the robot; (b) the human has to solve a special puzzle and the robot follows the solution and gives advice—here the denotatum is explicitly represented to both the human and the robot; (c) the robot listens to a story that is being told by the human, and therefore, the denotatum here is represented by the human. In all these conditions, the robot has to apply different gaze control modes to balance between the social (responsive) gaze, the side gaze, and the gaze to the physical object in problem space, if it exists. Although natural gaze behavior is rather compound and may differ in varying communicative situations, we iteratively extended the suggested model in order to cover these conceptually symmetric types of communicative situations.

In general, eye movements tend to have at least an implicit influence on the robot's attractiveness to a user. In our recent study [35], we investigated the contribution of the robot's effectors' movements—such as those of the hands, head, eyes, and mouth—to their attractiveness to the user. The subjects were students, teachers, and counselors of an educational camp ($n = 29$, 17 females, mean age of 19). During the experiment, the subjects were asked to listen to five short stories narrated by the robot. The subjects had to listen to each story twice in random order—with the robot using all active organs—*full-motion mode*, and without movements of a specific active organ (no movement of the eyes, head, hands, and no mouth animation)—deficit mode. After presenting two experimental conditions for each story, the subject chose their most preferred type of the robot's behavior. The subjects were also asked to describe the difference between the two modes of storytelling by the robot. As expected, the subjects in all cases chose the robot that used full-motion behavior. The experimental study revealed that users are more likely ($p < 0.01$, Mann–Whitney U-test) to prefer a robot that uses gestures, head movements, eyes, and mouth animation in its behavior compared to a robot for which some part of its body is stationary. The impact of the robot's eye movements on the user was quite implicit; the subjects significantly more often ($p < 0.05$, Mann–Whitney U-test) preferred the robot with eye movements to the robot with motionless eyes, but they seldom ($p < 0.05$, Mann–Whitney U-test) explicitly noticed the difference between these two modes. The implicit nature of the preference was supported by the verbal responses to a subsequent questionnaire; the subjects rather indicated the irrelevant differences between the robots, for example, they assumed that one of the robots spoke faster, said something wrong, or was in some way kinder and/or more interested.

3.1. Experiment 1: The Robot Tells a Story

To reproduce the pattern of complex gaze behavior on the robot, we developed a model in Python in which different scripts change their activation in time, and the script with the highest activation takes control of the robot's eyes. Figure 3 shows the functioning of the model with three scripts: (a) to think—this state changes sinusoidally over time; (b) to speak—the model received data about the starting time of the utterance and the phrase accent; (c) to pay attention to the person—this state increases linearly if the robot was not looking at the person and resets its activation as soon as the robot looks at the person. Every 40 milliseconds, the system calculates the leading script and allows it to set the direction of gaze. This model was used in one of the experimental conditions and was compared to some simple models of gaze management, which are described later. The eyelids and mouth did not move in this experiment. The robot detected the position of the human face within the environment and directed its gaze at the location of the human's eyes so that the subjects could freely move in the space in front of the robot.

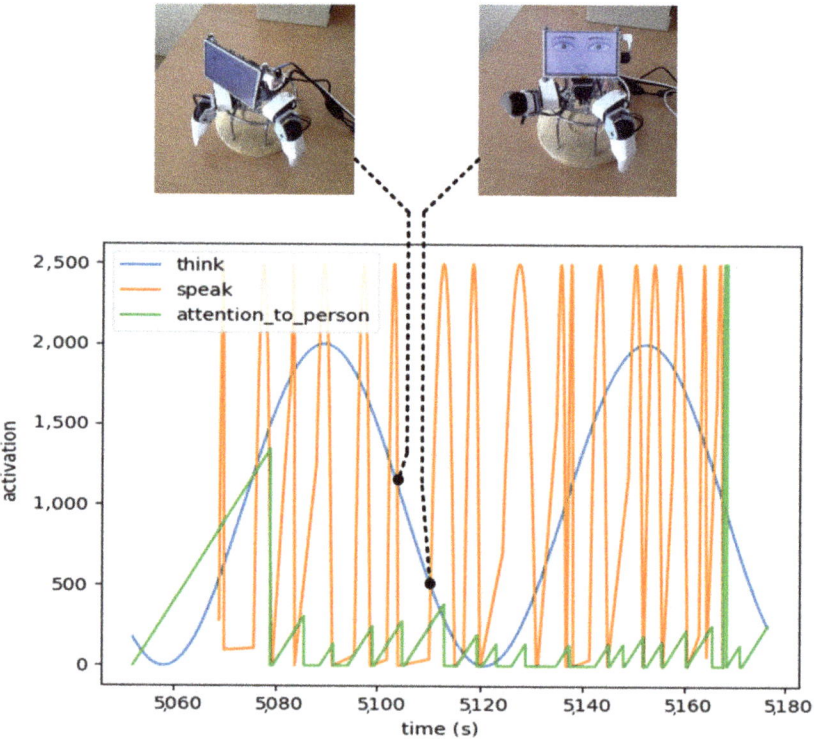

Figure 3. Gaze control model via the competition of control states; gaze direction is controlled by the leading state.

In the experiment, the participants ($n = 14$, six females, mean age of 27.2 years) evaluated the contribution of the gaze model to the robot's attractiveness (a preliminary report was given in a recent paper [36]). During the experiment, the robot narrated eight short stories. Each participant had to evaluate the robot on two modes of behavior (or conditions) according to Osgood's semantic differential scales. In Condition 1, the robot switched its gaze direction every 6 s from left to right, in turn, not looking at the user. In Condition 2, the gaze direction was controlled by the leading script, as mentioned above, and the robot could direct its gaze to the user. The activation of each of these scrips varied in time. When the leading state was changed, the new leader changed the gaze direction of the robot, as shown in Figure 3. The robot's utterances were accompanied by iconic gestures, which were similar in the two experimental conditions.

Results and a Preliminary Discussion

In the experimental setting, the subjects significantly more often ($p < 0.01$, Chi-Square) preferred the robot controlled by the balancing gaze model (Condition 2). This robot was significantly more often described as friendly, attractive, calm, emotional, and attentive ($p < 0.01$, Chi-Square). The results allow us to evaluate the contribution of the gaze direction to the positive perception of the robot and prove the effectiveness of the developed model of gaze behavior in the situation of human–machine interactions.

3.2. Experiment 2: The Robot Helps Humans in a Game

To further evaluate the influence of social gaze, we tested the model in a situation where the robot was helping a human solve a tangram puzzle. The goal of the experiment was to evaluate the influence of oriented gestures with directed gaze on a human ($n = 31$, 12 females, mean age of 27.4). The task of the participants was to arrange the elements within a given contour on a white sheet. During the experiment, the participant was to complete several figures, including a parallelogram, a fish, a triangle, and a ship. For each task, the game elements were placed in front of the participant on the left and right sides of the playing field. Two paired elements (two large triangles and two small triangles) were always placed on different sides of the playing field. The robot helped the human, indicating in speech which game element to take and where to place it (Figure 4). In half of the tasks, the robot used oriented communicative actions (pointing hand gestures, head movements, and gaze) to indicate the required game element, and then the correct position to place it (Condition 1). For the paired elements, the left or right element was chosen randomly. In the other half of the tasks, the robot used non-oriented, symmetric gestures while providing the same speech instruction (Condition 2). In the first condition, the direction of the robot's gaze was considered an indication. In its speech instructions, the robot always referred to an element by its shape and size, not by color. So, when the robot named paired elements, the reference was ambiguous and could indicate the element on either the right or the left side of the playing field.

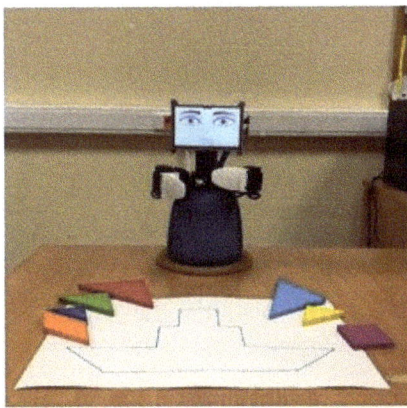 

Figure 4. Robot refers to the game element with speech (with ambiguous reference) or with speech and gesture.

Results and a Preliminary Discussion

According to the results of the experiment, most of the participants (64.5%) significantly ($p < 0.01$, Mann–Whitney U-test) preferred the robot pointing to the tangram elements and their locations with eyes and hands. At the same time, the difference between both experimental conditions was not obvious to the players; only half of them (15 people; 48.4% of the total group) noticed the difference between the robot's performance with and without the pointing gestures and gaze. Therefore, the evaluation of the robot can again be implicit because the participants did not distinguish between the two experimental conditions verbally, but simultaneously preferred the robot that indicated the necessary element and its location in the contour with the help of its head, eyes, and hand movements. This experiment is described in more detail elsewhere [37].

The results clearly show the importance of directional gaze for communication between a robot and a user. Directional gaze is perceived as a reflection of the robot's internal state. In the speech instructions, the robot prompts the user, and gaze is used to provide the reference for this prompt. It is this correspondence of mimicked expression to the robot's internal intention that is positively perceived by the subjects; however, it occurs at the implicit level.

3.3. Experiments 3: The Robot Listens to a Story

One of the key characteristics of social gaze is the ability to change the gaze direction following the addressee. To simulate the responsive gaze of the robot, we decided to examine a situation in which a person tells the robot a story and the robot acts as an active listener by demonstrating different modes of responsive gaze. The purpose of the new experiment was to study the effect of a robot's responsive gaze on its attractiveness to the user. In this experiment, two robots reacted to the communicative actions of a human in two different modes. In one case, the robot directed its gaze at the user; in the second case, it demonstrated side gaze, which is typical for the condition of thoughtfulness or when regulated by a politeness strategy to avoid face threatening acts aimed at the addressee [8,38]—video is provided within the Supplementary Materials. Within the experiment, subjects had to tell two robots stories following a list of pictures by Herluf Bidstrup (Figure 5; see more about his artwork at https://en.wikipedia.org/wiki/Herluf_Bidstrup, accessed on 1 October 2021). Each picture was represented as a stack of cards in random order. A total set of six stories was used; each subject had to tell three stories to each robot, switching to the other robot after each story.

A total of 46 subjects participated in the experiment (mean age of 27 years, 33 females). Before the experiment, the F-2 robots were introduced to the subjects. The respondents were told that the developers were currently trying to train the robot to follow the story narrated by a human. The main purpose of the study, which was to investigate the effect of the robot's responsive gaze, was not included in the introduction. We examined the following hypotheses: (a) the robot is perceived as more attractive if it establishes gaze contact with the user; (b) respondents with a high level of emotional intelligence better distinguish the behavioral patterns of the robots.

The orientation of the user's attention was roughly identified automatically by the orientation of the user's face and was recognized by a computer vision component based on OpenCV. This system was chosen as a possible "built-in" solution for emotional companion robots, allowing us to avoid the calibration typical for eye-tracker experiments, and thus, maintain more natural communication with the robots. To find a vector of the human face orientation that we interpret as a gaze vector, we built an approximate 3D model of human facial landmarks and developed the following steps. Human faces are detected within the camera video frames using a linear SVM classifier based on an advanced version of HoG features [39] implemented in the Dlib library. The image intensity histogram is normalized inside a box bounding the face. Then, landmarks of the found face are detected using an Ensemble of Regression Trees approach described in [40] and implemented in the Dlib library. These 2D landmarks' positions and 3D coordinates of the corresponding model points are used to solve a perspective-n-point (PnP) problem using the method suggested in [41] an implemented in the OpenCV library. Solving the PnP problem provides the orientation of the human face in 3D with respect to the camera. Next, a 3D gaze vector is built starting between the human's eyes and using the calculated face orientation, and the vector's coordinates are stabilized using a simple Kalman filter, assuming these coordinates to be independent. Then, a constant offset is added to the gaze vector's endpoint to make the vector point at the camera when the human is looking directly at the robot. The angle between the gaze vector and the direction from the camera center to the human's nose bridge provides information on if the human is looking at the robot.

Figure 5. Example of stimulus material—a graphic story by Herluf Bidstrup. Each story was represented as a stack of separate cards in random order.

The procedure of the experiment was as follows. A participant sat in a room in front of the two robots, identified by square and triangle marks on their bodies. The first robot said that it is ready to listen to a story. The subject took the cards from the stack and had to organize them into a story. While narrating, the subject was asked to show the robot the card corresponding to the current part of the story. At the end of the story, the robot expressed its gratitude and asked to tell the next story to the other robot. Within the setup of the experiment, three zones of attention were distinguished—a participant can organize the cards on the table and can show the cards to the left or the right robot. This setup ensured higher precision of the user recognition system, which was detecting the face orientation rather than the gaze itself. The subject could communicate with the left and right robots from the same position at the table, so the robots constantly maintained the corresponding gaze behavior. For example, each robot could respond to the user's gaze even if the user told a story to the other robot. Following the experiment, the subject had to express their preference for the triangle robot, the square robot, or both robots equally, as well as evaluate each robot on a five-point scale.

While "listening", the hands of the robots were controlled by the inactive component, which initiates permanent minor movements. Additionally, during the speech replies, the hands were controlled by more significant gestures that were coordinated to the utterances. At the same time, the head and eyes were constantly controlled by the two scripts responsible for the social gaze. When a person looked away from the robot (e.g., looked aside or looked at the table to follow the cards), each of the robots looked down—at the table with the cards. When a person looked at the robot, the first robot (marked with a square) looked back, raised its head, opened its eyelids, and raised its eyebrows (Table 1), while the second robot (marked with a triangle) demonstrated the side gaze, looking left or right and randomly changing the direction.

Table 1. Robots' reactions to human gaze: gaze responsive and gaze aversive behavior.

Title 1	Before or After the User's Gaze	During the User's Gaze
Gaze-responsive robot (marked with a square)		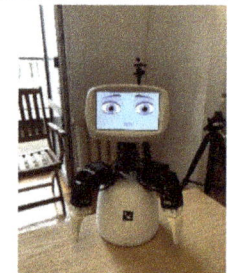
Gaze-aversive robot (marked with a triangle)		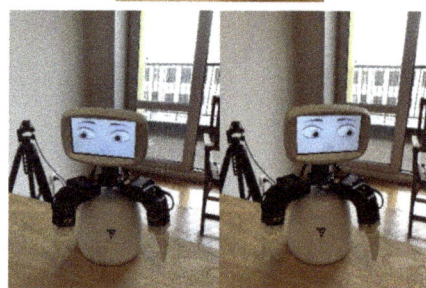
		Left- or right-side gazes are selected randomly

We used the Emotional Intelligence Test (EmIn) [42,43] to evaluate the level of emotional intelligence of the participants. The questionnaire is based on the interpretation of emotional intelligence as the ability to understand one's own and others' emotions, as well as to control one's own emotions. This understanding implies that a person (a) recognizes the presence of an emotional experience in oneself or another person, (b) identifies the expression and can find a verbal designation for one's own or another's emotion, and (c) understands the causes and effects of that emotional experience. Controlling emotions means that the person controls the intensity and external expression and can, if necessary, arbitrarily evoke a particular emotion. The test suggests an aggregated score of emotional intelligence, but we expected that the recognition of others' emotions would play a major role in the perception of the robots.

We also evaluated the accuracy of the recognition system. The automatically obtained data was compared with the observed gaze of the subjects during the experiment. The subject's gaze on the robot was defined as a movement of the gaze and head in the direction of the robot starting from 1 s. The number of reactions roughly corresponded to the number of gazes. The system was performing with a redundancy of 133% (for 1988 gazes, it demonstrated about 2650 responsive actions), the majority of redundant reactions were demonstrated during long human gazes (more than 30 s), when the system could execute several gaze responses. The statistics of real gazes were obtained by qualitative analysis of video recordings of the experiment. Taking into account the slight redundancy in performance for long gazes, the system was considered appropriate to execute gaze control in the actual communication.

Results and a Preliminary Discussion

According to the results of this experiment, if a person explicitly preferred a specific robot or evaluated it higher, they were assigned to the corresponding preference group. Within these groups, 26% of people ($n = 12$) preferred the gaze-avoiding robot (triangle), 28% of people ($n = 13$) preferred the gaze-responsive robot (square), and 46% ($n = 21$) evaluated the robots as equal. The subjects who preferred a specific robot demonstrated significantly different results on the scale of recognition of the emotions of others within the EmIn test. People who preferred the gaze-responsive robot had a level of emotional intelligence significantly higher than the group who preferred the gaze-avoiding robot (Figure 6). The subjects did not demonstrate any significant difference in any other scale of the emotional intelligence test.

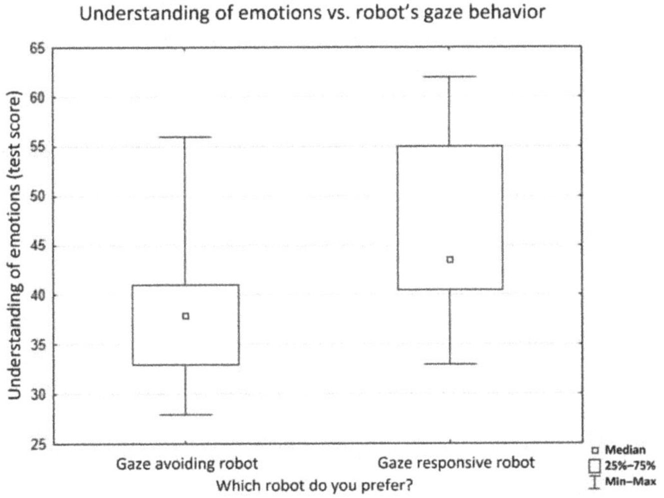

Figure 6. Level of emotional intelligence in our subjects vs. their preferences of the robots' social gaze behavior.

The results show that people with high sensitivity to emotions, measured with a dedicated scale of emotional intelligence, preferred the robot with a responsive gaze ($p < 0.05$, Mann–Whitney U-test). Several people in this group even described the gaze-avoiding robot as reacting to the user's gaze by looking sideways and characterized it as shy, thoughtful, or even female. Other respondents noted that they were eager to attract the attention of the gaze-avoiding robot, as they believed the robot was losing interest during the narrative. Several respondents noted that the triangle robot seemed to be simultaneously (a) paying attention to the human's narrative and (b) thinking about the events of the story or addressing some intrusive thoughts. For example, some people reported that "it is interesting how the robot seems to look at me while looking sideways". We can suggest the following interpretation of this phenomenon. The participants were perceiving the gaze-avoiding movement of the triangle robot as a compound pattern in which the robot (a) moves his gaze away immediately after the user's gaze, thus showing a type of responsive-gaze behavior, that is, paying attention to the user, and (b) looks sideways, thus showing a pattern typical for thinking about the events, according to the REC corpus. In other words, for these participants, the immediate start and the pattern of the movement (looking sideways) were invoked by two different internal states assigned to the robot. The immediate start of gaze movement was interpreted as expressed attention, and the looking sideways pattern was interpreted as thoughtfulness.

Although gaze-avoiding and gaze-responsive behaviors were designed as straightforward reactions to the user's gaze, several people noted different aspects of the movements. Some people who preferred the gaze-avoiding robot characterized it as more expressive and active since it was moving between three positions with high amplitude. People could also concentrate not on the responsive phase (when the robot looked up or sideways), but on the return movement to the inactive position (when the robot was looking down), which was sometimes interpreted as being attentive to the cards on the table or being upset about the events of the story.

Assuming these deviations in the descriptions, the evaluation of the group with the correct identification of the differences between the robots is rather speculative. As for the criteria, we selected people who explicitly indicated that the robot with a square was looking at me and/or the robot with a triangle was looking sideways. The rate of people who gave this response was 8% (1 of 12 persons) in the gaze-avoiding preference group, 23% (5 of 21 persons) in the neutral group, and 69% (9 of 13 persons) in the gaze-responsive preference group. It means that people who better recognized the difference between the gaze-responsive and gaze-aversive robots also significantly often ($p < 0.01$, Spearman correlation) preferred the robot with the gaze-responsive behavior. Thus, people with high sensitivity to the emotions of others better recognized the gaze behavioral types of the robots and preferred the robot with the responsive-gaze behavior.

4. General Discussion

Our results clearly evidence that one of the core features of the social gaze, when simulated by robots, is that it can provoke the human to assign cognitive and emotional properties to the robot. While this is not a new result in the field of human–robot interactions, some novel facts can be reported from our experiments. For example, although social gaze in robots makes their perception more anthropomorphic, this influence often remains only implicit. As a matter of fact, gaze can be perceived as a latent cue; people systematically react to it in special problem tasks, but half of them do not mention this feature in self-reports explicitly. Ordinary people without any noticeable neurological symptoms also demonstrated substantial individual differences. The responsive gaze was especially significant for people with a high level of emotional intelligence but not for all the users. This is the next novel result of our study. In particular, cases of aversive gazing can well be interpreted positively and increase mutual social rapport within an evolving joint activity, as in our experiments on problem solving and storytelling. Previous works have concentrated exclusively on momentary eye-to-eye contacts as the prime indication of the joint attention state ([5,22,44], among others). Therefore, researchers have ignored the necessary balancing of several behavioral acts by means of eye movements in a continuous interaction over time.

As perceived in its dynamics, gaze and the change of gaze direction can be described as a change of the object of interest (a more traditional approach) or as a change of the prevailing cognitive state, which controls the gaze. Thus, eye movements communicate at least two cognitive states—initial and final. We can consider this competition to be among the essential features of consciousness processes described by some authors as a competition and suppression of habitual strategies of behavior and cognitive states [45]. Not accidentally, people with high emotional intelligence can note and interpret minor features of the gaze (gaze response and gaze direction), assigning to it several cognitive states ("he heard me, but is thinking about something else"). The competitive gaze simulation can serve as a core method to communicate the conscious state of the agent by provoking listeners to attribute consciousness to the emotional robots during the communication.

Further perspectives in this line of research concern the imitation of phenomena such as the eye–voice span (eyes are ahead of the voice while reading aloud) and two main modes of eye movements and vision—one related to ambient vision (as in visual search) and one related to the focal mode of cognition and volition (as in identification and decision making). Although gaze was traditionally studied within the mechanism of attention of a subject, the modern approach to gaze activity as a means of communication can suggest new concepts and allow for creating a new generation of natural emotional interfaces. In further developments, we strongly hope for improvement in the registration of eye movements and social gaze beyond the current methodology. This will open the way to their broader use in engineering as interfaces for automated systems for data processing and control, such as the gaze–brain–computer interface to control the internet of robotic things or eye–hand coordination to replace the overwhelming, but not always precise (and sometimes even impossible), computer mouse movements. One example of a case in which the gaze is more precise and faster than the computer mouse movements is the localization of objects from within a speedy vehicle or in conditions of ordinary industrial production.

As can be seen particularly from the results of Experiments 2 and 3, several behavioral patterns could be executed on the robot at the same time. This was especially the case if their BMLs referred to complementary effectors, when, for example, a responsive gaze was performed by the eyes and head, speech production controlled the mouth and a pointing gesture was performed by a hand. An external stimulus can also invoke several scripts, each of which can suggest their behavioral pattern for the execution. This combination of patterns allows developers to create compound and rich behavior in the robot and even simulate some emotional blending when the activated scripts represent contradictory emotions [46]. This type of blending corresponds to the architecture of reflexive effects of consciousness, simulated earlier with the help of the scripts mechanism, where the first, most relevant script, is accompanied by secondary scripts that are seemingly less relevant but provide alternative views on the situation perceived and to be managed [47].

We do not currently have direct empirical evidence for this idea. However, indirectly, it is supported by the bulk of neurophysiological, neurolinguistic, and even neuromolecular data, disputing the old concept that the left cerebral hemisphere in humans "dominates" over the "subdominant" right hemisphere [48,49]. At least in a clinical context, disorders of spatial, corporeal, emotional, and self-related consciousness primarily result from lesions of the right brain hemisphere [50,51]. A neurolinguistic pendant to these data is the well-established knowledge that while the left hemisphere supports explicit linguistic functions in most humans, the right prefrontal cortex is important for the implicit understanding of specifically human communication pragmatics, metaphorical language, humor, irony, and sarcasm, as well as eye-to-eye contact [52–54]. Moreover, dedicated brain "machinery" seems to be behind these cognitive–affective effects with a specific course of neurodevelopment in early ontogenesis [55]. The role of gaze contact is different for children with brain-and-mind disturbances, such as autism spectrum disorders and Williams syndrome [56,57].

However, the pathway to reliable comprehension and modeling of this neurophysiological phenomenology is difficult and conceivably long. In terms of needed usability, computational solutions that have been delineated in the present article are more practical and feasible.

5. Conclusions

Social gaze is a compound phenomenon, as it can be controlled by different cognitive states and, thus, convey information on the rich internal organization of the subject. As we have shown, robots that control their gaze via a computational model of competition between several internal states are perceived as more emotional and invoke higher empathy in humans. In the situation of storytelling by the robot, this competition arose between the attention paid to the opponent (human), the politeness strategy (to avoid long gazes), and simulated thoughtfulness, thus manifesting courtesy and reflection on the part of the robot. In the situation of puzzle-solving, this conflict arose between the simulated attention to the game elements for the next successful move and the opponent who should understand the robot's instruction; this can manifest in the engagement of the robot to the process of problem-solving and be interpreted as its "intention" to cooperate with the human. In the situation of story listening, the attention to the user's face and gaze manifested the engagement of the robot to the story narrated by the user, while the preference of the gaze-responsive partner was correlated with the human ability to recognize others' emotions. In this respect, the social gaze system can be considered not only a system of visual perception attention and action but as a significant component to establishing fluent communication between robots and humans. Understandably, it is the key technology to make the users attribute human internal states to the robot in natural communication.

On a more conceptual level, one can consider essential features of consciousness as contrasted to the automated processes. Here, again, we used a minimalistic approach to try to find the minimal architecture of computation, demonstrating the essential features of reflexivity [47]. Interestingly, such architectures seem to be relatively simple, but they probably demand a kind of working memory extension to consider responses that seem to be not necessarily the most appropriate in the current context at the first sight. In our view, those are the "second-choice" resources for producing a robot's responses that will induce in humans an impression of being intelligent and even conscious. Though starting this endeavor with elementary hardware and processing architectures, we think that a computational solution to our fluent and partner interaction with artificial agents may be by far more feasible than in wet or in silico brain modeling.

Supplementary Materials: A video with the demonstration of the experiments is available online at www.youtube.com/channel/UCaJ7WUJb_NuqELyJyUrp45Q (accessed on 1 November 2021).

Author Contributions: Conceptualization, B.M.V. and A.K.; methodology, B.M.V., A.Z. and A.K.; software, N.A. and K.K.; formal analysis, A.K. and A.Z.; investigation, A.Z., L.Z. and A.K.; writing—original draft preparation, A.K., A.Z. and L.Z.; writing—review and editing, B.M.V. and A.K.; visualization, A.K.; supervision, B.M.V.; project administration, A.K. All authors have read and agreed to the published version of the manuscript.

Funding: The first two experiments were in part supported by the National Research Center "Kurchatov Institute" (decision 1057 from 2 July 2020). The development of the general model (part 2) and Experiment 3 (Section 3.3 and corresponding results) were supported by the Russian Science Foundation, grant number 19-18-00547.

Institutional Review Board Statement: The study was conducted according to the guidelines of the Declaration of Helsinki, and approved by the local Ethics Committee of National Research Center "Kurchatov Institute" (Protocol No.10 from the 1 August 2018).

Informed Consent Statement: Informed consent was obtained from all subjects involved in the study.

Data Availability Statement: Experimental data is available online at https://bit.ly/3giqV8b (accessed on 1 November 2021).

Acknowledgments: We appreciate the administrative and technical support from the Russian State University for the Humanities in conducting a series of human–robot experiments during the time of the pandemic.

Conflicts of Interest: The authors declare no conflict of interest. The funders had no role in the design of the study; in the collection, analyses, or interpretation of data; in the writing of the manuscript, or in the decision to publish the results.

References

1. Velichkovsky, B.M. Communicating Attention: Gaze Position Transfer in Cooperative Problem Solving. *Pragmat. Cognit.* **1995**, *3*, 199–223. [CrossRef]
2. Pagnotta, M.; Laland, K.N.; Coco, M.I. Attentional Coordination in Demonstrator-Observer Dyads Facilitates Learning and Predicts Performance in a Novel Manual Task. *Cognition* **2020**, *201*, 104314. [CrossRef]
3. Beyan, C.; Murino, V.; Venture, G.; Wykowska, A. Editorial: Computational Approaches for Human-Human and Human-Robot Social Interactions. *Front. Robot. AI* **2020**, *7*, 55. [CrossRef] [PubMed]
4. Li, M.; Guo, F.; Ren, Z.; Duffy, V.G. A visual and neural evaluation of the affective impression on humanoid robot appearances in free viewing. *Int. J. Ind. Ergon.* **2021**, 103159. [CrossRef]
5. Schrammel, F.; Pannasch, S.; Graupner, S.T.; Mojzisch, A.; Velichkovsky, B.M. Virtual friend or threat? The effects of facial expression and gaze interaction on psychophysiological responses and emotional experience. *Psychophysiology* **2009**, *46*, 922–931. [CrossRef] [PubMed]

6. Ito, J.; Yamane, Y.; Suzuki, M.; Maldonado, P.; Fujita, I.; Tamura, H.; Grün, S. Switch from ambient to focal processing mode explains the dynamics of free viewing eye movements. *Sci. Rep.* **2017**, *7*, 1082. [CrossRef] [PubMed]
7. Velichkovsky, B.M.; Korosteleva, A.N.; Pannasch, S.; Helmert, J.R.; Orlov, V.A.; Sharaev, M.G.; Velichkovsky, B.B.; Ushakov, V.L. Two Visual Systems and Their Eye Movements: A Fixation-Based Event-Related Experiment with Ultrafast fMRI Reconciles Competing Views. *STM* **2019**, *11*, 7–16. [CrossRef]
8. Brown, P.; Levinson, S.C. *Politeness: Some Universals in Language Usage (Studies in Interactional Sociolinguistics)*; Cambridge University Press: Cambridge, UK, 1987.
9. Ekman, P.; Friesen, W.V. *Facial Action Coding System: A Technique for the Measurement of Facial Movement*; Consulting Psychologists Press: Palo Alto, CA, USA, 1978; ISBN 0931835011.
10. Müller, C. Gesture and Sign: Cataclysmic Break or Dynamic Relations? *Front. Psychol.* **2018**, *9*, 1651. [CrossRef]
11. Iriskhanova, O.K.; Cienki, A. The Semiotics of Gestures in Cognitive Linguistics: Contribution and Challenges. *Vopr. Kogn. Lingvist.* **2018**, *4*, 25–36. [CrossRef]
12. Admoni, H.; Scassellati, B. Social Eye Gaze in Human-Robot Interaction: A Review. *J. Hum. Robot Interact.* **2017**, *6*, 25–63. [CrossRef]
13. Scassellati, B. Mechanisms of Shared Attention for a Humanoid Robot. *Embodied Cogn. Action: Pap. 1996 Fall Symp.* **1996**, *4*, 21.
14. Breazeal, C.; Scassellati, B. A Context-Dependent Attention System for a Social Robot. In Proceedings of the IJCAI International Joint Conference on Artificial Intelligence, San Francisco, CA, USA, 31 July–6 August 1999; Volume 2.
15. Kozima, H.; Ito, A. Towards Language Acquisition by an Attention-Sharing Robot. In Proceedings of the Joint Conferences on New Methods in Language Processing and Computational Natural Language Learning, Sydney, Australia, 11–17 January 1998; pp. 245–246.
16. Al Moubayed, S.; Skantze, G. Perception of Gaze Direction for Situated Interaction. In Proceedings of the 4th Workshop on Eye Gaze in Intelligent Human Machine Interaction, Gaze-In 2012, Santa Monica, CA, USA, 26 October 2012.
17. Normoyle, A.; Badler, J.B.; Fan, T.; Badler, N.I.; Cassol, V.J.; Musse, S.R. Evaluating Perceived Trust from Procedurally Animated Gaze. In Proceedings of the Proceedings-Motion in Games 2013, MIG 2013, Dublin, Ireland, 6–8 November 2013.
18. Lehmann, H.; Roncone, A.; Pattacini, U.; Metta, G. Physiologically Inspired Blinking Behavior for a Humanoid Robot. In *Lecture Notes in Computer Science*; Including subseries Lecture Notes in Artificial Intelligence and Lecture Notes in Bioinformatics; Springer: Cham, Switzerland, 2016; Volume 9979.
19. Yoshikawa, Y.; Shinozawa, K.; Ishiguro, H.; Hagita, N.; Miyamoto, T. The Effects of Responsive Eye Movement and Blinking Behavior in a Communication Robot. In Proceedings of the IEEE International Conference on Intelligent Robots and Systems, Beijing, China, 9–15 October 2006.
20. Yoshikawa, Y.; Shinozawa, K.; Ishiguro, H.; Hagita, N.; Miyamoto, T. Responsive Robot Gaze to Interaction Partner. In Proceedings of the Robotics: Science and Systems, Atlanta, GA, USA, 27–30 June 2007; Volume 2.
21. Huang, C.M.; Thomaz, A.L. Effects of Responding to, Initiating and Ensuring Joint Attention in Human-Robot Interaction. In Proceedings of the Proceedings-IEEE International Workshop on Robot and Human Interactive Communication, Atlanta, GA, USA, 31 July–3 August 2011.
22. Fedorova, A.A.; Shishkin, S.L.; Nuzhdin, Y.O.; Velichkovsky, B.M. Gaze Based Robot Control: The Communicative Approach. In Proceedings of the International IEEE/EMBS Conference on Neural Engineering NER, Montpellier, France, 22–24 April 2015; Volume 2015, pp. 751–754.
23. Minsky, M.L. *The Society of Mind*; Touchstone Book: New York, NY, USA, 1988.
24. Allen, S.R. Concern Processing in Autonomous Agents. Ph.D. Thesis, University of Birmingham, Birmingham, UK, 2001.
25. Kotov, A.; Zinina, A.; Filatov, A. Semantic Parser for Sentiment Analysis and the Emotional Computer Agents. In Proceedings of the AINL-ISMW FRUCT 2015, Saint Petersburg, Russia, 9–14 November 2015; pp. 167–170.
26. Baker, C.F.; Fillmore, C.J.; Lowe, J.B. *The Berkeley FrameNet Project*; Association for Computational Linguistics: Stroudsburg, PA, USA, 1998.
27. Lyashevskaya, O.; Kashkin, E. Framebank: A Database of Russian Lexical Constructions. In *Communications in Computer and Information Science*; Springer: Berlin/Heidelberg, Germany, 2015.
28. Fillmore, C.J. The Case for Case. In *Universals in Linguistic Theory*; Bach, E., Harms, R.T., Eds.; Holt, Rinehart &Winston: New York, NY, USA, 1968; pp. 1–68.
29. Wierzbicka, A. *Semantic Primitives*; Athenäum: Frankfurt, Germany, 1972; ISBN 376104822X.
30. Shvedova, N.Y. *Russian Semantic Dictionary. Explanatory Dictionary, Systematized by Classes of Words and Meanings*; Azbukovnik: Moscow, Russia, 1998. (in Russian)
31. Kotov, A.A.; Zaidelman, L.Y.; Arinkin, N.A.; Zinina, A.A.; Filatov, A.A. Frames Revisited: Automatic Extraction of Semantic Patterns from a Natural Text. *Comput. Linguist. Intellect. Technol.* **2018**, *17*, 357–367.
32. Vilhjálmsson, H.; Cantelmo, N.; Cassell, J.; Chafai, N.E.; Kipp, M.; Kopp, S.; Mancini, M.; Marsella, S.; Marshall, A.; Pelachaud, C.; et al. The Behavior Markup Language: Recent Developments and Challenges. In *Intelligent Virtual Agents*; Springer: Berlin/Heidelberg, Germany, 2007; pp. 99–111.
33. Kopp, S.; Krenn, B.; Marsella, S.; Marshall, A.; Pelachaud, C.; Pirker, H.; Thórisson, K.; Vilhjálmsson, H. Towards a Common Framework for Multimodal Generation: The Behavior Markup Language. In *Intelligent Virtual Agents*; Springer: Berlin/Heidelberg, Germany, 2006; pp. 205–217.

34. Kotov, A.A.; Budyanskaya, E. The Russian Emotional Corpus: Communication in Natural Emotional Situations. In *Computer Linguistics and Intellectual Technologies*; RSUH: Moscow, Russia, 2012; Volume 1, Issue 11 (18), pp. 296–306.
35. Zinina, A.; Zaidelman, L.; Arinkin, N.; Kotov, A.A. Non-Verbal Behavior of the Robot Companion: A Contribution to the Likeability. *Procedia Comput. Sci.* **2020**, *169*, 800–806. [CrossRef]
36. Tsfasman, M.M.; Arinkin, N.A.; Zaydelman, L.Y.; Zinina, A.A.; Kotov, A.A. Development of the oculomotor communication system of the F-2 robot based on the multimodal REC housing (in Russian). In Proceedings of the The Eighth International Conference on Cognitive Science: Abstracts of Reports, Svetlogorsk, Russia, 18–21 October 2018; pp. 1328–1330.
37. Zinina, A.; Arinkin, N.; Zaydelman, L.; Kotov, A.A. The Role of Oriented Gestures during Robot's Communication to a Human. *Comput. Linguist. Intellect. Technol.* **2019**, *2019*, 800–808.
38. Kotov, A.; Zinina, A.; Arinkin, N.; Zaidelman, L. Experimental study of interaction between human and robot: Contribution of oriented gestures in communication. In Proceedings of the XVI European Congress of Psychology, Moscow, Russia, 2–5 July 2019; 2019; p. 1158.
39. Felzenszwalb, P.F.; Girshick, R.B.; McAllester, D.; Ramanan, D. Object Detection with Discriminatively Trained Part-Based Models. *IEEE Trans. Pattern Anal. Mach. Intell.* **2010**, *32*, 1627–1645. [CrossRef]
40. Kazemi, V.; Sullivan, J. One millisecond face alignment with an ensemble of regression trees. *IEEE Conf. Comput. Vis. Pattern Recognit.* **2014**, 1867–1874. [CrossRef]
41. Terzakis, G.; Lourakis, M. A Consistently Fast and Globally Optimal Solution to the Perspective-n-Point Problem. In *Computer Vision–ECCV 2020. ECCV 2020. Lecture Notes in Computer Science*; Vedaldi, A., Bischof, H., Brox, T., Frahm, J.M., Eds.; Springer: Cham, Switzerland, 2020; Volume 12346. [CrossRef]
42. Lyusin, D.V. A New Technique for Measuring Emotional Intelligence: The EmIn Questionnaire (in Russian). *Psychol. Diagn.* **2006**, *4*, 3–22.
43. Lyusin, D.V. EMIN Emotional Intelligence Questionnaire: New psychometric data (in Russian). In *Social and Emotional Intelligence: From Models to Measurements*; Institute of Psychology of the Russian Academy of Sciences: Moscow, Russia, 2009; pp. 264–278.
44. Iwasaki, M.; Zhou, J.; Ikeda, M.; Koike, Y.; Onishi, Y.; Kawamura, T.; Nakanishi, H. "That Robot Stared Back at Me!": Demonstrating Perceptual Ability Is Key to Successful Human–Robot Interactions. *Front. Robot. AI* **2019**, *6*, 85. [CrossRef]
45. Posner, M.I. (Ed.) *Cognitive Neuroscience of Attention*; The Guilford Press: New York, NY, USA, 2004.
46. Ochs, M.; Niewiadomski, R.; Pelachaud, C.; Sadek, D. Intelligent Expressions of Emotions. In *International Conference on Affective Computing and Intelligent Interaction*; Tao, J., Tan, T., Picard, R.W., Eds.; ACII 2005, LNCS 3784; Springer: Berlin/Heidelberg, Germany, 2005; pp. 707–714.
47. Kotov, A.A. A Computational Model of Consciousness for Artificial Emotional Agents. *Psychol. Russia: State Art* **2017**, *10*, 57–73. [CrossRef]
48. Velichkovsky, B.M.; Krotkova, O.A.; Kotov, A.A.; Orlov, V.A.; Verkhlyutov, V.M.; Ushakov, V.L.; Sharaev, M.G. Consciousness in a Multilevel Architecture: Evidence from the Right Side of the Brain. *Conscious. Cogn.* **2018**, *64*, 227–239. [CrossRef] [PubMed]
49. Velichkovsky, B.M.; Nedoluzhko, A.; Goldberg, E.; Efimova, O.; Sharko, F.; Rastorguev, S.; Krasivskaya, A.; Sharaev, M.; Korosteleva, A.; Ushakov, V. New Insights into the Human Brain's Cognitive Organization: Views from the Top, from the Bottom, from the Left and, particularly, from the Right. *Procedia Comput. Sci.* **2020**, *169*, 547–557. [CrossRef]
50. Howard, I.P.; Templeton, W.B. *Human Spatial Orientation*; Wiley: London, UK, 1966.
51. Harrison, D.W. *Brain Asymmetry and Neural Systems: Foundations in Clinical Neuroscience and Neuropsychology*; Springer: Bern, Switzerland, 2015.
52. Shammi, P.; Stuss, D.T. Humour Appreciation: A Role of the Right Frontal Lobe. *Brain* **1999**, *122*, 657–666. [CrossRef] [PubMed]
53. Schilbach, L.; Helmert, J.R.; Mojzisch, A.; Pannasch, S.; Velichkovsky, B.M.; Vogeley, K. Neural Correlates, Visual Attention and Facial Expression during Social Interaction with Virtual Others. In Proceedings of the 27th Annual Conference of Cognitive Science Society, Stresa, Italy, 21–23 July 2005; Bara, B.G., Barsalou, L., Bucciarelli, M., Eds.; Lawrence Erlbaum: Mahwah, NJ, USA, 2005; pp. 74–86.
54. Kaplan, J.A.; Brownell, H.H.; Jacobs, J.R.; Gardner, H. The Effects of Right Hemisphere Damage on the Pragmatic Interpretation of Conversational Remarks. *Brain Lang.* **1990**, *38*, 315–333. [CrossRef]
55. Jones, W.; Klin, A. Attention to eyes is present but in decline in 2-6-month-old infants later diagnosed with autism. *Nature* **2013**, *504*, 427–431. [CrossRef] [PubMed]
56. Baron-Cohen, S. *Mindblindness: An Essay on Autism and Theory of Mind*; MIT Press: Cambridge, MA, USA, 1995.
57. Riby, D.M.; Hancock, P.J.; Jones, N.; Hanley, M. Spontaneous and cued gaze-following in autism and Williams syndrome. *J. Neurodev. Disord.* **2013**, *5*, 13. [CrossRef] [PubMed]

Article

Manipulating Stress Responses during Spaceflight Training with Virtual Stressors

Tor Finseth [1], Michael C. Dorneich [2,*], Nir Keren [3], Warren D. Franke [4] and Stephen B. Vardeman [2,5]

1. Department of Aerospace Engineering, Iowa State University, Ames, IA 50011, USA; tfinseth@iastate.edu
2. Department of Industrial and Manufacturing Systems Engineering, Iowa State University, Ames, IA 50011, USA; vardeman@iastate.edu
3. Department of Agricultural and Biosystems Engineering, Iowa State University, Ames, IA 50011, USA; nir@iastate.edu
4. Department of Kinesiology, Iowa State University, Ames, IA 50011, USA; wfranke@iastate.edu
5. Department of Statistics, Iowa State University, Ames, IA 50011, USA
* Correspondence: dorneich@iastate.edu

Abstract: Virtual reality (VR) provides the ability to simulate stressors to replicated real-world situations. It allows for the creation and validation of training, therapy, and stress countermeasures in a safe and controlled setting. However, there is still much unknown about the cognitive appraisal of stressors and underlying elements. More research is needed on the creation of stressors and to verify that stress levels can be effectively manipulated by the virtual environment. The objective of this study was to investigate and validate different VR stressor levels from existing emergency spaceflight procedures. Experts in spaceflight procedures and the human stress response helped design a VR spaceflight environment and emergency fire task procedure. A within-subject experiment evaluated three stressor levels. Forty healthy participants each completed three trials (low, medium, high stressor levels) in VR to locate and extinguish a fire on the International Space Station (VR-ISS). Since stress is a complex construct, physiological data (heart rate, heart rate variability, blood pressure, electrodermal activity) and self-assessment (workload, stress, anxiety) were collected for each stressor level. The results suggest that the environmental-based stressors can induce significantly different, distinguishable levels of stress in individuals.

Keywords: virtual reality; virtual environment; stress; spaceflight; training

1. Introduction

For decades, astronauts have used virtual reality environments (VRE) to practice extra vehicular activity, mass handling, and robotic arm manipulation [1–3]. These VR training scenarios have focused on tasks external to the International Space Station (ISS), rather than tasks on the interior of the station. Full-scale interior mock-ups of the ISS modules and visiting vehicles (e.g., Soyuz capsule) are used to train emergency procedures for fire, depressurization, or contaminant leaks [4,5]. While these training models are effective for task acquisition, the simulation of stressors is constrained by the resources available and physical infrastructure. This constraint may be detrimental as training environments that lack the magnitude of stress felt in a real situation may inadequately prepare individuals for coping [6]. Inadequate training to handle specific stress in the operational environment can result in the diversion of cognitive resources for managing emotional states (e.g., fear, distress, and anxiety), leaving less resources available for task problem solving.

When an individual perceives a situation as stressful, stress will consistently influence physiology and human performance [7,8]. The general impact of exposure to acute high stress is the reduction of cognitive resources available for processing task relevant information as well as reduction in cognitive regulation [9,10]. A deficiency in cognitive resources can affect many performance attributes, including attention, decision time, and short-term memory [7]. All these performance attributes may be critical for conducting an emergency procedure and avoiding harm. Developing strategies to train individuals for stress can reduce the impact on performance.

The ability to train with stressors in VREs may provide a safe, reliable, and operationally relevant environment for training resilience in hazardous situations. Stressors are defined as stimuli that are appraised as a threat/challenge and elicit a stress response in humans [11]. Past research on stressful operations in VR has included aerial firefighting [12], municipal firefighters [13], first-responders [14], surgeons [15], and military [16]. Emergency operations can be trained for specific simulated tasks or stressors to strengthen transfer to the real world. Some researchers have found that coping skills developed in simulated stressful task environments may be retained and transferred to environments with novel stressors and novel tasks [17]. However, the high rate of individual variability in perception of stress and lack of understanding the underpinning of stressors, make challenges in manipulating human stress responses via exposure to different stressor levels. While some research exists on virtual reality training for spaceflight emergency operations [18–20], more research is needed on the stressors present in the emergency and how manipulating stressors in VEs can approximate the physiological and subjective responses expected during a stress response.

The appraisal of a stressor cannot occur without experiencing a stressful event [21]. Therefore, task training that integrates stress management or graduated stress exposure often opt to experimentally induce stressful events [11]. This is advantageous with VR as the technology allows for experiencing a hazardous environment in a safe and controlled setting while measurements can be collected before the stress induction is divulged to the participant. However, the integration of VR with stress training has been limited by empirical research. As stated by [22], "Most of the studies, in fact, have VR-based stress management training programs only in theory, without providing data about trials conducted to test the effectiveness of the proposed approaches". The authors of [23] found that challenges exist in the selection of effective technologies for delivering stress training, methodological rigor of the training pedagogy, and multi-dimensional assessment of stress. Therefore, more research is needed to verify the efficacy for VR stress training, especially before integration with NASA skill training.

The objective of this study was to assess the extent to which operationally relevant VR stressor levels (low, medium, high) derived from existing emergency spaceflight procedures could evoke a reliable stress response. A panel of experts in spaceflight procedures and human stress response identified relevant stressors and created an experimental emergency procedure. The stressors and procedure were then used to design the VR International Space Station (VR-ISS) spaceflight environment. Participants were trained with the procedures and conducted a VR-ISS fire response for three stressor levels. Self-assessments of stress, workload, and anxiety were administered for each level. Physiological stress measures were used to distinguish varied stress responses. Reliable methods to induce different levels of stress in trainees in a virtual environment would enable training in operationally relevant scenarios such as spaceflight emergency procedures.

2. Background on Manipulating Stress in VR

The stress response induced by a VRE is a product of the stressors employed by the simulation and the cognitive appraisal process. A stressor refers to a specific stimulus within the environment that exert perceived demands on an individual with the outcome. Examples of stressors include noise, pain, task load, ambient temperature, and time pressure [7]. Stimuli are perceived as stressors through the cognitive appraisal process, which evaluates the meaning and significance to individual wellbeing. The appraisal process has two different categories: Primary appraisal, which evaluates the relevance and congruence of a situation with regards to the individual's goals and wellbeing, and secondary appraisal, which involves the evaluation of resources and options for coping with the perceived demand [21]. While the primary and secondary appraisals may be shaped by personal and situational factors, some researchers have focused on communal stressors related to self-preservation that have a higher likelihood of being appraised as threats [24,25]. Further, these stressors related to self-preservation are thought to be composed of underlying "thematic" stressor elements which may play a pivotal role in how stimuli are cognitively apprised as a stressor.

Of the known stressor elements, much attention has been paid to uncontrollability and unpredictability [24,25]. Uncontrollability manifests with situations or consequences that are outside one's control, difficult to change, and impede progress from attaining one's desired goal [26]. Unpredictability refers to the absence of knowing the conditions under which an event will occur or what an event will be like [27]. The context of the situation (information about outcome, time, stimuli, etc.) and the individual interpretation of the stimuli determine the amount that uncontrollability and unpredictability experienced. Thus, the same stressor manipulated in different contexts may have different amounts of unpredictability and uncontrollability.

Aside from stressor elements, the appraisal process also needs simulations to be immersive enough to recognize a stimulus as an existential threat or challenge. One remarkable aspect of simulated VREs is that individuals tend to respond realistically to virtual simulations [28]. Furthermore, exposure to VR simulations results in a "response-as-if-real" even when the visual fidelity is low, and the representation of the physical reality is significantly reduced. Three concepts have been reported as strong generators of immersive simulations: place illusion, plausibility, and virtual body ownership. Place illusion is the strong sensation of being present in the space generated by the virtual reality environment, even though participants know they are not there [29–32]. Plausibility is the component of presence that is the illusion that the perceived events in the virtual environment are really happening [33]. In comparison to place illusion, which is a static characteristic, plausibility is more concerned with the dynamics of events and the situation portrayed. Lastly, virtual body ownership means utilization of multisensory correlations to provide people the illusion that foreign objects are part of their body [34]. For example, when individuals with virtual hands touch a simulated spider, results show that they find that the virtual hand increased fear [35].

Insight into how to induce stress can be gained from standardized stress tests which have been used by researchers to reliably manipulate stress levels precisely and consistently in a laboratory setting. Several stress tests include public speaking, mental arithmetic, and cold-pressor [36,37]. These stress tests aim to use the most potent stressors possible to elicit a general physiological stress response, often unrelated to the environment the individual is in. However, these lab-based stress manipulations do not involve any operationally-relevant tasks. Potential stressors used for task training should be stressors that could occur in the operational environment in which the individual will perform tasks [38]. These stressors could be related to the environment (e.g., distractions), the task (e.g., increased difficulty), or that state of the human (e.g., fatigue). To identify stressors that are relevant to the operational environment, where the environment is more dynamic and it is harder to anticipate stressors, some researchers have used expert opinion to identify prominent stressors and measured the manipulation using subjective stress ratings and physiological indices of stress [39,40]. However, more research is needed on stressors for operational tasks and how they can be manipulated in VR.

3. Environment Design

To inform how stressors in VR should be designed, a workshop was held with a panel of subject matter experts to identify stressors that contribute to the stress in a spaceflight environment. Subject matter experts have been used in previous research to identify environmental stressors (e.g., noise, radiation) in a disabled submarine scenario [41]. Attending experts included a retired U.S. NASA astronaut, a retired NASA ISS flight director, and three experts on psychological and physiological stress. The panel was presented with existing ISS emergency fire procedures and layouts of the ISS [42]. The panel identified a series of potential stressors, how they are mapped onto the emergency response procedures, and what effects the stressors might have on astronauts. Each stressor's intensity was rated on how the subject was likely to feel, and what the effects on performance will be. In addition, it was assumed that certain stressors would increase stress levels without influencing task-based workload, so the stressors were categorized into task-related and environment-related stressors. In other words, the stressors during an emergency fire were categorized by their ability to change the environment without substantially changing how the individual would perform the procedure.

A list of stressors was compiled by the panel (Table 1). These stressors are related to an emergency fire procedure, and therefore may exclude peripheral stressors previously identified by subject-matter experts that may be associate with deep-space missions [43]. The stressors were categorized by the type of stress induction, manipulation type (environment or task manipulation), within or between experimental comparison. The process of categorizing environmental stressors was similar to the method proposed by [41]. Stressor levels and potential deviations from the selected emergency procedure were also listed. Three environmental stressors were selected for use in the study and the stressor intensity was varied among training scenarios: alarms, flickering lights, and degraded visibility from smoke. Priority was given to stressor manipulations that did not affect tasks requirements, to avoid later confounds in experimental design. The goal was to increase stress while keeping the task requirements the same to allow direct comparisons of stress training with and without stressor manipulation. By using only environmental stressors and not task stressors, any changes from such a comparison would not be confounded by differences in task requirements. For example, intensity of noise and visible smoke may be stressful, but will not change the task procedure for locating and extinguishing a fire. The selected stressors were then used to establish three stressor levels of low, medium, and high during the VR training scenarios. A few of the task stressors (e.g., rising atmospheric contaminants) were included in all the training scenarios to distribute task load equally.

Table 1. Stressors Identified from ISS Emergency Fire Procedure.

Stressor	How to Manipulate	Type of Stress Induction	Scale	Manipulation Type (Task/Envir.)	Within/Between Training Session
Fire Alarm	Magnitude	Divided Attention	Linear	Envir.	Within
	Noise type	Divided Attention	Binary	Envir.	Between
Warning Alarm	Magnitude	Divided Attention	Linear	Envir.	Within
	Noise type	Divided Attention	Binary	Envir.	Between
Caution Alarm	Magnitude	Divided Attention	Linear	Envir.	Within
	Noise type	Divided Attention	Binary	Envir.	Between
Lights	Flashing	Divided Attention	Linear	Envir.	Within
Visibility (smoke obscurity)	Magnitude	Task Difficulty, Time pressure	Linear	Task, Envir.	Within
	Rate of change	Task Difficulty, Time pressure	Linear	Task, Envir.	Within

Table 1. Cont.

Stressor	How to Manipulate	Type of Stress Induction	Scale	Manipulation Type (Task/Envir.)	Within/Between Training Session
Multiple Sources (fire, smoke, electrical trips)	Multiple sources in module(s)	Concurrent Task Mgmt.	Binary	Task	Between
Location of smoke/fire	Source in Soyuz	Task Difficulty	Binary	Task	Between
	Source separating crew	Task Difficulty, Empathy, Difficult Tradeoff	Binary	Task	Between
	Source in module(s)	Task Difficulty	Binary	Task	Between
Open Flame	Magnitude	Visual Threat, Task Difficulty, Time pressure	Linear	Task	Within
	Spread rate	Visual Threat, Task Difficulty, Time pressure	Linear	Task	Within
Oxygen/Emer. Mask	Limited oxygen supply	Time Pressure	Linear	Task	Within
	No oxygen, faulty	Task Difficulty	Binary	Task	Between
	Reduced peripheral vision	Task Difficulty	Linear	Task	Between
	Clarity of other crew's voices	Task Difficulty	Linear	Task	Within
Contaminants	Magnitude	Threat, Task Difficulty	Linear	Task	Within
	Rate	Time Pressure	Linear	Task	Within
Team Member	Language, accent	Task Difficulty	Linear	Task	Between
	Experience of crew	Task Difficulty	Linear	Task	Between
	Location of crew	Task Difficulty	Linear	Task	Between
	Sleeping crew	Task Difficulty	Binary	Task	Between
	Reaction of crew	Task Difficulty	Linear	Task	Within
Comm Operation	Comm delay	Task Difficulty	Linear	Task	Between
	No Comm	Task Difficulty	Binary	Task	Between
	No lights (power outage)	Concurrent Task Mgmt.	Binary	Task	Within
Power Outage	Compromised life support	Concurrent Task Mgmt., Time pressure	Binary	Task	Between
MCC Communication	Frequency of communication	Concurrent Task Mgmt.	Linear	Task	Between
	Inconsistent instructions	Task Difficulty	Linear	Task	Between
Fire Extinguishers	Limited PFEs available	Task Difficulty	Linear	Task	Between
	Limited PFE uses	Task Difficulty	Linear	Task	Between

The panel then developed a simplified fire procedure for trainees to follow. A flowchart of the selected emergency procedure that could be used for laboratory experiments is illustrated in Figure 1. This procedure was modified from the existing ISS Emergency Procedures [42] by shortening the duration of the procedure down to 5–10 min, eliminating communication with Mission Control Center (MCC) and crewmember induced stressors, having the smoke alarm indicate the presence of a fire (as opposed to being a false alarm), and minimizing tangential procedure steps that do not directly help the trainee locate the fire source. This procedure is also like the ISS fire strategy created by [18] for the purpose of task training. Further, this procedure was designed to have limited decision nodes to avoid task branches and maintain experimental replicability.

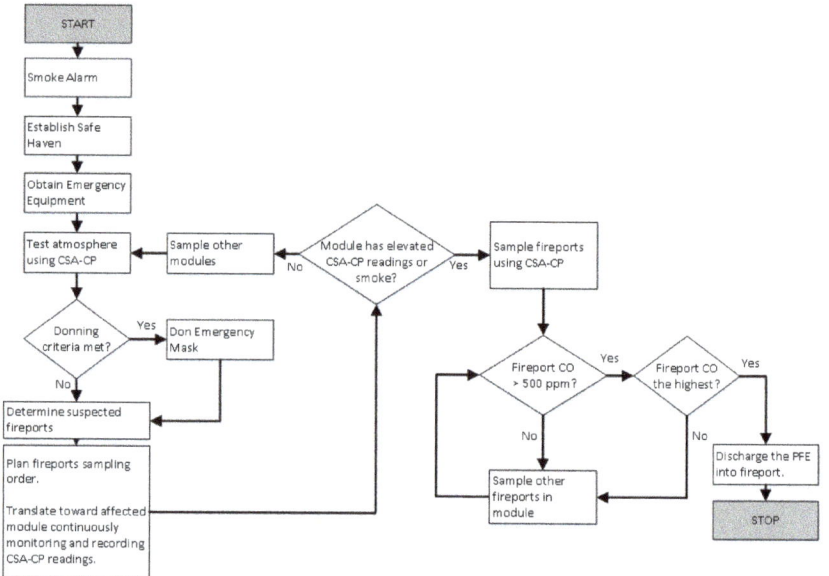

Figure 1. VR-ISS emergency fire procedure steps created by the workshop, modified from existing NASA ISS Emergency Fire Procedures.

This procedure begins with an auditory smoke alarm alerting crew to the presence of a fire. A safe haven is to be established in a location closest to the point of egress (e.g., Soyuz capsule) where crewmembers can formulate a response plan and assign duties. Emergency equipment vital to the fire response is obtained from storage compartments. Readings of contaminant levels are reviewed with a NASA Compound Specific Analyzer–Combustion Products (CSA-CP). The purpose of the CSA-CP onboard the ISS is to determine the level of atmospheric contaminants, including carbon monoxide (CO) in parts per million. If CO levels are rising, crewmembers should don oxygen masks. Based on CSA-CP readings, suspected fireports should be determined and crewmembers should translate from module to module, then narrow the search to sample local fireports as contaminant levels increase with proximity. If the sampled fireport CO > 500 and increasing, then discharge the portable fire extinguisher (PFE).

To integrate this procedure into VR, the stressors and procedure were then used to design the VR-ISS spaceflight environment. The VR-ISS environment is based on 3D models from [44] and [19] but has been largely modified from its former state to be used with VR head mounted display (HMD) and facilitate spaceflight procedure training. To simplify training participants, the VR-ISS consisted only of three of the existing U.S. Orbital Segment modules, Figure 2 illustrates the VR-ISS configuration used in the experiment: only Node 1, US Lab, and Node 2 were used.

An avatar was used to increase virtual body ownership, illustrated in Figure 3. The avatar's hands tracked VR hand controllers. Zero gravity mocking locomotion is integrated into the simulation to allow participants to float across the VE by grabbing and pushing against objects/surfaces.

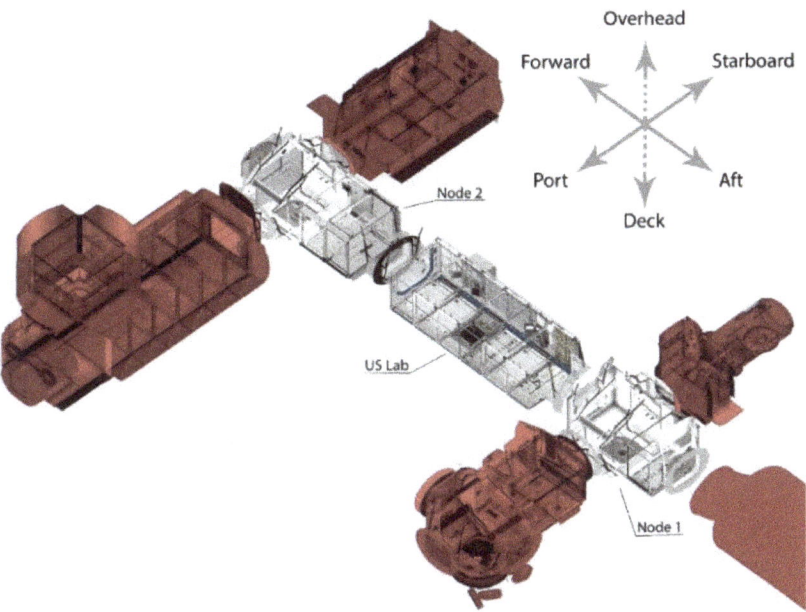

Figure 2. VR-ISS configuration. Some sections of the ISS U.S. and Russian segment were not included in the simulation reproduced with permission from [45], IEEE, 2020.

Figure 3. Third person view of participant's avatar in the VR-ISS.

Several dynamic interactions were included in the VR-ISS to aid detecting and locating the source of a fire. These dynamic functions were designed to enhance the immersion through increased place illusion and plausibility. Atmospheric contaminant levels rose as a function of time and distance from the virtual fire source. Since, the readings change as a function of time, the participant must not only remember previous readings at different locations to compare with the current reading, but also account of the fact that readings become less reliable the further back in time it was taken. Virtual smoke changed in density as a function of time and spread in a uniform pattern, consistent with expected smoke behavior in a microgravity; therefore, participants could not rely solely on visual smoke patterns to detect the location of the source.

Emergency equipment was also simulated for plausibility and place illusion. Readings of contaminant levels could be collected with a simulated NASA CSA-CP. The purpose of the CSA-CP onboard the ISS is to determine the level of atmospheric contaminants. Virtual CSA-CP displayed levels of oxygen, carbon monoxide (CO), hydrogen chloride (HCl), and hydrogen cyanide (HCN) in parts per million (Figure 4). Using voice commands, a floating CSA-CP appeared in front of the participant with the contaminant concentration values visible. The window disappeared after three seconds. Participants are expected to identify the location of the source by following the invisible path mentally established from recalling highest levels of contaminants in each VR-ISS module. A recording at the start of the simulation gave instructions to retrieve a Portable Fire Extinguisher (PFE) and Portable Breathing Apparatus (PBA) when the contaminate levels are excessive (Figure 4). The PFE is used to extinguish a fire source behind a rack fireport. The PFE has the capability for two uses before the canister is empty. Five PFEs are available in cabinets in the VR-ISS. PBAs are available in the same cabinets and can be done on the participant avatar's head. A Caution and Warning (C&W) panel displayed flashing lights to alert participants to a potential fire (Figure 4). Once the participants identified where the fire source was located, they began sampling fireports within the module to locate the "rack" that caused the fire. The VR-ISS included approximately 150 fireport labels, accurately placed on the racks throughout the ISS.

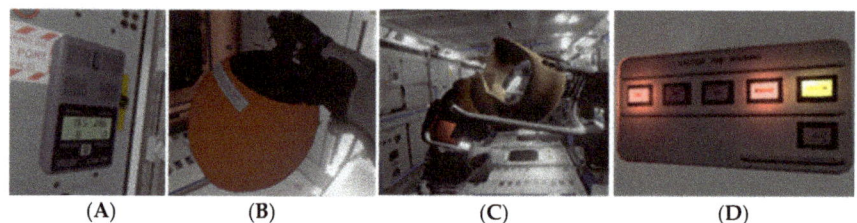

Figure 4. VR-ISS emergency fire equipment, (**A**) Compound Specific Analyzer–Combustion Products next to a fireport label, (**B**) Portable Fire Extinguisher, (**C**) Portable Breathing Apparatus, and (**D**) Caution & Warning Panel reproduced with permission from [45], IEEE, 2020.

4. Materials and Methods

4.1. Participants

Forty subjects participated (Male = 37, Female = 3). The study was reviewed and approved by Iowa State University Institutional Review Board. The participant mean age was 20.5 years (SD = 2.6). Thirty-eight of the participants in this study had a STEM background.

4.2. Task Environment

VR-ISS is the operational environment for this experiment. Participants were tasked with locating and extinguishing the location of a potential fire on the VR-ISS. Several dynamic interactions were included in the VR-ISS to aid detecting and locating the source of a fire. Atmospheric contaminant levels rose as a function of time and distance from the virtual fire source. Virtual smoke changed in density as a function of time and spread in a uniform pattern, consistent with expected smoke behavior in a microgravity; therefore, participants could not rely solely on visual smoke patterns to detect the location of the source.

Participants reviewed readings of contaminant levels with CSA-CP. When participants identified the fireport which had the highest level of contaminant and extinguished the fire with the PFE, the contaminate levels were reset and a new randomized fire source was created. The task ended five minutes after the beginning of the simulation.

4.3. Independent Variables

The VE was designed with three environmental stressors identified from the panel SMEs: alarms, flickering lights, and degraded visibility from smoke. The smoke filled the modules rapidly until the intensity level was achieved, but also moved dynamically through the module to increase plausibility. The lights for selected modules flickered randomly to simulate temporary power failure, presenting the participant with temporary near-complete darkness. A fire alarm and caution alarm were used for different levels.

The simulation had three different stress levels, each with a fire location randomized (Figure 5). The low stress level indicated a fire using increased CSA-CP contaminate values and C&W panel lights. The medium stress level indicated a fire using increased CSA-CP contaminate values, C&W panel lights, a continuous Caution alarm, and low levels of smoke (visibility of 6 ft). The high stress level indicated a fire using increased CSA-CP contaminate values, C&W panel lights, a continuous Caution alarm, a continuous Fire alarm, flickering lights, and high levels of smoke spread over time (visibility of 1 ft).

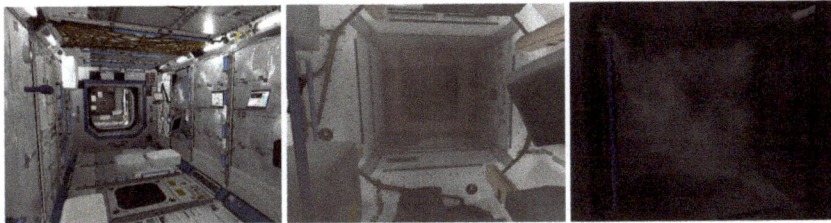

Figure 5. VR-ISS emergency fire (low, medium, high stressor scenarios). Reproduced with permission from [45], IEEE, 2020.

4.4. Dependent Variables

The study used both physiological and subjective indices of stress. The dependent variables are summarized in Table 2.

Table 2. Description of dependent variable sensor/metrics and frequencies.

Dependent Variable	Sensor/Metric	Features/Type	Description, Association	Frequency
Physiological stress response	Electrocardiogram (ECG) and Heart Rate Variability (HRV)	HR	Heart rate	Throughout trial
		RMSSD	The root mean of the sum of the squares of differences between beat intervals	
		pNN50	Proportion of the successive normal to normal beat intervals that differ more than 50 ms	
Physiological stress response	Continuous Non-Invasive Blood Pressure	SBP	Systolic Blood Pressure	Throughout trial
		DBP	Diastolic Blood Pressure	
Physiological stress response	Electrodermal Activity (EDA)	EDA	Tonic component (0–0.16 Hz)	Throughout trial
Subjective Stress	Post Stress Task Reaction Measure (PTSR)	9-point Likert scale		After trial
Subjective Stress	Free Stress Comparison	100-point scale		After experiment
Subjective Stress	Short State Stress Questionnaire (SSSQ)	24 items, 5-point Likert scale	Engagement, Distress, Worry	Before, after trial
Workload	NASA TLX	6 items, 21-point scale	Mental demand, Physical demand, Temporal demand, Performance, Effort, and Frustration	After trial
Anxiety	State-trait anxiety inventory (STAI)	20 items, 7-point Likert scale		After trial

4.4.1. Heart Rate and Heart Rate Variability

Heart rate increases with higher physiological stress response. Heart rate data were collected via electrocardiogram (ECG, modified CS5 lead configuration). The ECG was sampled at 2048 Hz using Biopac MP150 hardware and recorded using AcqKnowledge software (Version 3.8.2, Biopac Systems Inc.). Heart Rate is influenced by both sympathetic and parasympathetic branches of the autonomic nervous system [46]. Activity in these branches is indicative of the body's state of excitation vs. relaxation (vagal response).

Time domain analysis of the ECG was performed to quantify changes in Heart Rate Variability (HRV). The vagal response, which is indicative of relaxation, was assessed with two HRV metrics relating to the amount of variance in the inter-beat-interval through the root-mean-square difference of successive normal R-wave intervals (RMSSD) and the proportion of the number of pairs of successive normal R-waves that differ by more than 50 ms (pNN50). Increased pNN50 and RMSSD indicated a more relaxed state and greater ability to cope with increased stress. More specifically, both RMSSD and pNN50 represent vagal control within the time domain and are correlated to high frequency power [46]. The HRV time domain for each participant were calculated in 30-s intervals over the duration of each session.

4.4.2. Blood Pressure

Systolic blood pressure (SBP) and diastolic blood pressure (DBP) were collected as another measure of cardiovascular reactivity. Both SBP and DBP have been shown to increase in reaction to increased stress [47]. A finger cuff was placed on the participants' non-dominant hand over the middle phalanx of either the long or ring finger (CNAP Monitor 500, CNSystems Medizintechnik AG). The non-dominant arm was placed in an arm-sling to standardize the position of the hand relative to the heart between all participants. Data were recorded at 1024 Hz. To calibrate the finger cuff, an oscillometric non-invasive blood pressure cuff was placed on the participant's non-dominant upper arm. The CNAP blood pressure readings were calibrated before the start of each trial. The participant remained seated for the duration of the experiment to prevent orthostatic pressure changes.

4.4.3. Electrodermal Activity

The electrodermal activity (EDA) signal increases under higher acute stress [48]. EDA is sympathetically mediated, therefore can be used to verify autonomic influences on heart rate. The EDA signal was corrected with an IIR low pass 2nd-order Butterworth filter fixed at 5 Hz. The EDA tonic component, also called skin conductance level, was extracted by low pass filtering with a cut-off frequency of 0.16 Hz and used as an indicator of sympathetic activity [49]. Data were recorded in 30 s epochs for statistical analysis.

4.4.4. Subjective Stress

Two kinds of ratings were used to assess the perception of low, medium, and high stress: Post Task Stress Reaction (PTSR) [50], Free Stress Comparison of events. The PTSR and Free Stress Comparison questionnaires were modified from previous uses [40,47,51]. The PTSR asks participants to rate the ground truth simulations on a scale of "1" to "9" where a rating of "1" was used to represent experiencing "no stress", "5" was used to represent "medium stress", and "9" was used to represent experiencing "high stress". The PTSR is intended to measure the immediate retrospective stress after completing a trial.

In addition, the Free Stress Comparison has participants rate the relative stress level on a scale of 0 to 100 (least to most stressful) in comparison to other simulations. The stress appraisal process is continuous and relative, with reappraisals of the experience happening long after the stressful exposure [52]. The Free Stress Comparison was intended to measure the retrospective reappraisal after completing all the simulations, then relatively comparing their stressfulness.

4.4.5. Subjective Stress State

The Short Stress State Questionnaire (SSSQ) [53] assessed the subjective states pre- and post-trial to measure three state factors: task engagement, distress, and worry. Engagement refers to qualities of energetic arousal, motivation, and concentration. Distress is defined as feelings of tense arousal, hedonic tone, and confidence-control. Worry relates to self-focus, self-esteem, and cognitive interference [54]. The stress state acts as a mediator between the stressor and cognition or information processing, whereby the three aspects represent components of conscious experience during person–task–environment transactions [55].

4.4.6. Workload

The NASA Taskload Index (TLX) [56] was used to assess the subjective workload during exposure. The NASA TLX measures six dimensions of workload: mental demand, physical demand, temporal demand, performance, effort, and frustration level. NASA TLX was administered after the completion of a trial. Participant scores on the six numerical rating scales were computed in the 0 to 100 range and as an unweighted participant mean for each of the six-dimensional subscales [57].

4.4.7. Anxiety

To measure the anxiety during the experiment, State-Trait Anxiety Inventory (STAI) [58] Form Y-1 was given after the completion of a trial. The STAI-Y1 is a 20-item self-report scale for the assessment of state anxiety in adults. Based on the answers, the STAI score can be interpreted in the ranges: no stress <30, low <40, medium 40–55, high >55 [59].

4.5. Experiment Design

A within-subject 1 × 3 (trial) experiment was conducted. Each participant completed the same task of locating an onboard fire, but each trial had one of three different stressor levels (low, medium, and high). The order of stressor levels was assigned via Latin square to counterbalance and minimize the effect of training order's influence from differences among the treatment effects (i.e., each levels stressors) [60].

4.6. Hypothesis

It was hypothesized that manipulating different levels of VR environmental stressors (low, medium, and high) during an operational task would induce different levels of stress, with higher stressor levels resulting in greater increases in HR, decreases in RMSSD, decreases in pNN50, increases in SBP and DBP, increases in EDA, increases in PTSR, increase in Free Stress Comparison, no change in engagement, increases in distress, no change in worry, no change in workload, and increases in anxiety.

4.7. Procedure

The experiment was completed in a single laboratory visit, lasting approximately 60 min. At the beginning of the experiment, participants completed a series of pre-trial questionnaires, including demographic questions, a SSSQ to measure the stress in response immediately before the trials, and training on how to use the NASA TLX. To acclimate to VR before the data collection tasks, participants were trained on navigating, operating, and controlling the VR simulation (e.g., head-mounted display, hand controls, "play-area" boundaries represented but a visual blue-grid). Participants were asked to report cybersickness and were withdrawn if symptoms persisted.

For the VR-ISS, participants completed a VR interactive tutorial that included information about the ISS layout, how to navigate, fire equipment, and the appropriate emergency fire response. Participants practiced the procedure in the tutorial until memorized.

Participants then completed three trials: low, medium, and high stressor levels. After each trial, participants completed several questionnaires, including the post-trial SSSQ, NASA TLX, STAI, and PTSR. Participants were given 5–10 min between trials to complete questionnaires. At the end of the experiment, participants completed the Free Stress Comparison to compare the trials in retrospect.

4.8. Experiment Materials

The apparatus consisted of two parts: an HTC VIVE (professional version; HTC, 2016) consumer VR headset. The Unity (5.4.0f3, Unity Technologies, 2014) 3D game engine was used to facilitate all aspects of the VR-ISS as a virtual environment. The HTC VIVE setup consists of the HMD and two Lighthouse sensors that are responsible for tracking the headset position and orientation. For this experiment, the lighthouse sensors were positioned facing each other at opposite ends of our lab space, 8 ft high with 12×12 ft detectable play area.

4.9. Data Analysis

Data analysis was performed using SPSS software (Version 28.0; IBM Corp.). Distributions were tested for normality using skewness and kurtosis divided by the standard error and concluded to be normal if less than 1.96 [61]. Each dependent variable was visually inspected for errors (e.g., signal artifacts, miscalibration, electrode disconnect) and participant trials were omitted if they displayed erratic patterns. Alterations were made to discontinue some questionnaires during the experiment, resulting in analysis on only a subset of the sample. Repeated measure analysis of variance (RM-ANOVA) was used to calculate the fixed effect of stressor level. Physiological data were standardized against the first 90-s of the low stressor trial to emulate change from baseline, to which the 90-s were then omitted from the analysis. The first 90-s was also omitted from the medium and high stressor trials to account for the physiological transition in response to a stressor. To account for correlation of physiological data, autoregressive (AR1) models were used for the covariance matrix with a participant random effect. AR1 model fit was assessed with Akaike's Information Criteria (AIC). Significant differences were located using pairwise comparisons, and acceptance level was adjusted to control for type I errors (Bonferroni adjustment). Results were considered significant for $p \leq 0.05$ and marginally significant for $p < 0.10$ [62]. The effect sizes, given in partial eta squared for the models, were transformed, and reported as Cohen's d effect size under the conservative assumption the stress level means are separated with maximum variability [63]. Cohen's d was used for assessing effect size, where $0.2 < |d| < 0.5$ is considered small effect size, medium effect size when $0.5 < |d| < 0.8$, and large effect size for $|d| > 0.8$ [63].

The three-factor SSSQ scale scores for pre- and post-trial were calculated for each participant. The factor scores from both pre- and post-trial are standardized against normative means and standard deviation values from a large sample of British participants [64] and standardized using methods in [55]. Change scores were calculated for each factor using the z-score Formula (1) which has been used in previous studies [55]. The z-score then represents the change between pre- and post-trial in units of the deviation from the population mean.

$$z = (\text{standardized post-score} - \text{standardized pre-score}). \quad (1)$$

5. Results

A summary of the results is provided in Table 3, with greater description of the results in the following subsections.

5.1. Heart Rate and Heart Rate Variability

The main effect of stressor level on the change in heart rate ($N = 35$) was significant, $F(2, 63.7) = 4.34$, $p = 0.017$, with a large effect size, $d = 0.90$ (Figure 6a). Pairwise comparison indicated the change in heart rate was significantly higher ($p = 0.017$, $d = 0.58$) for participants in high stressor ($M = 1.41$, $SD = 3.05$) compared to low stressor ($M = 0.37$, $SD = 1.42$), but not significantly different ($p = 0.16$) for high stressor compared to medium stressor

(M = 0.58, SD = 1.59) and not significantly higher (p = 0.99) the for medium stressor compared to low stressor.

Table 3. Summary of Results.

Metric	N	Low M (SD)	Medium M (SD)	High M (SD)	Main Effect F (p)	Low vs. Medium p (d)	Low vs. High p (d)	Medium vs. High p (d)
Physiological Measures								
ΔHR	35	0.37 (1.42)	0.58 (1.59)	1.41 (3.05)	4.34 (0.017)	0.99	0.017 (0.58)	0.16
ΔRMSSD	35	0.63 (3.07)	0.48 (2.78)	−0.25 (2.44)	3.29 (0.04)	0.99	0.048 (0.37)	0.18
ΔpNN50	35	−0.02 (1.42)	0.13 (1.81)	−0.076 (1.51)	0.079 (0.92)	NA	NA	NA
ΔSBP	32	0.95 (2.44)	1.94 (4.11)	2.04 (6.09)	2.84 (0.066)	0.26	0.08 (0.43)	0.99
ΔDBP	32	1.09 (2.91)	2.05 (6.20)	0.82 (3.70)	0.69 (0.51)	NA	NA	NA
ΔEDA	32	−0.82 (2.45)	1.94 (4.42)	1.29 (3.97)	7.05 (0.002)	0.002 (0.71)	0.046 (0.42)	0.71
Subjective Measures								
PSTR	39	4.67 (1.61)	5.26 (1.6)	6.34 (1.68)	27.9 (<0.001)	.11	<0.001 (1.0)	<0.001 (0.66)
Free Stress	39	30.4 (20.5)	45.9 (17.6)	78.7 (18.8)	80.6 (<0.001)	<0.001 (0.81)	<0.001 (2.46)	<0.001 (1.8)
ΔEngage	23	−0.024 (0.51)	0.089 (0.61)	0.012 (0.65)	0.47 (0.63)	NA	NA	NA
ΔDistress	23	0.5 (0.79)	0.39 (0.59)	0.82 (0.94)	7.05 (0.002)	0.73	0.072 (0.37)	0.01 (0.55)
ΔWorry	23	−0.28 (0.73)	−0.38 (0.59)	−0.22 (0.8)	3.13 (0.054)	0.3	0.99	0.096 (0.23)
Workload	39	45.1 (17.8)	46.9 (17.4)	57.5 (17.6)	12.5 (<0.001)	0.99	0.002 (0.7)	<0.001 (0.61)
Anxiety	23	37.5 (9.76)	36.2 (9.58)	40.6 (10.7)	5.25 (0.009)	0.92	0.24	0.005 (0.43)

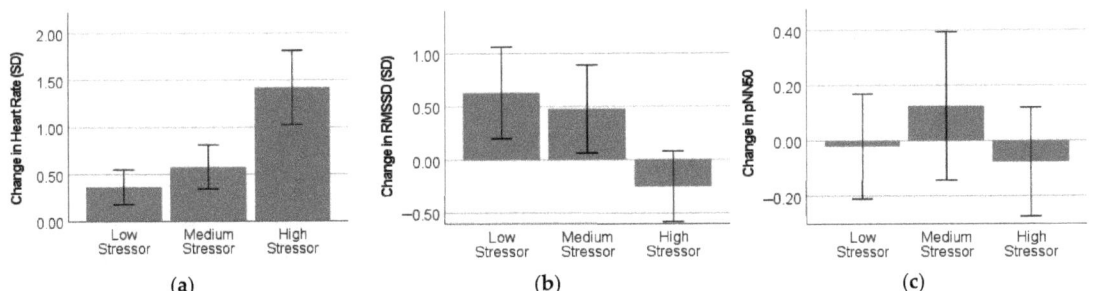

Figure 6. Heart rate metrics: (a) Heart Rate; (b) RMSSD; (c) pNN50. Error bars representing 95% confidence intervals.

The main effect of stressor level on the change in RMSSD (N = 35) was significant, $F(2,170)$ = 3.29, p = 0.04, with a medium effect size, d = 0.48 (Figure 6b). Pairwise comparison indicated the change in RMSSD was significantly lower (p = 0.048, d = 0.37) for participants in high stressor (M = −0.25, SD = 2.44) compared to low stressor (M = 0.63 SD = 3.07), but not significantly different (p = 0.18) for high stressor compared to medium stressor (M = 0.48, SD = 2.78) and not significantly higher (p = 0.99) the for medium stressor compared to low stressor.

The main effect of stressor level on the change in pNN50 ($N = 35$) was not significant, $F(2,185) = 0.079$, $p = 0.92$ (Figure 6c) for the low stressor ($M = -0.02$, $SD = 1.42$), medium stressor ($M = 0.13$, $SD = 1.81$), or high stressor ($M = -0.076$, $SD = 1.51$).

5.2. Blood Pressure

The main effect of stressor level on the change in SBP ($N = 32$) was marginally significant, $F(2,60.5) = 2.84$, $p = 0.066$, with a medium effect size, $d = 0.61$ (Figure 7a). Pairwise comparison indicated the change in blood pressure was marginally higher ($p = 0.08$, $d = 0.43$) for participants in high stressor ($M = 2.04$, $SD = 6.09$) compared to low stressor ($M = 0.95$, $SD = 2.44$), but not significantly higher ($p = 0.99$) for high stressor compared to medium stressor ($M = 1.94$, $SD = 4.11$) and not significantly higher ($p = 0.26$) for medium stressor compared to low stressor.

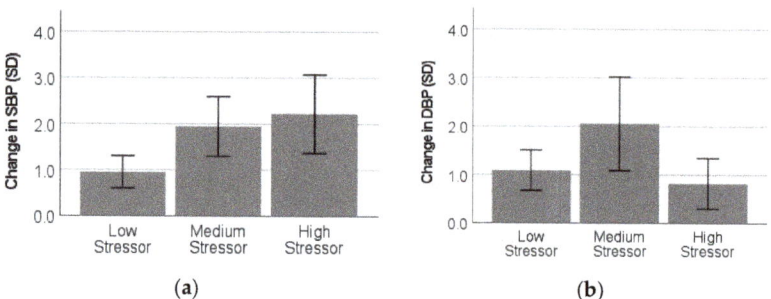

Figure 7. Blood pressure metrics: (**a**) systolic blood pressure (SBP); (**b**) diastolic blood pressure (DBP). Error bars representing 95% confidence intervals.

The main effect of stressor level on the change in DBP ($N = 32$) was not significant, $F(2,70.4) = 0.69$, $p = 0.51$ (Figure 7b) for the low stressor ($M = 1.09$, $SD = 2.91$), medium stressor ($M = 2.05$, $SD = 6.20$), or high stressor ($M = 0.82$, $SD = 3.70$).

5.3. EDA Tonic

The main effect of stressor level on the change in EDA tonic ($N = 32$) was significant, $F(2,64.6) = 7.05$, $p = 0.002$, with a large effect size, $d = 0.93$ (Figure 8). Pairwise comparison indicated the change in EDA was significantly higher ($p = 0.046$, $d = 0.42$) for participants in high stressor ($M = 1.29$, $SD = 3.97$) compared to low stressor ($M = -0.82$, $SD = 2.45$), and significantly higher ($p = 0.002$, $d = 0.71$) for the medium stressor (M = 1.94, SD = 4.42) compared to low stressor, but not significantly different ($p = 0.71$) for high stressor compared to medium stressor.

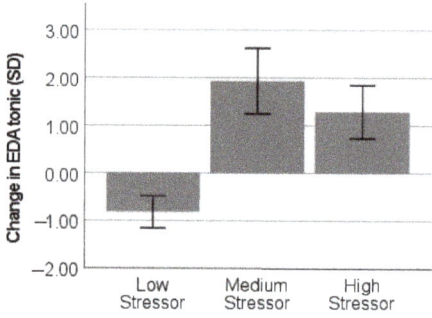

Figure 8. Electrodermal Activity tonic component. Error bars representing 95% confidence intervals.

5.4. Subjective Stress

The main effect of stressor level on subjective stress measured by PTSR ($N = 39$) was significant, $F(2,76) = 27.9$, $p < 0.001$, $d = 1.71$ (Figure 9a). Pairwise comparison indicated the subjective stress was significantly higher ($p < 0.001$, $d = 1.0$) for participants in high stressor ($M = 6.34$, $SD = 1.68$) compared to low stressor ($M = 4.67$, $SD = 1.61$), significantly higher ($p < 0.001$, $d = 0.66$) for high stressor compared to medium stressor ($M = 5.26$, $SD = 1.6$), but not significantly different ($p = 0.11$) for medium stressor compared to low stressor.

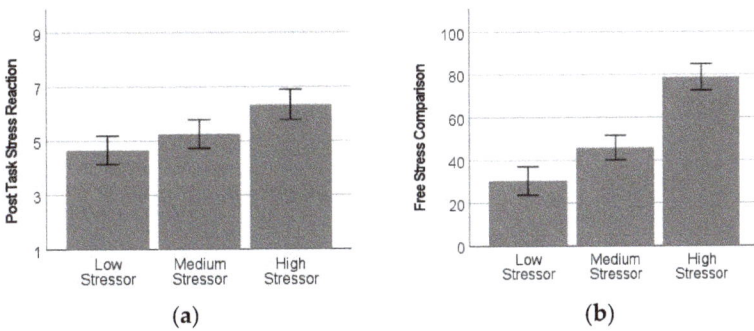

Figure 9. Subjective stress measures: (**a**) Post Task Stress Reaction (PTSR); (**b**) Free Stress Comparison. Error bars representing 95% confidence intervals.

The main effect of stressor level on subjective stress measured by Free Stress Comparison ($N = 39$) was significant, $F(2,76) = 80.6$, $p < 0.001$, $d = 2.91$ (Figure 9b). Pairwise comparison indicated the subjective stress was significantly higher ($p < 0.001$, $d = 2.46$) for participants in high stressor ($M = 78.7$, $SD = 18.8$) compared to low stressor ($M = 30.4$, $SD = 20.5$), significantly higher ($p < 0.001$, $d = 1.8$) for high stressor compared to medium stressor ($M = 45.9$, $SD = 17.6$); and significantly higher ($p < 0.001$, $d = 0.81$) for medium stressor compared to low stressor.

5.5. SSSQ: Stress State

The main effect of stressor level on the change in engagement ($N = 23$) was not significant, $F(2,44) = 0.47$, $p = 0.63$ (Figure 10a) for the low stressor ($M = -0.024$, $SD = 0.51$), medium stressor ($M = 0.089$, $SD = 0.61$), or high stressor ($M = 0.012$, $SD = 0.65$).

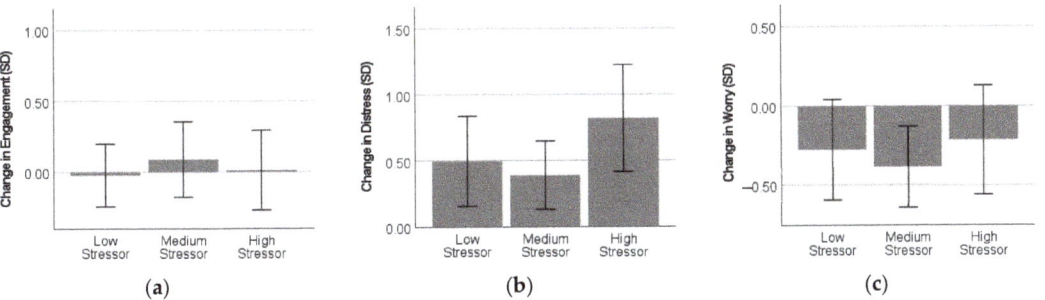

Figure 10. The change in stress state measured by the SSSQ: (**a**) Engagement; (**b**) Distress; (**c**) Worry. Error bars representing 95% confidence intervals.

The main effect of stressor level on the change in distress ($N = 23$) was significant, $F(2,44) = 7.05$, $p = 0.002$, with a large effect size, $d = 1.13$ (Figure 10b). Pairwise comparison indicated the change in distress was marginally significantly higher ($p = 0.072$, $d = 0.37$) for participants in high stressor ($M = 0.82$, $SD = 0.94$) compared to low stressor ($M = 0.50$,

$SD = 0.79$), significantly higher ($p = 0.01$, $d = 0.55$) for high stressor compared to medium stressor ($M = 0.39$, $SD = 0.59$), but not significantly different ($p = 0.73$) for medium stressor compared to low stressor.

The main effect of stressor level on the change in worry ($N = 23$) was marginally significant, $F(2,44) = 3.13$, $p = 0.054$, with a medium effect size, $d = 0.76$ (Figure 10c). Pairwise comparison indicated the change in worry was marginally significantly higher ($p = 0.096$, $d = 0.23$) for participants in medium stressor ($M = -0.38$, $SD = 0.59$) compared to high stressor ($M = -0.22$, $SD = 0.8$), but not significantly different ($p = 0.30$) for medium stressor compared to low stressor ($M = -0.28$, $SD = 0.73$), and not significantly different ($p = 0.99$) for the high stressor compared to low stressor.

5.6. NASA-TLX: Workload

The main effect of stressor level on workload ($N = 39$) was significant, $F(1.63,62.1) = 12.5$, $p < 0.001$, with a large effect size, $d = 1.15$ (Figure 11). Pairwise comparison indicated the workload was significantly higher ($p = 0.002$, $d = 0.7$) for participants in high stressor ($M = 57.5$, $SD = 17.6$) compared to low stressor ($M = 45.1$, $SD = 17.8$), significantly higher ($p < 0.001$, $d = 0.61$) for high stressor compared to medium stressor ($M = 46.9$, $SD = 17.4$), but not significantly ($p = 0.99$) different for medium stressor compared to low stressor.

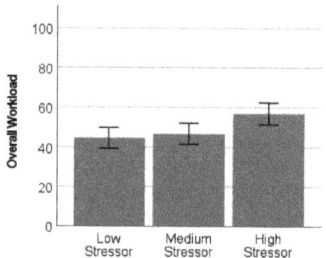

Figure 11. Overall workload for different levels of stressors obtained by NASA Task Load Index (TLX). Error bars representing 95% confidence intervals.

Within the NASA-TLX subscales, mental workload was significantly different, $F(1.7,66.2) = 4.28$, $p = 0.022$, $d = 0.67$, with the high stressor ($M = 55.6$, $SD = 21.9$) being marginally significantly higher than the low stressor ($M = 45.8$, $SD = 23.1$; $p = 0.064$, $d = 0.44$). Physical workload was significantly different, $F(2,90) = 7.78$, $p = 0.001$, with the high stressor ($M = 45.5$, $SD = 26.6$) being significantly higher than the low stressor ($M = 34.7$, $SD = 26.2$; $p = 0.01$, $d = 0.41$). Temporal workload was significantly different, $F(2,76) = 14.1$, $p < 0.001$, with the high stressor ($M = 68.2$, $SD = 26.6$) being significantly higher than the low stressor ($M = 47.8$, $SD = 23.8$; $p < 0.001$, $d = 0.81$), high stressor being significantly higher than the medium stressor ($M = 58.4$, $SD = 25.3$; $p = 0.011$, $d = 0.38$), and medium stressor being significantly higher than the low stressor ($p = 0.036$, $d = 0.43$). Performance was significantly different, $F(1.8,68.6) = 5.22$, $p = 0.01$, $d = 0.74$, with the high stressor ($M = 63.4$, $SD = 32.8$) being significantly higher than the medium stressor ($M = 43.7$, $SD = 32.9$; $p = 0.008$, $d = 0.6$). Effort was significantly different, $F(1.63,62.1) = 5.1$, $p = 0.013$, with the high stressor ($M = 67.8$, $SD = 22.8$) being significantly higher than the medium stressor ($M = 59.4$, $SD = 23.1$; $p = 0.001$, $d = 0.37$), and the high stressor being marginally significantly higher than the low stressor ($M = 59$, $SD = 21.4$; $p = 0.053$, $d = 0.40$). Frustration was significantly different, $F(1.77,67.3) = 6.53$, $p = 0.004$, with the high stressor ($M = 45.3$, $SD = 31.2$) being significantly higher than the medium stressor ($M = 32.4$, $SD = 28$; $p = 0.012$, $d = 0.43$) and low stressor ($M = 34.4$, $SD = 27.5$; $p = 0.043$, $d = 0.37$).

5.7. STAI: Anxiety

The main effect of stressor level on post-trial anxiety ($N = 23$) was significant, $F(2,44) = 5.25$, $p = 0.009$, with a large effect size, $d = 0.98$ (Figure 12). Pairwise comparison indicated the anxiety was significantly higher ($p = 0.005$, $d = 0.43$) for participants in high stressor ($M = 40.6$,

$SD = 10.7$) compared to medium stressor ($M = 36.2$, $SD = 9.58$), but not significantly different ($p = 0.92$) for the medium stressor compared to the low stressor ($M = 37.5$, $SD = 9.76$) or ($p = 0.24$) for the high stressor compared to the low stressor.

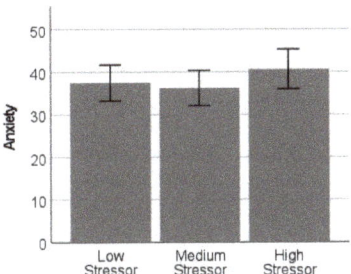

Figure 12. Anxiety obtained by the State Trait Anxiety Inventory (STAI). Error bars representing 95% confidence intervals.

6. Discussion

A spaceflight emergency procedure was conducted in VR-ISS with the aim at evaluating VR stressors. The hypothesis that manipulating different levels of VR environmental stressors during an operational task can induce different levels of stress was partially supported. The results demonstrated that levels of indicators of physiological stress (HR, RMSSD, EDA, SBP), subjective stress (PTSR, Free Stress Comparison), and ratings of distress, workload, and anxiety were significantly different for those training simulations. Some measures did not find significant differences, including pNN50 and DBP. Further, most measures showed differences between the low stressor and high stressors but could not discriminate the medium stressor level.

Several of the physiological stress indicators were found to be different for the stressor levels. The HR was different between the low and high stressor levels and similarly different for the RMSSD. The decrease in RMSSD for the high stress indicates parasympathetic inhibition and reflects an inability to relax. Further, there was an increase in EDA which indicates sympathetic activation, which is the likely cause for the increase in HR and SBP. Altogether, this suggests the high stressor simulation was effective at inducing a robust physiological stress response.

In contrast, the low and medium stressor levels were difficult to distinguish from each other and had varied physiological responses. The RMSSD remained elevated and the HR remained depressed for both levels. However, the medium level did have an increase in EDA, suggesting there was some sympathetic activation concurrent with parasympathetic attenuation. Research on cardiac autonomic balance versus cardiac regulatory capacity [65] suggests that the coactivity is still representative of a stress response, albeit a mild response due to the lack of support from other measures.

The subjective stress was found to be different for the stressor levels based on immediate ratings after each trial (PTSR), ratings after the experiment (Free Stress Comparison), and changes in distress during the trials (SSSQ). The results suggest that design of the VR simulations using environmental stressors was successful at manipulating trainee stress levels. However, the distress results show that low and medium were hard to distinguish between. The cause may be attributed to which stressors were selected for the simulation, the magnitude of each environmental stressors, or the combination of different stressors (e.g., noise, smoke). Previous research has shown that different stressors can elicit varying stress responses and can have a cumulative effect that may be greater than the individual effects of the stressors alone [66,67]. The present experiment used expert opinion to inform how the stressors training simulations but had little empirical support. By changing the stressors, magnitude, or combination, it may be possible to have the VR stressor scenarios result in physiological and distress that are more discriminable in future research. Nevertheless, the large effect of the stressor levels on the subjective stress measures demonstrates distinguishable training levels that can be used to induce stress in VR.

The results show that the workload was higher for the high stressor compared to the other stressor levels. Because the simulations were designed to only change environment stressors and not task load, it was expected that workload would not change between the three stressor levels because the procedure was the same for each. Since workload can be thought of as a stress derived from the stressor of task load, it was predicted that the TLX rating for frustration and performance may be different, but not enough to impact the overall workload. However, it is unexpected to find a difference for the high stressor level from the TLX mental workload, physical workload, temporal, and effort. A possible explanation can be discerned from the increase in the high stressor anxiety levels. The attentional control theory [10] suggests that an increased environmental stressor resulted in more stress, less resources for emotion regulation, and thereby, increases in the perceived demand of the workload. Similarly, previous research that found heightened stress reactivity and threat sensitivity when individuals are in conditions with high amount of stress versus a lower threat sensitivity and reactivity in conditions with no stress [68]. The implications are that that anxiety can have adverse effects on processing efficiency and central executive capacity, thus, a subjective workload may increase concurrently with increasing stress levels, even when the task load stays constant.

While VREs are a promising modality for training operations in stressful environments, questions still exist on how VRE designs effectively translate to the real world. The VR-ISS was created with focus on place illusion, plausibility, and virtual body ownership to enhance immersion and the potential to induce a stress response. However, a meta-analysis on stress biomarkers in VR by [69] found inconsistent responses across studies that induce a robust stress response with both high and low levels of immersion. Further, the individual difference may have a large impact on general biomarker responsivity; thus, new methods may be needed for induction to discriminate finer levels of stress.

This study had several limitations that should be considered with regards to interpretation of results and future work. First, the sample size may have been small and statistically underpowered. Several of the measures (SBP, change in worry) had large effects sizes but marginally significant results, such that there is a possibility the effect exists when tested with more subjects. Running more participants would increase the statistical power and confidence in the results. Second, the study recruited participants from the general population rather than astronauts. The general population is less familiar with the ISS layout and procedures, which could possibly lead to stress or confusion from being trained in a short period of time. Further, astronauts may have stronger associations between threat cues and spaceflight hazards, and simultaneously, possible development of coping skills to manage the threat appraisal. Third, while Latin squares was used to minimize the influence of training order, statistical analysis to verify that assumption for six training-order sequences would be underpowered. Therefore, analysis of the counterbalance was not conducted, and it is unknown to the extent that the effects of the training-order influenced the differences among effects in the stressor levels [60]. Future work will include evaluating the simulations with a participant sample similar to the age range, education level, and demographic of astronauts. Future investigation of other training effects would be beneficial and include memory consolidation/retention, task performance, and physiological habituation over multiple sessions.

7. Conclusions

This research found that stressors present in an emergency procedure could be manipulated in VR and approximate the physiological and subjective responses expected during a stress response. Results from this experiment were mixed. The high and low stressor levels had distinguishable results, but the medium level was harder to distinguish.

The findings complement past research that has investigated levels of stressors in VREs [16], which also found varying levels of stress when individuals are exposed to gradually increasing stressor. However, this experiment expands on those studies by adding the context of an operational task procedure and relevant environmental stressors.

While the measurement of stress is empirically straightforward, questions still remain about stressor underpinnings, association with immersion, and impact of VRE design on stressor appraisal. VR stressor scenarios may have more discernable physiological and distress by changing the stressors, magnitude, or combination. Future research should investigate how the combinations of stressors may influence the resulting stress response and how similar the stress is to that felt in the real environment.

Astronauts may benefit from training with VREs that simulate procedures on the interior of the ISS. Simulating stressors with a VRE allows for an immersive experience that is not constrained by resources (e.g., money, staff) or physical infrastructure. A training environment that can induce specific stress similar to the operational environment may help individual manage cognitive resources and emotional states with less consequence than in a real situation. By validating stressor levels, astronauts may someday use VR for training emergency fires and other stressful spaceflight procedures.

Author Contributions: Conceptualization, T.F., M.C.D., N.K., W.D.F. and S.B.V.; methodology, T.F., M.C.D., N.K., W.D.F. and S.B.V.; software, T.F.; validation, T.F.; formal analysis, T.F.; investigation, T.F.; resources, N.K. and W.D.F.; data curation, T.F.; writing—original draft preparation, T.F. and M.C.D.; writing—review and editing, M.C.D., N.K., W.D.F. and S.B.V.; visualization, T.F.; supervision, M.C.D.; project administration, M.C.D.; funding acquisition, M.C.D., N.K., W.D.F. and S.B.V. All authors have read and agreed to the published version of the manuscript.

Funding: This research was funded by National Aeronautics and Space Administration, grant number 80NSSC18K1572.

Institutional Review Board Statement: The study was conducted in accordance with the Declaration of Helsinki, and approved by the Institutional Review Board of Iowa State University (protocol code 19-184, date of approval: 5 September 2019).

Informed Consent Statement: Informed consent was obtained from all subjects involved in the study.

Data Availability Statement: Not applicable.

Acknowledgments: The authors thank Clayton C. Anderson, Tomas Gonzalez-Torres, and Elizabeth Shirtcliff for providing expertise on spaceflight procedures and the human stress response. The authors also thank Robin Gillund, Silvia Verhofste, Matthew Kreul, and Kelly Thompson for their laboratory assistance with research participants. For their help developing the VR-ISS and fire equipment models, the authors thank Pete Evans, Grant Leacox, Peter Carlson, and Robert Slezak.

Conflicts of Interest: The authors declare no conflict of interest.

References

1. Garcia, A.D.; Schlueter, J.; Paddock, E. Training astronauts using hardware-in-the loop simulations and virtual reality. In Proceedings of the AIAA Scitech 2020 Forum, Orlando, FL, USA, 6–10 January 2020; p. 167. [CrossRef]
2. Homan, D.; Gott, C. An integrated EVA/RMS virtual reality simulation, including force feedback for astronaut training. In Proceedings of the Flight Simulation Technologies Conference, San Diego, CA, USA, 29–31 July 1996; p. 3498. [CrossRef]
3. Cater, J.P.; Huffman, S.D. Use of the Remote Access Virtual Environment Network (RAVEN) for Coordinated IVA—EVA Astronaut Training and Evaluation. *Presence Teleoperators Virtual Environ.* **1995**, *4*, 103–109. [CrossRef] [PubMed]
4. Eichler, P.; Seine, R.; Khanina, E.; Schön, A. Astronaut Training for the European ISS Contributions Columbus Module and ATV. *Acta Astronaut.* **2006**, *59*, 1146–1152. [CrossRef]
5. Marciacq, J.B.; Bessone, L. Crew Training Safety: An Integrated Process. In *Safety Design for Space Systems*; Musgrave, G., Larsen, A., Sgobba, T., Eds.; Butterworth-Heinemann: London, UK, 2009; pp. 745–815.
6. Driskell, J.E.; Salas, E.; Johnston, J.H.; Wollert, T.N. Stress Exposure Training: An Event-Based Approach. In *Performance under Stress*; Hancock, P., Szalma, J.L., Eds.; Ashgate: London, UK, 2008; pp. 271–286.
7. Staal, M. *Stress, Cognition, and Human Performance: A Literature Review and Conceptual Framework*; NASA Technical Memorandum 212824; NASA Ames Research Center: Moffett Field, CA, USA, 2004.
8. Hockey, G.R.J. Compensatory Control in the Regulation of Human Performance under Stress and High Workload: A Cognitive-Energetical Framework. *Biol. Psychol.* **1997**, *45*, 73–93. [CrossRef]

9. Gaillard, A.W.K. Stress, workload, and fatigue as three biobehavioral states: A general overview. In *Stress, Workload, and Fatigue*; Hancock, P.A., Desmond, P.A., Eds.; Lawrence Erlbaum: Mahwah, NJ, USA, 2001; pp. 623–640.
10. Eysenck, M.W.; Derakshan, N.; Santos, R.; Calvo, M.G. Anxiety and Cognitive Performance: Attentional Control Theory. *Emotion* **2007**, *7*, 336. [CrossRef] [PubMed]
11. Kilby, C.J.; Sherman, K.A.; Wuthrich, V. Towards Understanding Interindividual Differences in Stressor Appraisals: A Systematic Review. *Personal. Individ. Differ.* **2018**, *135*, 92–100. [CrossRef]
12. Clifford, R.M.S.; Jung, S.; Hoermann, S.; Billinghurst, M.; Lindeman, R.W. Creating a Stressful Decision Making Environment for Aerial Firefighter Training in Virtual Reality. In Proceedings of the 2019 IEEE Conference Virtual Real 3d User Interfaces (VR), Osaka, Japan, 23–27 March 2019; pp. 181–189. [CrossRef]
13. Keren, N.; Bayouth, S.T.; Franke, W.D.; Godby, K.M. Examining the Effect of Level of Stress on Firefighters' Time-to-Decision in Virtual Reality. In Proceedings of the Human Factors and Ergonomics Society Annual Meeting, San Diego, CA, USA, 30 September–4 October 2013; Volume 57, pp. 299–303. [CrossRef]
14. Binsch, O.; Bottenheft, C.; Landman, A.; Roijendijk, L.; Vermetten, E.H. Testing the Applicability of a Virtual Reality Simulation Platform for Stress Training of First Responders. *Mil. Psychol.* **2021**, *33*, 182–196. [CrossRef]
15. Frederiksen, J.G.; Sørensen, S.M.D.; Konge, L.; Svendsen, M.B.S.; Nobel-Jørgensen, M.; Bjerrum, F.; Andersen, S.A.W. Cognitive Load and Performance in Immersive Virtual Reality versus Conventional Virtual Reality Simulation Training of Laparoscopic Surgery: A Randomized Trial. *Surg. Endosc.* **2020**, *34*, 1244–1252. [CrossRef]
16. Parsons, T.D.; Rizzo, A.A.; Courtney, C.G.; Dawson, M.E. Psychophysiology to Assess Impact of Varying Levels of Simulation Fidelity in a Threat Environment. In *Advances in Human-Computer Interaction*; Hindawi Limited: London, UK, 2012; pp. 1–9. [CrossRef]
17. Driskell, J.; Johnston, J.; Salas, E. Does Stress Training Generalize to Novel Settings? *Hum. Factors* **2001**, *43*, 99–110. [CrossRef]
18. Uhlig, T.; Roshani, F.-C.; Amodio, C.; Rovera, A.; Zekusic, N.; Helmholz, H.; Fairchild, M. ISS Emergency Scenarios and a Virtual Training Simulator for Flight Controllers. *Acta Astronaut.* **2016**, *128*, 513–520. [CrossRef]
19. Finseth, T.T.; Keren, N.; Dorneich, M.C.; Franke, W.D.; Anderson, C.C.; Shelley, M.C. Evaluating the Effectiveness of Graduated Stress Exposure in Virtual Spaceflight Hazard Training. *J Cogn. Eng. Decis. Mak.* **2018**, *12*, 248–268. [CrossRef]
20. Olbrich, M.; Graf, H.; Keil, J.; Gad, R.; Bamfaste, S.; Nicolini, F. Virtual reality based space operations–a study of ESA's potential for VR based training and simulation. In Proceedings of the International Conference on Virtual, Augmented and Mixed Reality (VAMR) 2018, Held as Part of HCI International 2018, Las Vegas, NV, USA, 15–20 July 2018; pp. 438–451. [CrossRef]
21. Lazarus, R.S.; Folkman, S. *Stress, Appraisal, and Coping*; Springer: New York, NY, USA, 1984.
22. Pallavicini, F.; Argenton, L.; Toniazzi, N. Virtual Reality Applications for Stress Management Training in the Military. *Aerosp. Med. Hum. Perform.* **2016**, *87*, 1021–1030. [CrossRef]
23. Serino, S.; Triberti, S.; Villani, D.; Cipresso, P.; Gaggioli, A.; Riva, G. Toward a Validation of Cyber-Interventions for Stress Disorders Based on Stress Inoculation Training: A Systematic Review. *Virtual Real.* **2014**, *18*, 73–87. [CrossRef]
24. Koolhaas, J.M.; Bartolomucci, A.; Buwalda, B.; de Boer, S.F.; Flügge, G.; Korte, S.M.; Meerlo, P.; Murison, R.; Olivier, B.; Palanza, P.; et al. Stress Revisited: A Critical Evaluation of the Stress Concept. *Neurosci. Biobehav. Rev.* **2011**, *35*, 1291–1301. [CrossRef] [PubMed]
25. Dickerson, S.S.; Kemeny, M.E. Acute Stressors and Cortisol Responses: A Theoretical Integration and Synthesis of Laboratory Research. *Psychol. Bull.* **2004**, *130*, 355–391. [CrossRef]
26. Blascovich, J.; Tomaka, J. The biopsychosocial model of arousal regulation. *Adv. Exp. Soc. Psychol.* **1996**, *28*, 1–51. [CrossRef]
27. Miller, S.M. Predictability and Human Stress: Toward a Clarification of Evidence and Theory. *Adv. Exp. Soc. Psychol.* **1981**, *14*, 203–256. [CrossRef]
28. Slater, M. Place Illusion and Plausibility Can Lead to Realistic Behaviour in Immersive Virtual Environments. *Philos. Trans. R. Soc. B Biol. Sci.* **2009**, *364*, 3549–3557. [CrossRef]
29. Held, R.M.; Durlach, N.I. Telepresence. *Presence Teleoperators Virtual Environ.* **1992**, *1*, 109–112. [CrossRef]
30. Sheridan, T.B. Musings on telepresence and virtual presence. *Presence Teleoperators Virtual Environ.* **1992**, *1*, 120–126. [CrossRef]
31. Barfield, W.; Weghorst, S. The sense of presence within virtual environments: A conceptual framework. In *Human-Computer Interaction: Software and Hardware Interfaces*; Salvendy, G., Smith, M., Eds.; Elsevier: Amsterdam, The Netherlands, 1993; Volume 19, pp. 699–704.
32. Slater, M.; Wilbur, S. A framework for immersive virtual environments (FIVE): Speculations on the role of presence in virtual environments. *Presence Teleoperators Virtual Environ.* **1997**, *6*, 603–616. [CrossRef]
33. Bergstorm, I.; Azevedo, S.; Papiotis, P.; Saldanha, N.; Slater, M. The plausibility of a string quartet performance in virtual reality. *IEEE Trans. Vis. Comput. Graph.* **2017**, *23*, 1332–1339. [CrossRef] [PubMed]
34. Botvinick, M.; Cohen, J. Rubber hands "feel" touch that eyes see. *Nature* **1998**, *391*, 756. [CrossRef] [PubMed]
35. Peperkorn, H.M.; Diemer, J.E.; Alpers, G.W.; Mühlberger, A. Representation of Patients' Hand Modulates Fear Reactions of Patients with Spider Phobia in Virtual Reality. *Front. Psychol.* **2016**, *7*, 268. [CrossRef] [PubMed]
36. Kirschbaum, C.; Pirke, K.; Hellhammer, D. The 'Trier Social Stress Test'—A Tool for Investigating Psychobiological Stress Responses in a Laboratory Setting. *Neuropsychobiology* **1993**, *28*, 76–81. [CrossRef]

37. Plarre, K.; Raij, A.; Hossain, S. Continuous Inference of Psychological Stress from Sensory Measurements Collected in the Natural Environment. In Proceedings of the 10th ACM/IEEE International Conference on Information Processing in Sensor Networks, Chicago, IL, USA, 12–14 April 2011; pp. 97–108.
38. Johnston, J.; Cannon-Bowers, J.A. Training for stress exposure. In *Stress and Human Performance*; Driskell, J., Salas, E., Eds.; Lawrence Erlbaum: Mahwah, NJ, USA, 1996; pp. 223–256.
39. Dorneich, M.C.; Mathan, S.; Ververs, P.M.; Whitlow, S.D. Cognitive State Estimation in Mobile Environments. In *Augmented Cognition: A Practitioner's Guide*; Schmorrow, D., Stanney, K., Reeves, L., Eds.; HFES Press: Santa Monica, CA, USA, 2008; pp. 75–111.
40. Healey, J.A.; Picard, R.W. Detecting stress during real-world driving tasks using physiological sensors. *IEEE Trans. Intell. Transp. Syst.* **2005**, *6*, 156–166. [CrossRef]
41. Chabal, S.; Bohnenkamper, A.; Reinhart, P.; Quatroche, A. *Stressors Present in a Disabled Submarine Scenario: Part 1. Identification of Environmental, MENTAL, and Physical Stressors*; Naval Submarine Medical Research Laboratory: Groton, CT, USA, 2019.
42. United States National Aeronautics and Space Administration (NASA). *International Space Station, Emergency Procedures 1a: Depress, Fire, Equipment Retrieval (No. JSC-48566)*; NASA Johnson Space Center: Houston, TX, USA, 2013.
43. Bartone, P.T.; Roland, R.R.; Bartone, J.V.; Krueger, G.P.; Sciarretta, A.A.; Johnsen, B.H. Human Adaptability for Deep Space Missions: An Exploratory Study. *J. Hum. Perform. Extreme Environ.* **2019**, *15*, 5. [CrossRef]
44. IGOAL, NASA/JSC, "ISS (Internal)," NASA 3D Resources. 27 March 2017. Available online: https://nasa3d.arc.nasa.gov/detail/iss-internal (accessed on 27 March 2017).
45. Finseth, T.; Dorneich, M.C.; Keren, N.; Franke, W.D.; Vardeman, S. Designing Training Scenarios for Stressful Spaceflight Emergency Procedures. In Proceedings of the 2020 AIAA/IEEE 39th Digital Avionics Systems Conference (DASC), San Antonio, TX, USA, 11–15 October 2020; pp. 1–10. [CrossRef]
46. McCraty, R.; Shaffer, F. Heart Rate Variability: New Perspectives on Physiological Mechanisms, Assessment of Self-Regulatory Capacity, and Health Risk. *Glob. Adv. Health Med.* **2015**, *4*, 46–61. [CrossRef]
47. Sharma, N.; Gedeon, T. Modeling Observer Stress for Typical Real Environments. *Expert Syst. Appl.* **2014**, *41*, 2231–2238. [CrossRef]
48. Mestanik, M.; Visnovcova, Z.; Tonhajzerova, I. The Assessment of the Autonomic Response to Acute Stress Using Electrodermal Activity. *Acta Med. Martiniana* **2014**, *14*, 5–9. [CrossRef]
49. Braithwaite, J.; Watson, D.; Jones, R. A Guide for Analysing Electrodermal Activity (EDA) & Skin Conductance Responses (SCRs) for Psychological Experiments. *Psychophysiology* **2013**, *49*, 1017–1034.
50. Penley, J.A.; Tomaka, J. Associations among the Big Five, Emotional Responses, and Coping with Acute Stress. *Personal. Individ. Differ.* **2002**, *32*, 1215–1228. [CrossRef]
51. Singh, R.R.; Conjeti, S.; Banerjee, R. A comparative evaluation of neural network classifiers for stress level analysis of automotive drivers using physiological signals. *Biomed. Signal Process.* **2013**, *8*, 740–754. [CrossRef]
52. Carpenter, R. A Review of Instruments on Cognitive Appraisal of Stress. *Arch. Psychiatr. Nurs.* **2016**, *30*, 271–279. [CrossRef] [PubMed]
53. Helton, W. Validation of a Short Stress State Questionnaire. In Proceedings of the Human Factors and Ergonomics Society Annual Meeting, New Orleans, LA, USA, 20–24 September 2004; Volume 48. [CrossRef]
54. Matthews, G.; Joyner, L.; Gilliland, K.; Campbell, S.; Falconer, S.; Huggins, J. Validation of a Comprehensive Stress State Questionnaire: Towards a State "Big Three". In *Personality Psychology in Europe*; Mervielde, I., Deary, I.J., De Fruyt, F., Ostendorf, F., Eds.; Tilburg University Press: Tilburg, The Netherlands, 1999; Volume 7, pp. 335–350.
55. Helton, W.; Näswall, K. Short stress state questionnaire. *Eur. J. Psychol. Assess.* **2015**, *31*, 20–30. [CrossRef]
56. Hart, S.G.; Staveland, L.E. Development of NASA-TLX (Task Load Index): Results of Empirical and Theoretical Research. *Adv. Psychol.* **1988**, *52*, 139–183. [CrossRef]
57. Nygren, T.E. Psychometric Properties of Subjective Workload Measurement Techniques: Implications for Their Use in the Assessment of Perceived Mental Workload. *Hum. Factors* **1991**, *33*, 17–33. [CrossRef]
58. Spielberger, C.D.; Gorsuch, R.L.; Lushene, R.; Vagg, P.R.; Jacobs, G.A. *Manual for the State-Trait Anxiety Inventory*; Consulting Psychologists Press: Palo Alto, CA, USA, 1983.
59. Tartarisco, G.; Carbonaro, N.; Tonacci, A.; Bernava, G.; Arnao, A.; Crifaci, G.; Cipresso, P.; Riva, G.; Gaggioli, A.; Rossi, D.D.; et al. Neuro-Fuzzy Physiological Computing to Assess Stress Levels in Virtual Reality Therapy. *Interact. Comput.* **2015**, *27*, 521–533. [CrossRef]
60. Reese, H.W. Counterbalancing and Other Uses of Repeated-Measures Latin-Square Designs: Analyses and Interpretations. *J Exp. Child Psychol.* **1997**, *64*, 137–158. [CrossRef]
61. Kim, H.Y. Statistical notes for clinical researchers: Assessing normal distribution (2) using skewness and kurtosis. *Restor. Dent. Endod.* **2013**, *38*, 52. [CrossRef]
62. Gelman, A. Commentary: P values and statistical practice. *Epidemiology* **2013**, *24*, 69–72. [CrossRef]
63. Cohen, J. *Statistical Power Analysis for the Behavioural Sciences*; Lawrence Earlbaum Associates: Hillside, NJ, USA, 1988.
64. Matthews, G.; Campbell, S.E.; Falconer, S.; Joyner, L.A.; Huggins, J.; Gilliland, K.; Grier, R.; Warm, J.S. Fundamental Dimensions of Subjective State in Performance Settings: Task Engagement, Distress, and Worry. *Emotion* **2002**, *2*, 315–340. [CrossRef] [PubMed]

65. Berntson, G.G.; Norman, G.J.; Hawkley, L.C.; Cacioppo, J.T. Cardiac Autonomic Balance versus Cardiac Regulatory Capacity. *Psychophysiology* **2008**, *45*, 643–652. [CrossRef] [PubMed]
66. Abdelall, E.S.; Eagle, Z.; Finseth, T.; Mumani, A.A.; Wang, Z.; Dorneich, M.C.; Stone, R.T. The Interaction Between Physical and Psychosocial Stressors. *Front. Behav. Neurosci.* **2020**, *14*, 63. [CrossRef] [PubMed]
67. Pedrotti, M.; Mirzaei, M.A.; Tedesco, A.; Chardonnet, J.-R.; Mérienne, F.; Benedetto, S.; Baccino, T. Automatic Stress Classification with Pupil Diameter Analysis. *Int. J. Hum.-Comput. Int.* **2014**, *30*, 220–236. [CrossRef]
68. Akinola, M.; Mendes, W.B. Stress-Induced Cortisol Facilitates Threat-Related Decision Making among Police Officers. *Behav. Neurosci.* **2012**, *126*, 167–174. [CrossRef]
69. van Dammen, L.; Finseth, T.; McCurdy, B.H.; Barnett, N.P.; Conrady, R.; Leach, A.G.; Deick, A.F.; Stennis, A.V.; Gardner, R.; Smith, B.; et al. Evoking stress reactivity in virtual reality: A systematic review and meta-analysis. *Neurosci. Biobehav. Rev.* **2022**; submitted.

Emulating Cued Recall of Abstract Concepts via Regulated Activation Networks

Rahul Sharma [†,‡], Bernardete Ribeiro [‡], Alexandre Miguel Pinto [‡] and Amílcar Cardoso [*,‡]

Center of Informatics and Systems (CISUC), DEI, Polo-II University of Coimbra, Pinhal de Marrocos, 3030-290 Coimbra, Portugal; rahul@dei.uc.pt (R.S.); bribeiro@dei.uc.pt (B.R.); ampinto@dei.uc.pt (A.M.P.)
* Correspondence: amilcar@dei.uc.pt
† Current address: Centre for Informatics and Systems of the University of Coimbra Polo II, Pinhal de Marrocos, 3030-290 Coimbra, Portugal.
‡ Dr. Rahul Sharma is the main contributor under the supervision of Dr. Bernardete Ribeiro, Dr. Alexandre Miguel Pinto, and Dr. Amílcar F. Cardoso.

Abstract: Abstract concepts play a vital role in decision-making or recall operations because the associations among them are essential for contextual processing. Abstract concepts are complex and difficult to represent (conceptually, formally, or computationally), leading to difficulties in their comprehension and recall. This contribution reports the computational simulation of the cued recall of abstract concepts by exploiting their learned associations. The cued recall operation is realized via a novel geometric back-propagation algorithm that emulates the recall of abstract concepts learned through regulated activation network (RAN) modeling. During recall operation, another algorithm uniquely regulates the activation of concepts (nodes) by injecting excitatory, neutral, and inhibitory signals to other concepts of the same level. A Toy-data problem is considered to illustrate the RAN modeling and recall procedure. The results display how regulation enables contextual awareness among abstract nodes during the recall process. The MNIST dataset is used to show how recall operations retrieve intuitive and non-intuitive blends of abstract nodes. We show that every recall process converges to an optimal image. With more cues, better images are recalled, and every intermediate image obtained during the recall iterations corresponds to the varying cognitive states of the recognition procedure.

Keywords: computational psychology; computational cognitive modeling; machine learning; concept blending; conceptual combinations; recall; computational creativity

1. Introduction

Concepts are an important object of research in cognitive and psychological research. Usually, the conceptual representations are process-oriented, symbolic or distributed, and knowledge-based [1–3]. In general, a hierarchical structure defines an organization of concepts where the concrete concepts are placed in the lower level, and the abstract Concepts occupy the higher levels (see the example of the vehicle and cars in Figure 1). Therefore, abstract concepts are also seen as the generalization of concrete concepts [4,5]. Abstract concepts are studied mathematically [6] and theoretically [7,8], but computational studies are scarce [1]. This article uses a computational model, regulated activation network (RAN) [9–11], capable of building representation of convex abstract concepts, which are later used in recall simulations.

The prime aspect of this article is to emulate the recall procedure, that can be viewed as the cognitive process of remembering. Context plays an eminent role in the recall of concrete concepts (such as the word "table"), which is often termed as concreteness effect. The concreteness effect is expressed through the dual-coding theory [12,13] and the context availability hypothesis [14]. According to this theory, individually, abstract concepts are abstruse in context retrieval when compared with concrete concepts; therefore, their recall procedure is complex. However, an interesting work suggests that context retrieval of abstract concepts is possible when response pairs of abstract concepts are related to one another, thus providing context for the abstract stimuli [15]. In RAN modeling we can learn associations among concepts (including the abstract concepts). These learned associations provide adequate context relations among the abstract concepts. In this work, we exploit these learned associations among the abstract concepts to simulate a regulated recall operation.

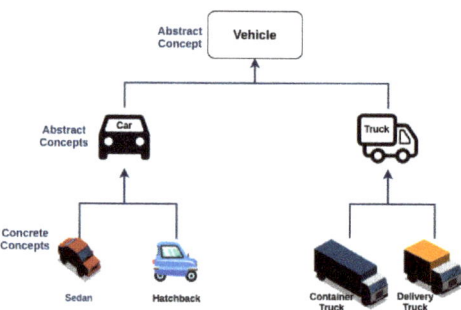

Figure 1. Hierarchy abstraction of concrete and abstract concepts.

In recent years, advances in technologies have played an essential role in cognitive and psychological research, e.g., use of devices like GP3 [16], TheEyeTribe [17], and electroencephalography (EEG) [18,19] to study visual attention. In this article, we focus on representations of computational models that are also very useful in understanding the psychological and cognitive phenomena, validating the existing cognitive theories, and helping to formulate fresh ideas related to cognition [20–23]. The representations of these computational approaches are either symbolic (amodal), distributed (multimodal), or hybrid [24], which helps in simulating or understanding various cognitive phenomena. Below are a few examples of computational models or architecture used to study psychological and cognitive phenomena based upon their representation:

- Symbolic: Adaptive Control of Thought–Rational (ACT-R) [25] is a symbolic architecture intended to model memory [26] and simulate attention [25,27], decision-making [28], recognition [29], and forgetting [29].
- Distributed: Multimodal approaches such as artificial neural networks (ANNs), including the restricted Boltzmann machine (RBM) [30], deep neural networks [31], stacked auto-encoders [32], and convolution neural networks (CNN) [33], have significant contributions in feature recognition [34] and distributed memory representation [35]. Methods like Random Forests have also been used in studies related to visual attention [36].
- Hybrid: Cognitive architectures like Connectionist Learning with Adaptive Rule Induction On-line (CLARION) [37] simulate scenarios related to cognitive and social psychology.

This article makes use of the RAN's hybrid nature and modeling to build the representation of convex abstract concepts and further simulate recall of abstract concepts. F First, the model generation takes place with four basic steps of the RAN approach [9], i.e., concept identification, concept creation, inter-layer learning, and upward activation propagation. An intra-layer procedure also takes place at all the layers to identify the association among the concepts at the same level. Further, these learned associations are uniquely interpreted to determine whether the impact of their learned weights is inhibitory, excitatory, or neutral. Later, these impacts are applied to obtain a regulatory effect on peer concepts (abstract concepts or input layer concepts) during recall operation. A Toy-data problem was used for modeling with RAN and demonstrating the novel geometric back-propagation algorithm for the simulation-cued recall operation. The benchmark dataset of the image domain, MNIST, is also used to demonstrate the cued recall experiment. These experiments also show how blends of abstract concepts can be recalled. To summarize, the following are the main contributions of the article: first, the impact factor calculation to determine the inhibitory, excitatory, or neutral effect of one node over other; second, the novel intra-layer regulation algorithm for the use of the impact factor in order to regulate the activation of other concepts; third, the novel geometric back-propagation algorithm and recall simulations using the geometric back-propagation algorithm.

The remainder of this article is organized in the following way: Section 2 puts forward the state of the art related to recall operations; RAN modeling, the intra-layer regulation algorithm, and the geometric back-propagation algorithm are detailed in Section 3 using a Toy-data problem; the cued recall demonstration with the MNIST dataset is reported in Section 4; Section 5 concludes the article.

2. Related Work

Recall or retrieval is a cognitive process [38] of remembering a thing or an event. While recalling, the brain activates a neural assembly that was created when the original event occurred [38]. In psychology, there is a plethora of articles studying the recall process. Psychologists used free-recall, cued-recall, and serial-recall as tools to investigate memory processes [39]. Recall has been used to study the effect of cognitive strategies, such as chunking and the use of mnemonics for memorization of things (such as large numbers) [40]. One interesting study reported the benefits of subsequent recall in retrieval operations where memories are related or competing [41]. The proverb "practice makes a man perfect" relates to the fortification of memory, and an investigation shows how retrieval plays an important role in this memory strengthening [42]. Technologies such as functional magnetic resonance imaging (fMRI), magnetic resonance imaging (MRI), positron emission tomography (PET), and electroencephalography (EEG) played an active role in validating many recall related hypotheses [41,43–45].

Notable contributions to the modeling of memory recall procedures are observed. Based on the temporal context model [46,47] of human behavior, human memory performance was modeled using a probabilistic approach during free-recall experiments [48]. A computational model of interaction between the prefrontal cortex and medial temporal lobe in memory usage was designed to study the prefrontal control in a recall process [49]. The model was a simple neural network with quick and flexible reinforcement learning exhibiting strategic recall. Another computational model differentiates recall from the recognition process depending upon the number of cues involved in the retrieval procedure [50]. For encoding, the model used an inference-based model of memory [51], and retrieval was carried out using a Bayesian observer model [52]. A large number of computational psychology contributions examining the recall process and recognition using the neural networks are available [53–56].

An interesting study simulated the free-recall process using the ACT-R architecture [57], showing that the classical effect of primacy and recency can be recreated through rehearsal theory based upon ACT-R and Baddeley's phonological Loop [58]. ACT-R architecture was also used to propose a new theory of memory retrieval to predict for intricate serial and free-recall operations [59]. This research also focused on the prospects of associative learning by introducing a strengthening and decaying mechanism depending upon the similarity of the input stimulus. The serial recall has been modeled in a scientific contribution using ACT-R architecture to explain the processes involved while recalling a list of words [60]. The traditional ACT-R recall operations had a limitation: here, the memory access depends upon limiting the capacity of the activation process, consequently inducing errors in the contents being recalled. This theory overcomes the limitations by predicting the latency and errors in a serial recall process.

The free recall process was also modeled using CLARION to determine the role of distractions in an incubation task [61]. This study made a striking observation that rehearsals play an important role in memory consolidation during the free recall procedure, and distractions can hinder the free recall and eventually effect memory strengthening. CLARION was also used to emulate, acquire, and expound human-centric data relevant to incubation and insight through free recall, lexical decision, and problem-solving tasks [62].

This article introduces a novel algorithm named geometric back-propagation, which enable us to simulate the recall simulation using RAN modeling. The main objective of the experiments in this paper is to demonstrate the role cue (activation) on abstract concepts (nodes) in recall operations. An additional goal is to show that when larger number abstract concepts participate (i.e., more cues are available) in recall operations, then better recall is observed in the experiment.

3. RAN Methodology to Simulate Recall Operations

Here, we describe the emulation of the recall process using RAN modeling. For background understanding in Section 3.2 we describe RAN modeling along with two learning mechanisms, i.e., inter-layer and intra-layer learning. Having explained RAN modeling, the two contributions of this article are elucidated: first, the regulation mechanism is described, the biological inspiration behind this operation is descibed in Section 3.1. Secondly, a novel geometric back-propagation algorithm is proposed that propagates activations from abstract level to input Level. RAN methodology and the article's contributions are illustrated using a Toy-data set. At the end of this section, the experiments of RAN modeling with the Toy-data are also reported, demonstrating the recall operation.

3.1. Biological Inspiration of Regulation Operation on RAN's Modeling

The nerve cell (neuron) consists of several main components: the dendrites, the cell body, and the axon, as shown in Figure 2. When an electric signal traverses the whole axon and reaches one of its terminations, it releases chemicals called neurotransmitters, which diffuse across the synaptic gap and are absorbed by the receptive neuron's dendrite. Depending on the neurotransmitter, this absorption can either enhance or inhibit the receptive neuron's activation.

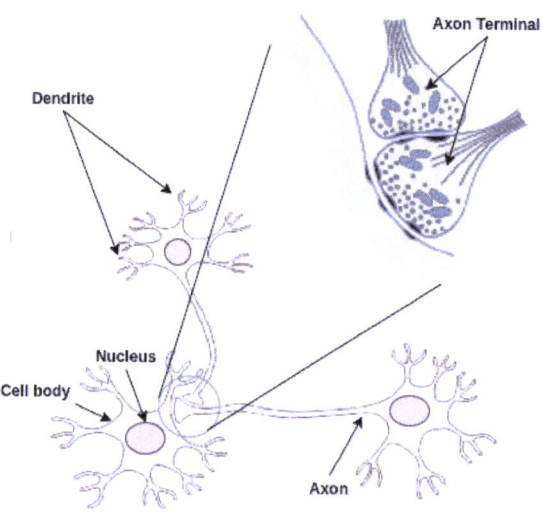

Figure 2. Biological neuron showing the axoaxonic synapses.

Other kinds of synapses occur in biological brains, such as axoaxonic synapses, as shown in Figure 2. These synapses occur when the axon of a neuron connects to the axon of another neuron instead of to its dendrites. Such configuration usually plays a regulatory role by mediating presynaptic inhibition and presynaptic facilitation [63]. By virtue of artificial axoaxonic synapses, this contribution realizes the inhibitory, excitatory, and neutral activation propagation phenomenon in RAN modeling, which is used to induce a regulatory effect on the activation at nodes during recall operation.

3.2. Abstract Concept Modeling with RAN

This section is dedicated to describing convex abstract concept modeling with RAN [11]. To demonstrate RAN methodology, a Toy-data problem is used—see Figure 3. The Toy-data are synthetically produced by generating a 2D dataset with five classes. In Figure 3, we can see that out of the five clusters, three are far apart from one another; however, two clusters are very close to each other. This arrangement of clusters was introduced into the Toy-data problem to demonstrate the excitatory and inhibitory impact of concepts, representing each cluster at an abstract level. The dataset consists of 1800 data instances with an equal distribution in all of the classes. RAN modeling is performed using the four basic steps, where step 1 and step 4 consist of two concurrent operations.

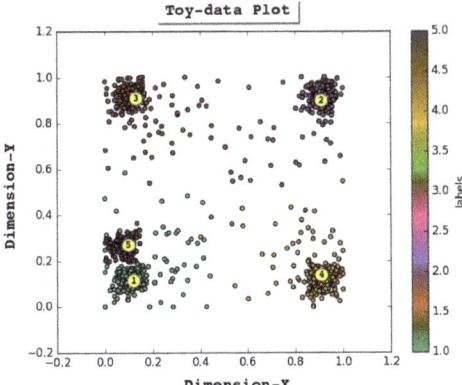

Figure 3. Graph of 2D Toy-data with five clusters, along with their respective cluster centers (1, ..., 5).

3.2.1. Step 1a: Convex Concept Identification (CCI)

CCI is a method to determine convex groups in a given dataset. In RAN, each data instance is considered a point in n-dimensional geometric feature space, inspired by the theory of conceptual spaces [64]. In this method, we also determine the cluster centers that are used in the inter-Layer learning operation (see Section 3.2.4). Step 1a in Figure 3 shows input Layer 0; here, nodes S_1 and S_2 correspond to the dimensions of the input Toy-data. To identify the five convex groups, the K-mean [65] clustering algorithm is chosen as the concept identifier (CI), and the value of K is set to 5 to determine five clusters. Five cluster centers are also identified in this process, as shown in Figure 3, as cluster representative data points (CRPD). Any clustering algorithm can act as a concept identifier provided that it enables the identification of convex regions along with their respective cluster centers.

3.2.2. Step 1b and Step 4b: Concept Similarity Relation Learning (CSRL)

CSRL is an intra-layer operation in RAN modeling explained in [66]. The main purpose of this process is to determine the alikeness among the concepts and associate them through a similarity relation. This relation also simulates the behavior of activation found in biological neurons; i.e., affine neurons are activated concurrently upon receiving input stimuli, whereas unrelated neurons remain relatively inactive for the same input stimulus. This phenomenon is expressed mathematically through Equation (1) to calculate a pair-wise relation/weight $w_{m \to n}$ between node m, and node n at a layer. The numerator $(1 - |A_m^I - A_n^I|)$ calculates the similarity of activation (A_m^I is the activation of Ith instance of propagated data at node C_m in a layer, and similarly, A_n^I is the activation of Ith instance of propagated data at node C_n. $m \neq n$ and m, n are integers.) of node m w.r.t. node n, and the product $(1 - A_m^I) * (1 - A_n^I)$ is used to reduce the impact of similarity on weight $w_{m \to n}$

when both activations (i.e., A_m^I, and A_n^I) are very close to 0, though similar. Consequently, we obtain a symmetric $k \times k$ matrix as learned concept similarity relation weights (CSRW) among the nodes within the layer.

$$w_{m \to n} = \frac{\sum_I [(1 - |A_m^I - A_n^I|) - (1 - A_m^I) * (1 - A_n^I)]}{\sum_I [1 - (1 - A_m^I) * (1 - A_n^I)]} \quad (1)$$

where $m \in 1, ..., k$; $n \in 1, ..., k$; and $m \neq n$.

This learning mechanism is performed two times while modeling with Toy-data: First, in Step 1b at the input layer 0, (see Figure 4, step 1b), and second, when the input data is propagated upward to the convex concept layer 1. The learning at layer 0 has a size of 2×2, as the input layer has two nodes. In contrast, the learning at layer 1 has a size of 5×5 (see Figure 5 for the CSRL weights).

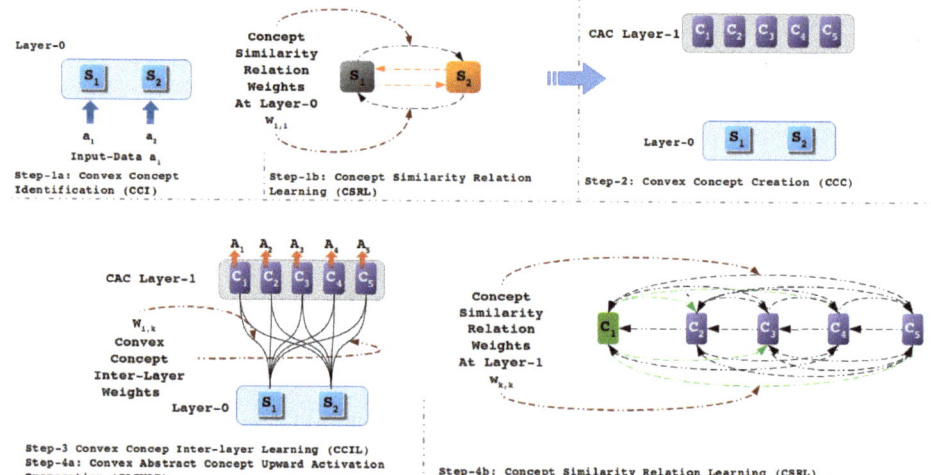

Figure 4. Regulated activation network (RAN) convex abstract concept modeling process. The procedure displays the four steps in RAN modeling. This figure shows the three learning procedures, i.e., two similarity relation learning procedures at two layers, and one inter-layer learning procedure between layer 1 and layer 0. In step 1, similarity relation learning (step 1a) is performed along with the concept identification process (step 1b). Similarly, in step 4, similarity relation learning (step 4b) is performed together with the upward activation propagation method (step 4a).

3.2.3. Step 2: Convex Abstract Concept Creation (CACC)

Convex abstract concept creation is a method of dynamically creating a layer that consists of nodes as an abstract representative of categories determined in the convex concept identification process (see Section 3.2.1). For instance, step 2 in Figure 4 depicts the creation of new a convex abstract concept (CAC) layer 1 with five nodes (C_1, ..., C_5). These five nodes (C_1, C_2, C_3, C_4, and C_5) represent clusters 3, 1, 4, 2 and 5 in Figure 3, respectively. The count of the CAC layer nodes depends upon the number of clusters identified in the CCI operation at the input layer; i.e., if k clusters were determined in the CCI mechanism, then in step 2, a new layer is created consisting of k nodes.

3.2.4. Step 3: Convex Concept Inter-Layer Learning (CCILL)

Besides intra-Layer learning in Steps 1b and 4b, the second learning mechanism in RAN modeling is used to identify the association among the nodes at the CAC layer and input layer. Since the nodes are the abstract representative of the clusters identified in the CCI process, and according to the theory of prototype, the cluster center (i.e., the CRPDs) is the most probable representative of a cluster [67–69]. Therefore, the cluster centers are understood as an association among the CAC layer nodes and input layer nodes, as depicted by Equation (2).

$$W_{k,i} = \begin{bmatrix} W_{1,1}, W_{1,2}, \ldots, W_{1,n_a} \\ \ldots \\ W_{t,1}, W_{t,2}, \ldots, W_{t,n_a} \\ \ldots \\ W_{n_A,1}, W_{n_A,2}, \ldots, W_{n_A,n_a} \end{bmatrix} = \begin{bmatrix} C_1 \\ \ldots \\ C_t \\ \ldots \\ C_{n_A} \end{bmatrix} \quad (2)$$

where $k = 1, 2, \ldots, n_A$, and $i = 1, 2, \ldots, n_a$

In Equation (2) we can see that the coordinates (feature values of data) of centers identified in the CCI process are assigned as learning W, i.e., convex concepts inter-layer wiights (CCILW). In the experiment with Toy-data, we learned a 5 × 2 weight matrix between the five nodes at CAL layer 1 and two nodes of input layer 0, as shown in Figure 4, step 3. Having completed step 3, a basic RAN model is obtained consisting of input layer 0, CAC layer 1, learning between two layers, and learning among the nodes at input layer 0.

CSRL at Layer-0

	S_1	S_2
S_1	0	0.56
S_2	0.56	0

σ at Layer-0

	S_1	S_2
S_1	0	0.002
S_2	0.002	0

CSRL at Layer-1

	C_1	C_2	C_3	C_4	C_5
C_1	0	0.5	0.55	0.54	0.52
C_2	0.5	0	0.48	0.46	0.86
C_3	0.55	0.48	0	0.56	0.46
C_4	0.54	0.46	0.56	0	0.45
C_5	0.52	0.86	0.46	0.45	0

σ at Layer-1

	C_1	C_2	C_3	C_4	C_5
C_1	0	0	0.0012	0.0007	0.0001
C_2	0	0	-0.0001	-0.0007	0.3663
C_3	0.0012	-0.0001	0	0.0014	-0.0004
C_4	0.0007	-0.0007	0.0014	0	-0.001
C_5	0.0001	0.3663	-0.0004	-0.001	0

Figure 5. The concept similarity relation learning (CSRL) weight matrices learned with Toy-data and their corresponding impact dactor (σ) at layer 0 and layer 1. σ is calculated using Equation (5).

3.2.5. Step 4a: Convex Abstract Concept Upward Activation Propagation (CACUAP)

This step is used to propagate i-dimensional input data vector a_i to the CAC layer and to obtain k-dimensional data vector A_k. This mechanism is used in two stages: in the first stage, Euclidean distance is calculated among the input data a_i and all the CCILWs $W_{k,i}$. This distance is further normalized (using the denominator in Equation (3)) to obtain distance in the range [0, 1](in RAN's modeling, the activation values of are, by definition, real values in the [0, 1] interval—and in such a setting, in an n-dimensional space, the maximum possible Euclidean distance between any two points is $\sqrt{\sum_{i=1}^{n}(a_i - 0)^2} = \sqrt{n}$, where $a_i = 1$.).

$$d_k = \frac{\sqrt{\sum_{i=1}^{n_a}(W_{k,i} - a_i)^2}}{\sqrt{n_a}} \quad (3)$$

In the second stage, the normalized distance obtained from Equation (3) is transformed non-linearly, establishing a similarity relation conforming to the following three conditions: (1) $f(d = 0) = 1$, i.e., when distance is 0, similarity is 100%; (2) $f(d = 1) = 0$ i.e., when distance is 1, similarity is 0%; and (3) $f(d = x)$ is continuous, monotonous, and differentiable in the [0, 1] interval.

$$f(x) = (1 - \sqrt[3]{x})^2 \quad (4)$$

This similarity relation equation (Equation (4)) transforms the distance values observed at each node in the CAC layer into its similarity value. These similarity values act as degree of confidence (DoC) values to recognize the category being represented by the nodes at the CAC layer. Upon propagating all input values to the CAC Layer, the observed outputs A_k are used to perform concept similarity relation learning (CSRL), as shown in Figure 4, step 4b. After completing step 4b, the RAN modeling procedure terminates, and a model is obtained, as shown in step 3 in Figure 4.

In order to build more than one layer, all of the steps are repeated iteratively, and the output of all the intermediate CAC layers is pipelined as input to the new layer being built.

3.3. Regulation Mechanism

The regulation operation in RAN modeling is performed in three steps: first, an impact factor of the CSRL matrix is deduced; second, the intra-layer (IL) contribution of activation at a node by another node in the same layer is determined; third, activation at a node by a function of self-activation and intra-layer activation induced by other nodes on the latter are obtained.

3.3.1. Impact Factor (σ) Construction and Interpretation

The impact factor is a function that interprets the CSRL weight values (in the range [0, 1]) as excitatory, inhibitory, or neutral weights. The purpose of CSRL weights is to determine how concurrently two nodes (e.g., S_1 and S_2) are active. If the CSRL weight is intermediate, i.e., 0.5, it signifies that the two nodes are 50% concurrently active (depicting a state of confusion). Therefore, these nodes do not have an impact on each other in the same layer. If the CSRL weight of the two nodes were "0", then the two nodes were never active simultaneously. This also indicates that the two nodes are inhibitors of each other. Finally, if the CSRL weights of the two nodes were "1", then the nodes were always active conjointly, exciting the activation of each other.

$$\sigma_{m \to n} = [2 * (W_{m \to n} - 0.5)]^3 \tag{5}$$

The aforementioned comprehension of CSRL weights W is exhibited by a mathematical Equation (5), where $\sigma_{m \to n}$ is the impact of node m over node n. Figure 6 shows a graphical view of the impact factor σ (Equation (5)), depicting the excitatory, inhibitory, or neutral interpretations of CSRL weights. Figure 5 shows the CSRL and their respective σ weights for both layers. At layer 0, the nodes S_1 and S_2 have a very minimal excitatory impact on each other (in Figure 3, every node at layer 1 can be related to clusters C_1, ..., C_5 serially). However, at layer 1, node C_1 has no impact on node C_2 (and vice versa). There are many negative weights in the σ matrix of layer 1, indicating that these nodes inhibit each other. In Figure 3, we see that the clusters C_2 and C_5 are very close, and the activations observed at both of the nodes must be very similar. Hence, high CSRL weight is learned between nodes C_2 and C_5. Notably, both exhibit good excitatory behavior towards each other.

3.3.2. Intra-Layer Activation

The objective of calculating intra-layer (IL) activation is to determine the amount of activation a node n receives from all the other m nodes of the same Layer. To obtain the intra-layer activation at node n, the approach must address three prospects. First, intra-layer activation must consider the impact (σ) of excitatory, inhibitory, or neutral effects of all m nodes over node n. Second, the current activation of m nodes and their CSRL weight ($Wm \to n$) to node n should be considered in calculating the activation of node n. Third, the intra-layer activation computed for node n must be in the range [0, 1]. Equation (8) conforms to all three requirements.

$$\chi_m = (a_m * W_{m \to n}) \tag{6}$$

$$\neg \chi_m = (1 - a_m) * (1 - W_{m \to n}) \tag{7}$$

$$IL(a_n) = \frac{\sum_m \sigma_{m \to n}(\chi_m + \neg\chi_m)}{\sum_m \sigma_{m \to n}} \quad (8)$$

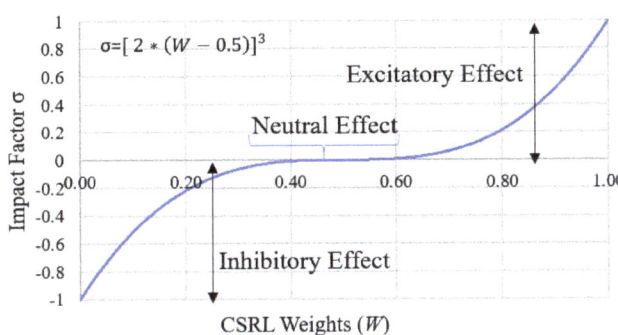

Figure 6. Excitatory, inhibitory, and neutral effect of CSRL weights (W) when transformed using the impact factor σ.

3.3.3. Intra-Layer Regulation

To identify the actual activation $AA(a_n)$ at node n, this operation uses a regulation factor (ρ) to decide the share of self-contribution of activation by node n and intra-layer activation at node n, i.e., $IL(a_n)$. Equation (9) shows the mathematical function for the intra-layer regulation operation. From Equation (9), we can observe if the ρ is '0'; i.e., without any regulation, only the activation of node n contributes to the actual activation.

$$AA(a_n) = (1 - \rho) * a_n + \rho * IL(a_n) \quad (9)$$

Algorithm 1 presents the intra-layer regulation operation in an algorithmic form. This regulation operation has its importance when propagating the activation from an abstract concept layer to the input layer, as described in Section 3.4.

Algorithm 1: Intra-Layer Regulation

Input: current activation a_n at node n at layer L.
Input: CSRL W at layer L
Input: impact matrix σ at layer L
Initialization: regulation factor ρ, between [0, 1];
foreach a_n in L **do**
| Calculate $IL(a_n)$, using Equation (8);
| Calculate actual activation $AA(a_n)$, using Equation (9);
end
return $AA(a_n)$

3.4. Geometric Back-Propagation Operation

Geometric back propagation (GBP) is a downward propagation mechanism in RAN modeling. This method enables us to determine an activation vector a_m ($< a_1, .., a_i, .., a_m >$) at layer L-1, for an expected activation (E-A) vector A'_n ($< A'_1, .., A'_j, .., A'_n >$) at layer L. This operation is a window operation that takes place between two adjacent layers, i.e., layers L and L-1. For instance, if the RAN model has three layers, L_0 (input layer), L_1, and L_2 (output layer), then two GBP operations take place, first between L_2 and L_1, and then between L_1 and L_0. Figure 7 shows the single-window operation between two layers.

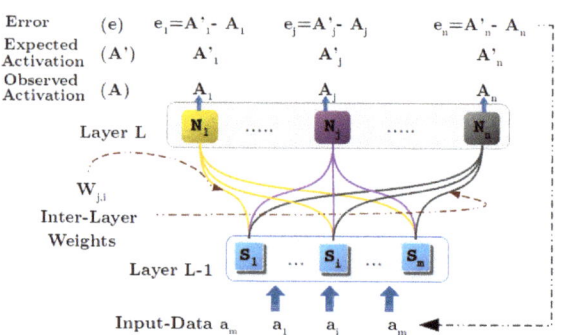

Figure 7. Single-window operation in geometric back-propagation operation. The figure also shows the error calculation and propagation.

The GBP mechanism commences with an expected activation (A'_n) vector at layer L. Next, a starting input vector a_m ($< a_1, .., a_i, .., a_m >$) is injected in layer L-1. Now, we enter into a cycle where we propagate the activation of the nodes in layer L-1 upwards to layer L and determine the observed activation (O-A) vector A_n ($< A_1, .., A_j, .., A_n >$) at layer L. Furthermore, an error vector e is calculated using A'_n and A_n (expected and observed activation vector) through Equation (10). The error vector e is used to determine an accumulated delta value $\triangle_{a_i}$ (see Equation (12)) based upon the function expressed by Equation (11). This $\triangle_{a_i}$ value is then added to the activation vector at layer L-1 (see Equation (13)) to obtain a new input vector a_m^{new}. Additionally, the cycle is repeated with the new a_m^{new} input at layer L-1 until the error is minimized or the cycle equals the user-defined maximum iteration threshold. Algorithm 2 presents the detailed geometric back-propagation algorithm. As mentioned earlier, the GBP operation takes place between two consecutive layers. However, if the hierarchy has more than two layers, then with the window operation, it is possible to propagate down the injected E-As at the nodes of the top-most layer L, to input layer 0.

$$e_j = A'_j - A_j \tag{10}$$

$$\triangle_{a_i, A_j} = (W_{j,i} - a_i) * (e_j) \tag{11}$$

$$\triangle_{a_i} = \sum_{j=1,..,i} \triangle_{a_i, A_j} \tag{12}$$

$$a_i^{new} = a_i + \triangle_{a_i} \tag{13}$$

3.5. Recall Demonstration with Toy-Data

There are two types of experiments performed in this section: first, the single-cue recall (SCR) operation, where the recall is performed based upon the expected activation by one node in an abstract concept layer and, second, the multiple-cue recall (MCR) mechanism, where the recall procedure is carried out for the expected activation at all the nodes in the abstract concept layer. The experiments demonstrated in this section use the two-layered model generated with RAN methodology (see the two-layered model obtained at step 3 in Figure 4 in Section 3.2). In the generated RAN model, the five abstract concept nodes (C_1, C_2, C_3, C_4, and C_5) correspond to clusters 3, 1, 4, 2 and 5, respectively (see Figure 3 for the clusters). In both SCR and MCR operations, the five abstract concept nodes at layer 1 are injected with an expected activation set. Furthermore, the geometric back-propagation algorithm (Algorithm 2) performs one thousand iterations of downward propagation of activation to obtain appropriate values at input layer 0 as recalled activation. In all of the recall simulations, the GBP operation is initialized with values 0 and 0.60 as *Starting-Point*, and *maxIter* is set for 1000 iterations. The expected activation varies with the experiment, and two sets of the regulation factor were determined empirically to demonstrate both recall procedures.

3.5.1. Single-Cue Recall (SCR) Experiment

In SCR experiments, the objective was to determine the recalled activation at input layer 0 by injecting binary activation values as expected activation in abstract concept layer 1. The expected activation vector contains value 1 for only one abstract concept node, and for the remaining nodes, 0 is assigned. In all SCR experiments, six regulation factors ρ (0%, 0.5%, 0.75%, 1%, 1.25% and 1.5%) were used. Table 1 logs the E-A and the thousandth iteration value of O-A for the five SCR experiments. The following are the five experiments to demonstrate the SCR operation along with observations:

- **Exp-1:** This is the first experiment in which we injected an E-A vector *[0, 0, 0, 0, 1]* of activation at abstract concept layer 1. The objective was to recall activations at input layer 0 for which a very high activation was observed at node C_5 at layer 1 and comparatively lower activation for the other four nodes at layer 1. The GBP algorithm was executed six times with an E-A of *[0, 0, 0, 0, 1]* for the six different regulation factors (ρ). The observation for Exp-1 (see Table 1) shows that with a ρ of 0.75%, the maximum activation of 0.85 was observed at node C_5. As expected, the node C_2 received good activation because nodes C_2 and C_5 represent clusters 1 and 5 (see Figure 3), which are close to one another. Figure 8a shows the six trajectories for the six regulation factors; each trajectory is formed by one thousand iterations. In Figure 8a, the yellow marker shows the CRDP of cluster C_5 and the trajectory with ρ of 0.75% converge closest to this CRPD. Thus, an activation vector *[0.1, 0.24]* is recalled at input nodes [S_1, S_2] for the given E-A vector *[0, 0, 0, 0, 1]*.
- **Exp-2:** In this experiment, the E-A provided to the GBP algorithm was *[0, 0, 0, 1, 0]* to recall activation at layer 0, which is strongly represented by node C_4. For each regulation factor, the GBP algorithm was run; the O-As obtained at layer 1 are listed in Table 1, and the corresponding recalled activation at layer 0 is shown in Figure 8b. From the observations, it can be deduced that the experiment with ρ of 0.75% produced the best outcome and recall activation *[0.9, 0.9]* for input layer 0.
- **Exp-3:** In this experiment, the GBP algorithm was supplied with an E-A vector of *[0, 0, 1, 0, 0]* to recall the input layer 0 vector, which is represented by node C_3 at layer 1. Figure 8c and Table 1 shows the recall trajectories at layer 0 and the O-A vector at layer 1, respectively. The experiment with the regulation of 0.75% displayed the best representation. A vector *[0.92, 0.11]* was recalled at layer 0 for the injected E-A vector.
- **Exp-4:** The aim of this experiment was to recall an input vector that closely represents the abstract concept node C_2 by feeding the GBP algorithm with an E-A vector of *[0, 1, 0, 0, 0]*. After applying the six regulation factors to each GBP operation, it was observed that the experiment with ρ of 0.75% displayed the best result. Table 1 shows the O-A for the E-A. Figure 8d shows the trajectories of the recalled values and shows the best outcome with ρ of 0.75% that converges to an activation vector *[0.14, 0.07]*.
- **Exp-5:** This experiment shows the recall vector obtained by initializing an E-A vector *[1, 0, 0, 0, 0]* with all GBP experiments with the six regulation values. Unlike the previous experiments, the best O-A was obtained with a regulation factor of 1%. Figure 8d shows the recall outcome for all six regulation factors. At input layer 0, the recall operation with ρ of 1% results into an activation vector *[0.15, 0.91]* for the given E-A.

Algorithm 2: Geometric Back-Propagation Operation

Input: A $ExpectActivation$ activation A'_n ($< A'_1, .., A'_j, .., A'_n >$) at layer L, with n nodes

Input: Desired Maximum Iteration $maxIter$

Output: An activation pattern a_m ($< a_1, .., a_i, .., a_m >$) at layer L-1, with m nodes

Set Regulation Factor ρ between [0, 1];
Set $currentActivation = Starting$-$Point$ (a vector of activation $< a_1, \ldots, a_m >$);
Set $previousActivation = currentActivation$;
Set $PropagateActivation$= CCUAP of $currentActivaiton$ to layer L (see Section 3.2.5);
Set $ObservedlActivation$ (A)= Regulate $PropagateActivation$ via Algorithm 1;
Calculate error vector (e) at layer L using Equation (10);
Set $iter = 0$;
repeat
  Set $iter = iter + 1$;
  foreach a_i in $previousActivation$ **do**
    Calculate the delta ($\triangle_{a_i, A_j}$) using Equation (11);
    Calculate the sum of delta for a_i, i.e., $\triangle_{a_i}$, using Equation (12);
    if $\triangle_{a_i} > 0$ **then**
      $a_{temp} = a_i + \triangle_{a_i} * (1 - a_i)$;
      if $a_{temp} > 1$ **then**
        Assign $a_i = 1$;
      end
      else if $a_{temp} < 0$ **then**
        Assign $a_i^{new} = 0$;
      end
      else
        Assign $a_i^{new} = a_{temp}$;
      end
    end
    else
      $a_{temp} = a_i + \triangle_{a_i} * (a_i)$;
      if $a_{temp} > 1$ **then**
        Assign $a_i^{new} = 1$;
      end
      else if $a_{temp} < 0$ **then**
        Assign $a_i^{new} = 0$;
      end
      else
        Assign $a_i^{new} = a_{temp}$;
      end
    end
  end
  Set $currentActivation = < a_1^{new}, .., a_i^{new}, .., a_m^{new} >$ (new activation vector at layer L-1);
  Set $previousActivation = currentActivation$;
  Set $PropagateActivation$= CCUAP of $currentActivaiton$ to layer L (see Section 3.2.5);
  Set $ObservedlActivation$ (A)= Regulate $PropagateActivation$ via Algorithm 1;
  Calculate error vector (e) at layer L using Equation (10);
until $iter = maxIter$;
return $currentActivation$

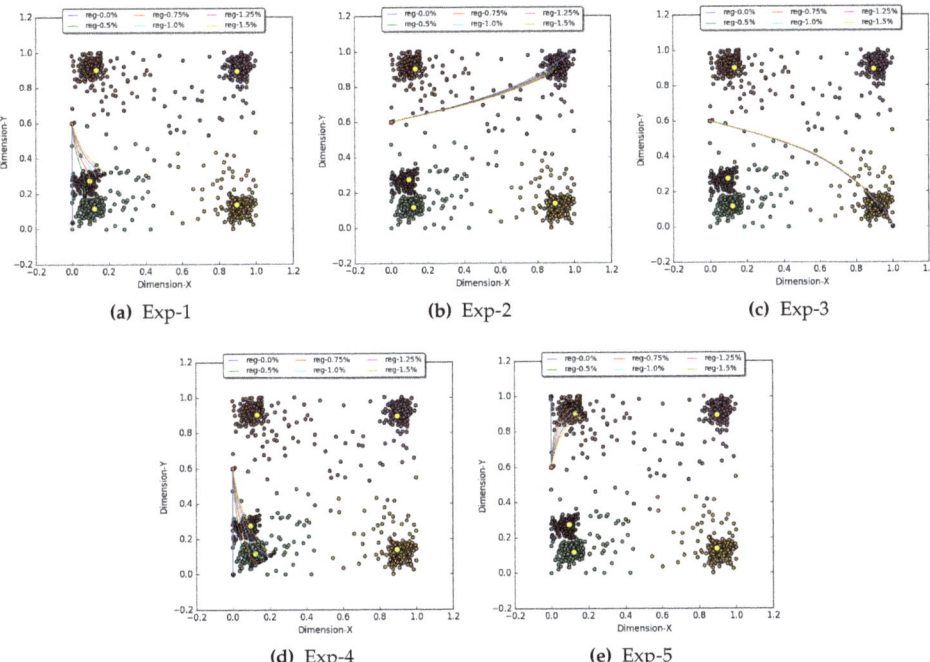

Figure 8. The trajectories of activation observed at input layer 0 while carrying out one thousand iterations of the geometric back-propagation (GBP) algorithm. The red circle is the starting point of trajectory, and the black circle is the activation value after the thousandth iteration. The graphs also depict the trajectories observed at input layer 0 with six regulation factors ρ (0%, 0.5%, 0.75%, 1%, 1.25%, and 1.5%). Each graph visualizes the recalled activation for five single-cue recall (SCR) experiments.

Table 1. Observations of activations at abstract concept layer 1 for SCR experiments.

Regulation	Experiment	E-A at Layer 1					O-A at Layer 1				
%		C_1	C_2	C_3	C_4	C_5	C_1	C_2	C_3	C_4	C_5
0	Exp-1	0	0	0	0	1	0.08	0.61	0.07	0.01	0.47
0.5	Exp-1	0	0	0	0	1	0.14	0.70	0.09	0.03	0.72
0.75	**Exp-1**	0	0	0	0	1	0.16	**0.64**	0.10	0.04	**0.85**
1	Exp-1	0	0	0	0	1	0.19	0.56	0.11	0.05	0.82
1.25	Exp-1	0	0	0	0	1	0.19	0.54	0.12	0.06	0.75
1.5	Exp-1	0	0	0	0	1	0.20	0.52	0.12	0.06	0.71
0	Exp-2	0	0	0	1	0	0.08	0.01	0.08	0.61	0.02
0.5	Exp-2	0	0	0	1	0	0.10	0.02	0.11	0.83	0.03
0.75	**Exp-2**	0	0	0	1	0	0.11	0.02	0.11	**0.93**	0.04
1	Exp-2	0	0	0	1	0	0.12	0.03	0.12	0.90	0.04
1.25	Exp-2	0	0	0	1	0	0.12	0.03	0.13	0.83	0.05
1.5	Exp-2	0	0	0	1	0	0.13	0.03	0.13	0.78	0.05
0	Exp-3	0	0	1	0	0	0.01	0.07	0.57	0.07	0.06
0.5	Exp-3	0	0	1	0	0	0.02	0.09	0.76	0.10	0.08
0.75	**Exp-3**	0	0	**1**	0	0	0.02	0.10	**0.84**	0.11	0.09
1	Exp-3	0	0	1	0	0	0.03	0.11	0.93	0.11	0.09
1.25	Exp-3	0	0	1	0	0	0.03	0.11	0.91	0.12	0.10
1.5	Exp-3	0	0	1	0	0	0.03	0.12	0.84	0.13	0.10
0	Exp-4	0	1	0	0	0	0.06	0.57	0.07	0.01	0.42
0.5	Exp-4	0	1	0	0	0	0.08	0.75	0.10	0.02	0.50
0.75	**Exp-4**	0	**1**	0	0	0	0.09	**0.80**	0.12	0.02	**0.52**
1	Exp-4	0	1	0	0	0	0.10	0.77	0.13	0.03	0.53
1.25	Exp-4	0	1	0	0	0	0.10	0.73	0.14	0.03	0.54
1.5	Exp-4	0	1	0	0	0	0.11	0.69	0.15	0.04	0.54

Table 1. Cont.

Regulation %	Experiment	E-A at Layer 1					O-A at Layer 1				
		C_1	C_2	C_3	C_4	C_5	C_1	C_2	C_3	C_4	C_5
0	Exp-5	1	0	0	0	0	0.58	0.07	0.01	0.07	0.13
0.5	Exp-5	1	0	0	0	0	0.78	0.09	0.02	0.10	0.15
0.75	Exp-5	1	0	0	0	0	0.85	0.10	0.02	0.11	0.16
1	**Exp-5**	**1**	**0**	**0**	**0**	**0**	**0.88**	**0.10**	**0.03**	**0.12**	**0.17**
1.25	Exp-5	1	0	0	0	0	0.84	0.11	0.03	0.13	0.17
1.5	Exp-5	1	0	0	0	0	0.79	0.11	0.04	0.13	0.18

E-A, expected activation, O-A, observed activation.

3.5.2. Multiple-Cue Recall (MCR) Experiment

The MCR experiments were carried out to determine the recall vector at input layer 0 for an E-A vector at layer 1. The constituents of the E-A vector are degree of confidence (DoC) values that define the expected representation of each abstract concept node at layer 1. To demonstrate MCR, five experiments were performed, and in every experiment, six regulation factors ρ (0%, 0.1%, 0.2%, 0.3%, 0.4%, and 0.5%) were used to make inferences. Table 2 lists the observations of O-A for the respective E-A in each MCR experiment. Figure 9 displays the trajectories of recalled activation in layer 0 for six regulation factors with respect to each experiment. The E-A vectors used in the MCR experiments are vectors obtained by propagating an input activation from layer 0 to layer 1. Hence, in MCR simulation, we also have an expected recall (E-R) vector to perform evaluations. The following are the MCR experiment descriptions along with observations:

- **Exp-6:** In this experiment, an E-A vector of *[0.57, 0.16, 0.06, 0.15, 0.25]* was provided to the GBP algorithm. With this E-A, we wanted to recall activation at input layer 0, that is 57%, 16%, 06%, 15% and 25% represented by nodes C_1, C_2, C_3, C_4, and C_5, respectively. Table 2 lists the O-A observed for the six regulation factors. The results with ρ of 0% and 0.1% show the outcome, which is almost identical to that of E-A. The E-R vector of this experiment was *[0.2256, 0.7610]*. With ρ of 0% and 0.1%, the observed recall was *[0.2260781, 0.7647118]* and *[0.2343844, 0.7602842]*, respectively, which are also similar to the expected recall vector. Figure 9a shows the trajectories of all the recalled activation vectors at layer 0 for the E-A vector w.r.t. their six regulation factors.
- **Exp-7:** For this experiment, the GBP algorithm was injected with an E-A vector of *[0.07, 0.04, 0.23, 0.46, 0.05]* for recalling an activation vector at input layer 0. The six O-A vectors obtained for the six regulation factors can be seen in Table 2. The O-A vector for ρ of 0% and 0.1% was almost the same as the E-A vector. The recalled activations regulation factor 0% was *[0.9875402, 0.6551013]*, which is almost similar to the E-R vector *[0.9896, 0.6568]*. Figure 9b shows all the recalled trajectories for this experiment.
- **Exp-8:** In this experiment, the GBP algorithm was initialized with the E-A vector of *[0.09, 0.5, 0.22, 0.05, 0.52]*. The E-R vector for this experiment was *[0.3458, 0.1157]*, and two similar vectors, *[0.3444873, 0.1032499]* and *[0.3489568, 0.1284956]*, were recalled in this experiment using regulation of 0% and 0.1%, respectively. Figure 9c shows all recalled trajectories. The O-A vectors obtained with the regulation of 0% and 0.1% were also identical to the E-A vector of this experiment—see Table 2.
- **Exp-9:** The recall simulation in this experiment was instantiated with an E-A vector of *[0.09, 0.40, 0,28, 0.07, 0.35]*, and a recall vector of *[0.4410, 0.1341]* was expected at input layer 0. Upon using the GBP algorithm with six regulation factors, the recall operation without regulation, i.e., ρ of 0%, produced the most similar recall vector *[0.4370989, 0.1308456]* and the corresponding O-A vector at layer 1. However, the outcome with 0.1% regulations was also similar to a recalled vector *[0.4392012, 0.1518065]* at input layer 0—see Figure 9d.
- **Exp-10:** This experiment used an E-A vector *[0.07, 0.09, 0.44, 0.26, 0.10]* in order to obtain an E-R vector *[0.8813, 0.4145]* at input layer 0. The six simulations were carried

out with different regulation factors, and it was observed that the results with ρ of 0% and 0.1% produced results very near those of the E-R, i.e., [0.8873921, 0.4137484] and [0.8702277, 0.4153301], respectively—see Figure 9e for all trajectories. The same observations were made at the O-A vectors for ρ of 0% and 0.1% at layer 1—see Table 2.

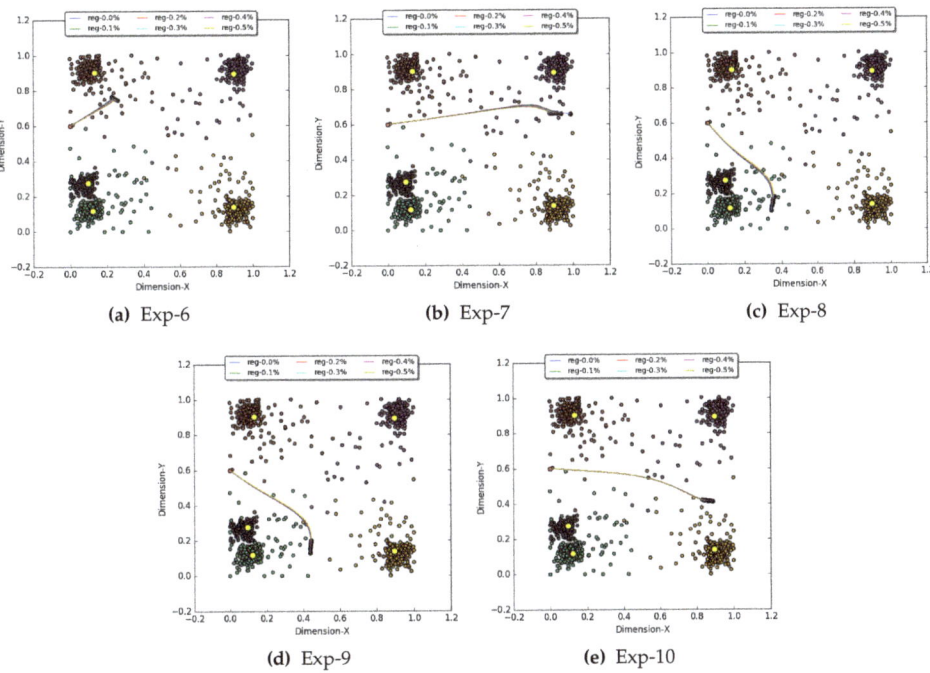

(a) Exp-6 (b) Exp-7 (c) Exp-8

(d) Exp-9 (e) Exp-10

Figure 9. The trajectories of activation observed at input layer 0 while carrying out one thousand iterations of the GBP algorithm. The red circle is the starting point of trajectory, and the black circle is the activation value after the thousandth iteration. The graphs also depict the trajectories observed at input layer 0 with six regulation factors ρ (0%, 0.1%, 0.2%, 0.3%, 0.4%, and 0.5%). Each graph visualizes the recalled activation for five multiple-cue recall (MCR) experiments.

Table 2. Observations of activations at abstract concept layer 1 for MCR experiments.

Regulation	Experiment	E-A at Layer 1					O-A at Layer 1				
%		C_1	C_2	C_3	C_4	C_5	C_1	C_2	C_3	C_4	C_5
0	Exp-6	0.57	0.16	0.06	0.15	0.25	0.579	0.162	0.063	0.148	0.246
0.1	Exp-6	0.57	0.16	0.06	0.15	0.25	0.567	0.163	0.066	0.151	0.247
0.2	Exp-6	0.57	0.16	0.06	0.15	0.25	0.556	0.164	0.068	0.154	0.249
0.3	Exp-6	0.57	0.16	0.06	0.15	0.25	0.546	0.166	0.070	0.156	0.250
0.4	Exp-6	0.57	0.16	0.06	0.15	0.25	0.537	0.167	0.072	0.159	0.251
0.5	Exp-6	0.57	0.16	0.06	0.15	0.25	0.529	0.168	0.074	0.161	0.252
0	Exp-7	0.07	0.04	0.23	0.46	0.05	0.072	0.039	0.233	0.465	0.051
0.1	Exp-7	0.07	0.04	0.23	0.46	0.05	0.086	0.048	0.235	0.482	0.061
0.2	Exp-7	0.07	0.04	0.23	0.46	0.05	0.093	0.052	0.235	0.486	0.067
0.3	Exp-7	0.07	0.04	0.23	0.46	0.05	0.099	0.055	0.236	0.487	0.071
0.4	Exp-7	0.07	0.04	0.23	0.46	0.05	0.103	0.058	0.236	0.487	0.074
0.5	Exp-7	0.07	0.04	0.23	0.46	0.05	0.107	0.061	0.236	0.485	0.077
0	Exp-8	0.09	0.50	0.22	0.05	0.42	0.091	0.502	0.218	0.051	0.414
0.1	Exp-8	0.09	0.50	0.22	0.05	0.42	0.099	0.497	0.221	0.056	0.424
0.2	Exp-8	0.09	0.50	0.22	0.05	0.42	0.105	0.492	0.223	0.061	0.430
0.3	Exp-8	0.09	0.50	0.22	0.05	0.42	0.110	0.486	0.224	0.064	0.434
0.4	Exp-8	0.09	0.50	0.22	0.05	0.42	0.114	0.481	0.225	0.067	0.437
0.5	Exp-8	0.09	0.50	0.22	0.05	0.42	0.117	0.476	0.226	0.070	0.439
0	Exp-9	0.09	0.40	0.28	0.07	0.35	0.090	0.401	0.280	0.070	0.350
0.1	Exp-9	0.09	0.40	0.28	0.07	0.35	0.096	0.397	0.281	0.076	0.355
0.2	Exp-9	0.09	0.40	0.28	0.07	0.35	0.101	0.394	0.282	0.080	0.358
0.3	Exp-9	0.09	0.40	0.28	0.07	0.35	0.105	0.391	0.282	0.084	0.360

Table 2. Cont.

Regulation %	Experiment	E-A at Layer 1					O-A at Layer 1				
		C_1	C_2	C_3	C_4	C_5	C_1	C_2	C_3	C_4	C_5
0.4	Exp-9	0.09	0.40	0.28	0.07	0.35	0.109	0.388	0.282	0.087	0.362
0.5	Exp-9	0.09	0.40	0.28	0.07	0.35	0.112	0.385	0.281	0.090	0.363
0	**Exp-10**	0.07	0.09	0.44	0.26	0.10	0.068	0.093	0.440	0.263	0.099
0.1	**Exp-10**	0.07	0.09	0.44	0.26	0.10	0.073	0.098	0.437	0.264	0.105
0.2	Exp-10	0.07	0.09	0.44	0.26	0.10	0.076	0.103	0.434	0.264	0.110
0.3	Exp-10	0.07	0.09	0.44	0.26	0.10	0.079	0.106	0.431	0.264	0.114
0.4	Exp-10	0.07	0.09	0.44	0.26	0.10	0.082	0.109	0.428	0.265	0.117
0.5	Exp-10	0.07	0.09	0.44	0.26	0.10	0.084	0.112	0.425	0.265	0.120

E-A, expected activation, O-A, observed activation.

3.5.3. Discussion

The experiments in Sections 3.5.1 and 3.5.2 demonstrate a notable behavior of RAN by simulating the cued recall operation through a Toy-data problem. The intra-Layer learning (i.e., CSRL) is uniquely utilized by RAN modeling to interpret the association among the concepts as inhibitory, excitatory, or neutral. Furthermore, the intra-layer regulation (Algorithm 1) uses intra-layer learning (CSRL) and its interpretations to produce a regulatory effect over the activation of the concepts (at the same layer). The geometric back-propagation operation (Algorithm 2) is a method analogous to remembering something learned in an abstract form and recalling its concrete features. For example, while remembering the abstract concept "house", we recall concrete features related to the house, such as "mother", "father", "wife", and "pets".

In the graphs in Figures 8 and 9, we can see that all the trajectories commence from a starting point (red dot) and converge to a point after one thousand iterations. Each point in a trajectory represents a temporal mental state while recalling a concrete concept. Every time a concrete concept (activation vector in layer 0) is recalled, its corresponding abstract concept (at layer 1) is compared with the expected abstract concepts. The difference between expected and observed activation is propagated back as the error to the previously recalled activations at layer 0. In the next instance, the corrected recalled activation at layer 0 repeats the process until one thousand iterations are completed.

It was observed that without regulation, i.e., 0% ρ, the trajectory converges to a point but with a minimal amount of regulation, and the result improves. For instance, in the graphs in Figure 8, only one abstract concept was being recalled and the results improved when the regulation was introduced. In the two experiments (SCR and MCR), we can see that the two different sets of regulation factors are considered. These sets were obtained empirically, but we can see that the set of the regulation factors for the SCR experiment has a higher value. This is because the GBP algorithm strives to minimizes the error at each abstract concept node at layer 1, and in the geometrical context, similarity cannot be the same for more than one abstract concept. Thus, the trajectory converges to a point, but the result improves when a minimal amount of regulation is induced. In the MCR experiment, the best outcome is observed with little or no regulation because the expected similarity (DoC and E-A) is a non-zero value. The other reason is that these are possible expected similarity vectors, unlike those in the SCR experiments.

4. Cued Recall Demonstration with MNIST Data

The MNIST [70] dataset is a collection of handwritten images of digits (0, 1, 2, 3, 4, 5, 6, 7, 8 and 9), where each image is black and white in color and has a 28 × 28 pixel size. This dataset of image domains is used to demonstrate the Ccued recall operation of learned abstract concepts representing different digits. Two types of investigations were conducted with this dataset: first, multiple binary valued cue recall (MBVCR), where the E-A vector is a binary value ([0, 1]) vector, and second, multiple-cue recall (MCR).

For this experiment, one thousand images were selected randomly from the MNIST dataset. The 28 × 28 image was transformed in a single vector of 784 attributes, where each attribute corresponds to a pixel of the image. Additionally, the attribute values of the data were normalized between 0 and 1 using min–max normalization (black pixel is min, i.e., 0, and white pixel is max, i.e., 255). Having preprocessed the data, the RAN modeling procedure was instantiated by selecting the K-mean clustering algorithm as the concept identifier. K was initialized with 30 to determine thirty categories in the input space. The model was configured to grow one level deep and build convex abstract concept (CAC) layer 1. After carrying out all four steps of RAN modeling (see Section 3.2), a model was obtained—see Figure 10.

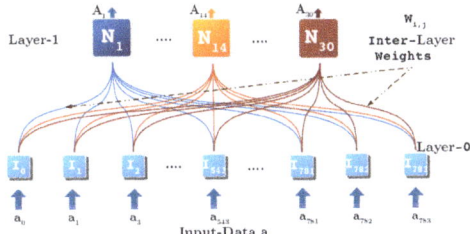

Figure 10. RAN model generated with MNIST dataset.

In Figure 10, layer 0 has 784 nodes representing each pixel; CAC layer 1 has 30 nodes representing the thirty categories identified during the CCI process in RAN modeling. The interlayer weights (ILWs) are the cluster centers (CRDPs) of the thirty clusters. Figure 11 shows the ILWs reconstructed in image form of 28 × 28 pixels. In RAN modeling, a CRDP is the optimum representative of an input level category at CAC layer 1. Therefore, Figure 11a–ad are the best represented by CAC node N_1, ..., N_{30}, respectively. In Figure 11 it is noticeable that each digit is represented by at least two CAC node of layer 1. The digit 9 is represented by the largest number of nodes, i.e., N_2, N_{15}, N_{18}, and N_{24}. In contrast, digit 4 is represented by two nodes, N_1 and N_{10}. Figure 11d,m,u,x show that the CAC nodes N_4, N_{13}, N_{21}, and N_{24} do not represent an individual digit. Node N_4 and N_{13} jointly represent digits 3 and 8; node N_{21} looks like two digits, 3 and 5; and N_{24} depicts digits 7 and 9.

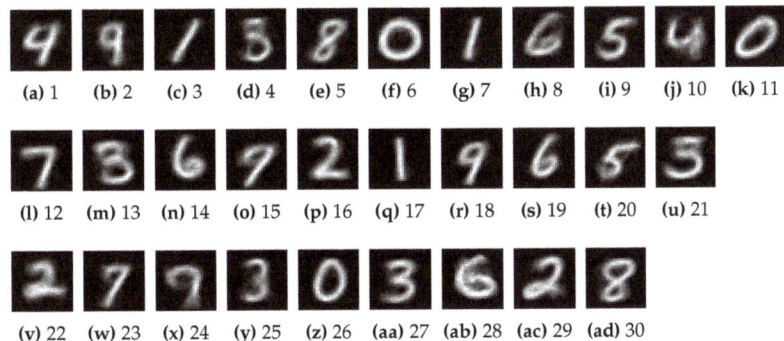

Figure 11. The thirty CRDPs (cluster centers). Each node in layer 1 of Figure 10 acts as the abstract representative of each CRDP.

For simplicity, Figure 10 shows only inter-layer learning; intra-Layer learning (CSRL weights) was also performed on both the input layer 0 and CAC layer 1. The CSRL weights at input layer 0 were a 784 × 784 matrix, and at CAC layer 1, a 30 × 30 matrix was learned. These two intra-layer learning procedures were utilized by the GBP algorithm to simulate the recall operations. In all of the experiments, the GBP algorithm was configured to iterate five hundred times. The GBP algorithm was initialized with a vector with activation 1 for all 784 nodes of input layer 0. The image at Iter-0 (see Tables 3 and 4) is white because activation 1 corresponds to pixel value of 255 depicting white color.

Table 3. Intuitive multiple binary valued cue recall (MBVCR) observations with RAN model of MNIST data.

Digit	æ	Iter ⇒	0	3	5	8	11	19	25	35	41	71	81	91	101	151	201	251	301	351	401	451	501
0	0%																						
0	0.009%																						
1	0%																						
1	0.009%																						
2	0%																						
2	0.009%																						
3	0%																						
3	0.009%																						
4	0%																						
4	0.009%																						
5	0%																						
5	0.009%																						
6	0%																						
6	0.009%																						
7	0%																						
7	0.009%																						
8	0%																						
8	0.009%																						
9	0%																						
9	0.009%																						

Table 4. Observations of multiple-cue recall operation with RAN model of MNIST data.

Digit	æ	Iter ⇒	0	3	5	8	11	19	25	35	41	71	81	91	101	151	201	251	301	351	401	451	501
0	0%																						
0	0.009%																						
1	0%																						
1	0.009%																						
2	0%																						
2	0.009%																						
3	0%																						
3	0.009%																						
4	0%																						
4	0.009%																						
5	0%																						
5	0.009%																						
6	0%																						
6	0.009%																						
7	0%																						
7	0.009%																						
8	0%																						
8	0.009%																						
9	0%																						
9	0.009%																						

In each experiment, the two cued recall demonstrations, MBVCR and MCR, use the expected activating (E-A) vector as listed in Table 5. The experiments of single digits and combined digits for the MBVCR operation used an E-A vector of binary values, where binary 1 at a node N was assigned w.r.t the digit(s) being recalled. For instance, the E-A vector of digit 2 was formed by initializing the E-A vector with binary 1 for nodes N_{16}, N_{22}, and N_{29} and binary 0 for the remaining 27 CAC nodes (see Table 5). The E-A vectors of MCR experiments are the actual activation values obtained by propagating the inter-layer weights upwards (CRDPs, see Figure 11) as input. The weights represented by Figure 11a–c,e,f,i,l–n,p were provided as input to the CCUAP operation to observe their respective activation at CAC layer 1. These observed activation vectors were used as E-A for each digit recall operation (see last ten MCR E-As in Table 5).

Table 5. Expected activation injected at thirty convex abstract concept (CAC) nodes in layer 1 of RAN model for MNIST data.

Digit	Exp	Expected Activation (E-A)
0	MBVCR	[0,0,0,0,0,1,0,0,0,0,1,0,0,0,0,0,0,0,0,0,0,0,0,0,0,1,0,0,0,0]
1	MBVCR	[0,0,1,0,0,0,1,0,0,0,0,0,0,0,0,1,0,0,0,0,0,0,0,0,0,0,0,0,0,0]
2	MBVCR	[0,0,0,0,0,0,0,0,0,0,0,0,0,0,0,1,0,0,0,0,0,1,0,0,0,0,0,0,1,0]
3	MBVCR	[0,0,0,1,0,0,0,0,0,0,0,0,1,0,0,0,0,0,0,0,0,0,0,0,1,0,1,0,0,0]
4	MBVCR	[1,0,0,0,0,0,0,0,0,1,0,0,0,0,0,0,0,0,0,0,0,0,0,0,0,0,0,0,0,0]
5	MBVCR	[0,0,0,0,0,0,0,1,0,0,0,0,0,0,0,0,0,0,1,1,0,0,0,0,0,0,0,0,0,0]
6	MBVCR	[0,0,0,0,0,0,0,1,0,0,0,0,0,0,1,0,0,0,0,1,0,0,0,0,0,0,0,1,0,0]
7	MBVCR	[0,0,0,0,0,0,0,0,0,0,0,1,0,0,0,0,0,0,0,0,0,1,0,0,0,0,0,0,0,0]
8	MBVCR	[0,0,0,0,1,0,0,0,0,0,0,0,0,0,0,0,0,0,0,0,0,0,0,0,0,0,0,0,0,1]
9	MBVCR	[0,1,0,0,0,0,0,0,0,0,0,0,0,0,1,0,0,1,0,0,0,0,0,0,1,0,0,0,0,0]
2 and 5	MBVCR	[0,0,0,0,0,0,0,0,0,1,0,0,0,0,0,0,0,1,0,0,0,1,1,1,0,0,0,0,0,1,0]
3 and 5	MBVCR	[0,0,0,1,0,0,0,0,0,1,0,0,0,1,0,0,0,0,0,0,0,1,1,0,0,0,1,0,1,0,0,0]
0 and 1	MBVCR	[0,0,1,0,0,1,1,0,0,0,0,1,0,0,0,0,0,1,0,0,0,0,0,0,0,0,1,0,0,0,0]
0	MCR	[0.26,0.25,0.22,0.30,0.27,1.00,0.23,0.30,0.28,0.30,0.35,0.26,0.32,0.26,0.25,0.28,0.22,0.23,0.25,0.33,0.28,0.25,0.29,0.28,0.39,0.28,0.31,0.27,0.30]
1	MCR	[0.29,0.37,1.00,0.35,0.43,0.22,0.49,0.33,0.34,0.32,0.26,0.31,0.31,0.28,0.34,0.39,0.38,0.35,0.34,0.39,0.31,0.35,0.43,0.31,0.42,0.27,0.36,0.24,0.31,0.33]
2	MCR	[0.28,0.35,0.39,0.38,0.40,0.28,0.38,0.36,0.33,0.31,0.26,0.29,0.36,0.35,0.29,1.00,0.34,0.31,0.37,0.34,0.36,0.43,0.34,0.29,0.47,0.32,0.37,0.33,0.37,0.34]
3	MCR	[0.31,0.38,0.31,0.46,0.35,0.32,0.32,0.31,0.35,0.33,0.26,0.32,1.00,0.33,0.31,0.36,0.31,0.33,0.31,0.30,0.40,0.34,0.33,0.35,0.38,0.34,0.46,0.35,0.27,0.37]
4	MCR	[1.00,0.46,0.29,0.32,0.38,0.26,0.30,0.38,0.37,0.44,0.31,0.42,0.31,0.39,0.44,0.28,0.31,0.48,0.36,0.37,0.29,0.34,0.37,0.42,0.34,0.34,0.31,0.32,0.40]
5	MCR	[0.37,0.42,0.34,0.44,0.47,0.28,0.36,0.41,1.00,0.29,0.39,0.36,0.35,0.34,0.37,0.33,0.36,0.42,0.43,0.40,0.42,0.33,0.38,0.35,0.43,0.37,0.45,0.29,0.33,0.41]
6	MCR	[0.39,0.40,0.28,0.38,0.33,0.26,0.31,0.45,0.34,0.38,0.28,0.31,0.33,1.00,0.34,0.35,0.32,0.36,0.42,0.32,0.32,0.40,0.33,0.35,0.35,0.36,0.42,0.33,0.33]
7	MCR	[0.42,0.45,0.31,0.34,0.35,0.26,0.34,0.32,0.36,0.34,0.28,1.00,0.32,0.31,0.41,0.29,0.35,0.47,0.33,0.35,0.29,0.32,0.43,0.47,0.38,0.30,0.34,0.25,0.28,0.34]
8	MCR	[0.38,0.41,0.43,0.39,1.00,0.27,0.40,0.38,0.47,0.32,0.33,0.35,0.35,0.33,0.40,0.40,0.37,0.44,0.42,0.44,0.38,0.39,0.45,0.32,0.44,0.34,0.44,0.29,0.39,0.48]
9	MCR	[0.46,1.00,0.37,0.40,0.41,0.25,0.39,0.38,0.42,0.38,0.29,0.45,0.38,0.40,0.40,0.35,0.39,0.51,0.40,0.39,0.32,0.40,0.43,0.44,0.43,0.32,0.44,0.32,0.30,0.40]

4.1. Multiple Binary Valued Cue Recall (MBVCR) Operation

For the MBVCR operation, the RAN model generated with MNIST data (see Figure 10) was used in order to obtain the recalled activation at input layer 0 for a given expected activation vector at CAC layer 1. As described earlier, the E-A vector for MBVCR is a vector of binary values, which is provided as input to the GBP algorithm to perform the recall operation. The experiments themselves are divided into two categories, i.e., intuitive and non-intuitive recall.

4.1.1. Intuitive MBVCR Experiment

In this experiment, by intuition, we hypothesize that if all CAC nodes (representing a digit) are activated with value 1, then its recall at layer 0 must depict that digit. For example, if the CAC nodes N_6, N_{11}, and N_{26} (see Figure 11) are activated with a value of 1 (and 0 for others), then we should obtain an image depicting a blend of zero digits after the recall operation. We performed this intuitive recall experiment for all ten digits. The binary E-A vector of all ten digits for the intuitive MBVCR operation is listed in Table 5. Table 3 displays the recalled images of all twenty experiments. For every digit, two investigations were made; the first without regulation, i.e., $\rho = 0$; the second with a regulation of 0.009%.

The first observation is that there is a very insignificant difference between the images recalled with and without regulation. After the second iteration, the digit being recalled begins to appear. Beyond the 80th iteration, no significant change is observed in the recalled images. The recalled images of digits 0, 1, 2, 3, 7, and 8 are recognizable after the 500th iteration. However, the digits 4, 5, and 9 are not very discernible in their last iteration; this is because these digits are cross-represented by CAC nodes (see Figure 11). All the images recalled in this experiment contain noise (i.e., the gray shades), because the E-A vector has two values, either 0 or 1, and a node can be 100% similar to only one other node. Therefore, the GBP algorithm adjusts the activation at the CAC node such that the best representation of the E-A is achieved.

4.1.2. Non-Intuitive MBVCR Experiment

In these experiments, the E-A vector contains an activation value of 1 for CAC nodes representing two different digits. The objective of the experiment was to determine what is recalled at the input layer 0 when the CAC nodes, representing two different digits, expect high activation. The three E-As used in this experiment are a combination of activation 2^s-with-5^s, 3^s-with-5^s, and 0^s-with-1^s (see Table 5 for E-A vectors with the coupled digits). The observations without regulation and with regulation are similar—see Table 6. The blend of the 2^s-with-5^s recalls an image that looks like the letter x. The fusion of 3^s-with-5^s recalls an image similar to the digit 3. The combination of 0^s-with-1^s in the beginning looked like the symbol Φ, but this was distorted later. It is also observed that the images obtained after all the iterations had less noise when compared to the those of the intuitive MBVCR experiments. This is probably because a number of CAC nodes were expecting activation, i.e., more cues were provided.

Table 6. Non-intuitive MBVCR observations with RAN model of MNIST data.

Digit	æ	Iter ⇒ 0	3	5	8	11	19	25	35	41	71	81	91	101	151	201	251	301	351	401	451	501
2 and 5	0%																					
2 and 5	0.009%																					
3 and 5	0%																					
3 and 5	0.009%																					
0 and 1	0%																					
0 and 1	0.009%																					

4.2. Multiple-Cue Recall (MCR) Experiment

This experiment is the same as the experiment discussed in Section 3.5.2. The E-A vectors are the activation values observed at CAC nodes by propagating the inter-layer weights using the CCUAP operation of RAN modeling.

Figure 11 shows the images re-constructed for each inter-layer weight. The E-As corresponding to Figure 11a–c,e,f,i,l–n,p are listed in Table 5 and are used in MCR demonstrations of this section.

The objective of this experiment was the same as that of MBVCR experiments, i.e., obtaining an activation vector at input layer 0 that corresponds to an E-A vector. However, in this experiment, an expected recall (E-R) was already known. Therefore, the E-As of ten digits (see MCR E-As in Table 5) were expected to recall the images in Figure 11a–c,e,f,i,l–n,p.

In this experiment, the observations with and without regulation are identical. It is also worth noting that after the 500th iteration, the recalled images of all ten digits were similar to the E-R images of each digits.

4.3. Discussion

There are a few things worth mentioning in the recall demonstrations of RAN modeling with the MNIST dataset. First, we can reconstruct cognizable images of a digit by activating the CAC nodes representing that digit. Second, it is possible to recall both an intuitive and non-intuitive blend of learned abstract concepts (in these experiments, the abstract concepts are a generic representations of digits). Third, the recalled activations, with and without regulation, are similar for a complex dataset like MNIST. Last, the more cues we provide in the E-A vector, the more accurate the recall operation becomes. The recall capability of RAN modeling was applied to the reconsctruction of an encoded image, where the image was encoded using RAN convex concept modeling and reconstruction was performed via the geometric back-propagation (or recall) operation [71].

5. Conclusions

Recall is a cognitive process that can also be seen as an act of remembering a concept. Concepts are normally perceived in a hierarchical form, where the concrete concepts occupy the lower level, and the abstract concepts take up the relatively higher level in the hierarchy. According to context availability theory, the context among the concrete concepts is easily determined when compared to abstract concepts; hence, their comprehension and recall are also difficult. However, if we can relate abstract concepts to one another, it is possible to deduce a contextual relationship among them. In this article, we exploited the intra-layer associations learned among the concepts (including abstract concepts) using RAN modeling to establish context among the concepts. We use this context-related information to induce a regulatory effect on the concepts and further to simulate the recall operations.

To demonstrate the effect of regulation of the recall process, a Toy-data problem was considered. First, we modeled with Toy-data to identify five abstract concepts. The proposed regulation algorithm utilized the learned intra-layer weight to determine the excitatory, neutral, and inhibitory impact induced by peer nodes on one another. Two types of cued recall experiments were performed using the unique geometric back-propagation algorithm: first, single-cue recall (SCR), simulation where the recall was simulated by activating only one abstract concept; second, multiple-cue recall (MCR) operation to retrieve the activation vector at the input level by injecting multiple cues at the abstract nodes. In SCR experiments, the regulation induced by peer nodes improved the recalled values. However, the observations with MCR operations were promising because they retrieved identical activation, as expected.

The benchmark MNIST dataset was used to exhibit cued recall as blends of learned abstract concepts. A two-layered model was generated with RAN to obtain thirty abstract concepts generically representing digits. In the multiple binary valued cue recall (MBVCR) experiment, multiple abstract nodes were injected with high activation to recall as blends of digits. Interestingly, it was observed in all the experiments that the blend of abstract nodes recalled an image of the digit that they represent at the abstract Level. The blend of different digits also produced some intriguing outcome; for exmaple, a blend of 2 and 5 recalled x, and a blend of 0 with 1 looked like a Φ symbol. The MCR operations were interesting as upon injecting the multiple cues, the recalled image was very similar to the expected recalled image.

Both the experiments displayed how oncepts can be contextually associated and impact each other's activation through regulation. Furthermore, with cue recall operations, it can be concluded that the more cues injected to an abstract concept, the better the obtained recall results. For future work, we intend to perform conceptual combination experiments and study the aspects of creative concept retrievals with the geometric back-propagation algorithm.

Author Contributions: R.S. performed state of the art, developed and implemented the methodology, carried out data selection and methodology validation, and prepared the original draft of the article. B.R. supervised the research work performed the formal analysis, review and editing, took care of funding. A.M.P. conceived the study plan and methodology, supervised the investigation, methodology development and implementation. A.C. supervised the research work, performed formal analysis, review and edition, managed funding. All authors have read and agreed to the published version of the manuscript.

Funding: The work presented in this paper was partially carried out in the scope of the SOCIALITE Project (PTDC/EEI-SCR/2072/2014), co-financed by COMPETE 2020, Portugal 2020—Operational Program for Competitiveness and Internationalization (POCI), European Union's ERDF (European Regional Development Fund), and the Portuguese Foundation for Science and Technology (FCT). This work was also partially funded by project ConCreTe. ConCreTe acknowledges the financial support of the Future and Emerging Technologies (FET) program within the Seventh Framework Programme for Research of the European Commission, under FET grant number 611733.

Data Availability Statement: In this work we used two datasets: first is the Toydata it is availabel at R.S. GitHub account at (https://github.com/rahulsharma-rs/datasets/blob/master/synthetic-data/Toy-dataRANstestclusterchapter6.csv accessed on 30 January 2021); second dataset is an image dataset named MNIST which is available in public domain (http://yann.lecun.com/exdb/mnist/ accessed on 30 January 2021).

Acknowledgments: I would like to express special thanks to Christine Zhang, Assistant Editor of the Applied Science Journal for her help in finding funds to support this article's publication cost.

Conflicts of Interest: The authors declare no conflict of interest.

Abbreviations

The following abbreviations and Notations that are used in this manuscript:

Abbreviations	Description
CAC	Convex Abstract Concept
CACC	Convex Abstract Concept Creation
CACUAP	Convex Abstract Concept Upward Activation Propagation
CCI	Convex Concept Identification
CCILL	Convex Concept Inter-Layer Learning
CCILW	Convex Concepts Inter-Layer Weights
CI	Concept Identifier
CRPD	Cluster Representative Data Points
CSRL	Concept Similarity Relation Learning
CSRW	Concept Similarity Relation Weights
E-A	Expected Activation
GBP	Geometric Back Propagation
IL	Intra-Layer
ILWs	Inter-Layer Weights
MBVCR	Multiple Binary Valued Cue Recall
MCR	Multiple Cue Recall
RAN	Regulated Activation Network
SCR	Single Cue Recall
Notations	**Description**
W	Convex Concept Inter-layer weight matrix
w	Similarity Relation weight matrix
C	Cluster center or Centroids
A	Output Activation
a	Input Activation
i, k, j	Variables to represent node index for 0th, 1st and 2nd *layer* respectively
m, n	Arbitrary node indexes for any layer
I	Ith instance of input data
$f(x)$	Transfer function to obtain similarity relation
t	Variable used to depict intermediate index
n_a	Size of input Vector at Layer-0
n_A	Size of Convex Abstract Concept vector at Layer-1

References

1. Kiefer, M.; Pulvermüller, F. Conceptual representations in mind and brain: Theoretical developments, current evidence and future directions. *Cortex* **2012**, *48*, 805–825. [CrossRef] [PubMed]
2. Bechtel, W.; Graham, G.; Balota, D.A. *A Companion to Cognitive Science*; Wiley-Blackwell: Hoboken, NJ, USA, 1998.
3. Xiao, P.; Toivonen, H.; Gross, O.; Cardoso, A.; Correia, J.A.; Machado, P.; Martins, P.; Oliveira, H.G.; Sharma, R.; Pinto, A.M.; et al. Conceptual Representations for Computational Concept Creation. *ACM Comput. Surv.* **2019**, *52*, 1–33. [CrossRef]
4. Rosch, E.; Mervis, C.B.; Gray, W.D.; Johnson, D.M.; Boyes-Braem, P. Basic objects in natural categories. *Cogn. Psychol.* **1976**, *8*, 382–439. [CrossRef]
5. Tversky, B.; Hemenway, K. Objects, parts, and categories. *J. Exp. Psychol. Gen.* **1984**, *113*, 169–193. [CrossRef]
6. Saitta, L.; Zucker, J.D. Semantic abstraction for concept representation and learning. In Proceedings of the Symposium on Abstraction, Reformulation and Approximation, Horseshoe Bay, TX, USA, 26–29 July 2000; pp. 103–120.
7. Borghi, A.M.; Barca, L.; Binkofski, F.; Tummolini, L. Varieties of abstract concepts: Development, use and representation in the brain. *Philos. Trans. R. Soc. Lond. Ser. B Biol. Sci.* **2019**, *373*, 20170121. [CrossRef]
8. Borghi, A.M.; Binkofski, F.; Castelfranchi, C.; Cimatti, F.; Scorolli, C.; Tummolini, L. The challenge of abstract concepts. *Psychol. Bull.* **2017**, *143*, 263–292. [CrossRef]
9. Sharma, R.; Ribeiro, B.; Pinto, A.M.; Cardoso, F.A. Modeling Abstract Concepts For Internet of Everything: A Cognitive Artificial System. In Proceedings of the 13th APCA International Conference on Automatic Control and Soft Computing (CONTROLO), IEEE, Ponta Delgada, Portugal, 4–6 June 2018; pp. 340–345.

10. Sharma, R.; Ribeiro, B.; Pinto, A.M.; Cardoso, A.F.; Raposo, D.; Marcelo, A.R.; Silva, J.S.; Boavida, F. *Computational Concept Modeling for Student Centric Lifestyle Analysis: A Technical Report on SOCIALITE Case Study*; Technical Report; Center of Information Science University of Coimbra: Coimbra, Portugal, 2017.
11. Sharma, R.; Ribeiro, B.; Miguel Pinto, A.; Cardoso, F.A. Exploring Geometric Feature Hyper-Space in Data to Learn Representations of Abstract Concepts. *Appl. Sci.* **2020**, *10*, 1994. [CrossRef]
12. Paivio, A. *Mental Representations: A Dual Coding Approach*; Oxford University Press: Oxford, UK, 1990.
13. Paivio, A. *Imagery and Verbal Processes*; Holt, Rinehart & Winston: New York, NY, USA, 1971.
14. Schwanenflugel, P.J.; Akin, C.; Luh, W.M. Context availability and the recall of abstract and concrete words. *Mem. Cogn.* **1992**, *20*, 96–104. [CrossRef] [PubMed]
15. Bransford, J.D.; McCarrell, N.S. A sketch of a cognitive approach to comprehension: Some thoughts about understanding what it means to comprehend. In *Thinking Reading in Cognitive Science*; Cambridge University Press: Cambridge, UK, 1974.
16. Costescu, C.; Rosan, A.; Brigitta, N.; Hathazi, A.; Kovari, A.; Katona, J.; Demeter, R.; Heldal, I.; Helgesen, C.; Thill, S.; et al. Assessing Visual Attention in Children Using GP3 Eye Tracker. In Proceedings of the 2019 10th IEEE International Conference on Cognitive Infocommunications (CogInfoCom), Naples, Italy, 23–25 October 2019; pp. 343–348. [CrossRef]
17. Ujbanyi, T.; Katona, J.; Sziladi, G.; Kovari, A. Eye-tracking analysis of computer networks exam question besides different skilled groups. In Proceedings of the 2016 7th IEEE International Conference on Cognitive Infocommunications (CogInfoCom), Wroclaw, Poland, 16–18 October 2016; pp. 000277–000282. [CrossRef]
18. Katona, J.; Ujbanyi, T.; Sziladi, G.; Kovari, A. Examine the effect of different web-based media on human brain waves. In Proceedings of the 2017 8th IEEE International Conference on Cognitive Infocommunications (CogInfoCom), Debrecen, Hungary, 11–14 September 2017; pp. 000407–000412.
19. Katona, J.; Ujbanyi, T.; Sziladi, G.; Kovari, A. Speed control of Festo Robotino mobile robot using NeuroSky MindWave EEG headset based brain-computer interface. In Proceedings of the 2016 7th IEEE International Conference on Cognitive Infocommunications (CogInfoCom), Wroclaw, Poland, 16–18 October 2016, pp. 000251–000256. [CrossRef]
20. Rolls, E.; Loh, M.; Deco, G.; Winterer, G. Computational models of schizophrenia and dopamine modulation in the prefrontal cortex. *Nat. Rev. Neurosci.* **2008**, *9*, 696–709. [CrossRef]
21. Kyaga, S.; Landén, M.; Boman, M.; Hultman, C.M.; Långström, N.; Lichtenstein, P. Mental illness, suicide and creativity: 40-Year prospective total population study. *J. Psychiatr. Res.* **2013**, *47*, 83–90. [CrossRef] [PubMed]
22. Braver, T.; Barch, D.; Cohen, J. Cognition and control in schizophrenia: A computational model of dopamine and prefrontal function. *Biol. Psychiatry* **1999**, *46*, 312–328. [CrossRef]
23. O'Reilly, R.C. Biologically based computational models of high-level cognition. *Science* **2006**, *314*, 91–94. [CrossRef] [PubMed]
24. Hayes, J.C.; Kraemer, D.J. Grounded understanding of abstract concepts: The case of STEM learning. *Cogn. Res. Princ. Implic.* **2017**, *2*, 7. [CrossRef] [PubMed]
25. Anderson, J.R.; Matessa, M.; Lebiere, C. ACT-R: A theory of higher level cognition and its relation to visual attention. *Hum. Comput. Interact.* **1997**, *12*, 439–462. [CrossRef]
26. Lovett, M.C.; Daily, L.Z.; Reder, L.M. A source activation theory of working memory: Cross-task prediction of performance in ACT-R. *Cogn. Syst. Res.* **2000**, *1*, 99–118. [CrossRef]
27. Anderson, J.R.; Bothell, D.; Byrne, M.D.; Douglass, S.; Lebiere, C.; Qin, Y. An integrated theory of the mind. *Psychol. Rev.* **2004**, *111*, 1036. [CrossRef] [PubMed]
28. Marewski, J.N.; Mehlhorn, K. Using the ACT-R architecture to specify 39 quantitative process models of decision making. *Judgm. Decis. Mak.* **2011**, *6*, 439–519.
29. Schooler, L.J.; Hertwig, R. How forgetting aids heuristic inference. *Psychol. Rev.* **2005**, *112*, 610. [CrossRef] [PubMed]
30. Hinton, G. A practical guide to training restricted Boltzmann machines. In *Neural Networks: Tricks of the Trade: Second Edition*; Springer: Berlin/Heidelberg, Germany, 2012; pp. 599–619.
31. Collobert, R.; Weston, J. A unified architecture for natural language processing: Deep neural networks with multitask learning. In Proceedings of the 25th International Conference on Machine Learning, ACM, Helsinki, Finland, 5–9 July 2008; pp. 160–167.
32. Vincent, P.; Larochelle, H.; Lajoie, I.; Bengio, Y.; Manzagol, P.A. Stacked denoising autoencoders: Learning useful representations in a deep network with a local denoising criterion. *J. Mach. Learn. Res.* **2010**, *11*, 3371–3408.
33. Krizhevsky, A.; Sutskever, I.; Hinton, G.E. ImageNet Classification with Deep Convolutional Neural Networks. *Commun. ACM* **2017**, *60*, 84–90. [CrossRef]
34. Babić, B.R.; Nešić, N.; Miljković, Z. Automatic feature recognition using artificial neural networks to integrate design and manufacturing: Review of automatic feature recognition systems. *Artif. Intell. Eng. Des. Anal. Manuf.* **2011**, *25*, 289–304. [CrossRef]
35. Gritsenko, V.; Rachkovskij, D.; Frolov, A.; Gayler, R.; Kleyko, D.; Osipov, E. Neural distributed autoassociative memories: A survey. *Cybern. Comput. Eng.* **2017**, *22*, 5–35.
36. Frutos-Pascual, M.; Garcia-Zapirain, B. Assessing visual attention using eye tracking sensors in intelligent cognitive therapies based on serious games. *Sensors* **2015**, *15*, 11092–11117. [CrossRef]
37. Sun, R.; Peterson, T. Learning in reactive sequential decision tasks: The CLARION model. In Proceedings of the IEEE International Conference on Neural Networks, Washington, DC, USA, 3–6 June 1996; Volume 2, pp. 1073–1078.
38. Buzsáki, G. Neural syntax: Cell assemblies, synapsembles, and readers. *Neuron* **2010**, *68*, 362–385. [CrossRef]

39. Bower, G.H. A brief history of memory research. In *The Oxford Handbook of Memory*; Oxford University Press: Oxford, UK, 2000; pp. 3–32.
40. Bermingham, D.; Hill, R.D.; Woltz, D.; Gardner, M.K. Cognitive strategy use and measured numeric ability in immediate-and long-term recall of everyday numeric information. *PLoS ONE* **2013**, *8*, e57999.
41. Rafidi, N.S.; Hulbert, J.C.; Brooks, P.P.; Norman, K.A. Reductions in Retrieval Competition Predict the Benefit of Repeated Testing. *Sci. Rep.* **2018**, *8*, 11714. [CrossRef] [PubMed]
42. Antony, J.W.; Ferreira, C.S.; Norman, K.A.; Wimber, M. Retrieval as a fast route to memory consolidation. *Trends Cogn. Sci.* **2017**, *21*, 573–576. [CrossRef]
43. Polyn, S.M.; Natu, V.S.; Cohen, J.D.; Norman, K.A. Category-specific cortical activity precedes retrieval during memory search. *Science* **2005**, *310*, 1963–1966. [CrossRef] [PubMed]
44. Hulbert, J.; Norman, K. Neural differentiation tracks improved recall of competing memories following interleaved study and retrieval practice. *Cereb. Cortex* **2014**, *25*, 3994–4008. [CrossRef]
45. Kapur, S.; Craik, F.I.; Jones, C.; Brown, G.M.; Houle, S.; Tulving, E. Functional role of the prefrontal cortex in retrieval of memories: A PET study. *Neuroreport* **1995**, *6*, 1880–1884. [CrossRef] [PubMed]
46. Sederberg, P.B.; Howard, M.W.; Kahana, M.J. A context-based theory of recency and contiguity in free recall. *Psychol. Rev.* **2008**, *115*, 893. [CrossRef] [PubMed]
47. Howard, M.W.; Kahana, M.J. A distributed representation of temporal context. *J. Math. Psychol.* **2002**, *46*, 269–299. [CrossRef]
48. Socher, R.; Gershman, S.; Sederberg, P.; Norman, K.; Perotte, A.J.; Blei, D.M. *Advances in Neural Information Processing Systems*; Curran Associates, Inc.: Red Hook, NY, USA, 2009; pp. 1714–1722.
49. Becker, S.; Lim, J. A computational model of prefrontal control in free recall: Strategic memory use in the California Verbal Learning Task. *J. Cogn. Neurosci.* **2003**, *15*, 821–832. [CrossRef]
50. Srivastava, N.; Vul, E. A simple model of recognition and recall memory. In *Advances in Neural Information Processing Systems*; Curran Associates, Inc.: Red Hook, NY, USA, 2017; pp. 293–301.
51. Gillund, G.; Shiffrin, R.M. A retrieval model for both recognition and recall. *Psychol. Rev.* **1984**, *91*, 1. [CrossRef]
52. Wei, X.X.; Stocker, A.A. A Bayesian observer model constrained by efficient coding can explain 'anti-Bayesian' percepts. *Nat. Neurosci.* **2015**, *18*, 1509. [CrossRef] [PubMed]
53. Ruppin, E.; Yeshurun, Y. An attractor neural network model of recall and recognition. In *Advances in Neural Information Processing Systems*; Morgan Kaufmann Publishers Inc.: Burlington, MA, USA, 1990; pp. 642–648.
54. Biggs, D.; Nuttall, A. *Neural Memory Networks*; Technical Report; Stanford University: Stanford, CA, USA, 2015.
55. Ruppin, E.; Yeshurun, Y. Recall and recognition in an attractor neural network model of memory retrieval. *Connect. Sci.* **1991**, *3*, 381–400. [CrossRef]
56. Recanatesi, S.; Katkov, M.; Romani, S.; Tsodyks, M. Neural network model of memory retrieval. *Front. Comput. Neurosci.* **2015**, *9*, 149. [CrossRef]
57. Taatgen, N. A model of free-recall using the ACT-R architecture and the phonological loop. In *Proceedings of Benelearn-96, Citeseer*; Herik, I.H.J., Weijters, T., Eds.; Universiteit Maastricht: Maastricht, The Netherlands, 1996.
58. Baddeley, A. Working memory. *Science* **1992**, *255*, 556–559. [CrossRef]
59. Thomson, R.; Pyke, A.; Hiatt, L.M.; Trafton, J.G. An Account of Associative Learning in Memory Recall. In Proceedings of the 37th Annual Conference of the Cognitive Science Society, Pasadena, CA, USA, 22–25 July 2015; pp. 2386–2391.
60. Anderson, J.R.; Matessa, M. A production system theory of serial memory. *Psychol. Rev.* **1997**, *104*, 728. [CrossRef]
61. Hélie, S.; Sun, R.; Xiong, L. Mixed effects of distractor tasks on incubation. In Proceedings of the 30th Annual Meeting of the Cognitive Science Society, Washington, DC, USA, 23–26 July 2008; Cognitive Science Society: Austin, TX, USA, 2008; pp. 1251–1256.
62. Hélie, S.; Sun, R. Incubation, insight, and creative problem solving: A unified theory and a connectionist model. *Psychol. Rev.* **2010**, *117*, 994–1024. [CrossRef]
63. Garrett, B. *Study Guide to Accompany Bob Garrett's Brain & Behavior: An Introduction to Biological Psychology*; Sage Publications: New York, NY, USA, 2014.
64. Gärdenfors, P. *Conceptual Spaces: The Geometry of Thought*; MIT Press: Cambridge, MA, USA, 2004.
65. Hartigan, J.A.; Wong, M.A. Algorithm AS 136: A k-means clustering algorithm. *J. R. Stat. Soc. Ser. C (Appl. Stat.)* **1979**, *28*, 100–108. [CrossRef]
66. Sharma, R.; Ribeiro, B.; Pinto, A.M.; Cardoso, F.A. Learning non-convex abstract concepts with regulated activation networks. *Ann. Math. Artif. Intell.* **2020**, *88*, 1207–1235. [CrossRef]
67. Rosch, E. Cognitive representations of semantic categories. *J. Exp. Psychol. Gen.* **1975**, *104*, 192–233. [CrossRef]
68. Mervis, C.B.; Rosch, E. Categorization of natural objects. *Annu. Rev. Psychol.* **1981**, *32*, 89–115. [CrossRef]
69. Rosch, E. Prototype classification and logical classification: The two systems. In *New Trends in Conceptual Representation: Challenges to Piaget's Theory*; University of California, Berkeley: Berkeley, CA, USA, 1983; pp. 73–86.

70. LeCun, Y.; Cortes, C.; Burges, C.J. The MNIST DATABASE of Handwritten Digits. 1999. Available online: http://yann.lecun.com/exdb/mnist/ (accessed on 10 July 2015).
71. Sharma, R.; Ribeiro, B.; Miguel Pinto, A.; Amílcar Cardoso, F. Reconstructing Abstract Concepts and their Blends Via Computational Cognitive Modeling. In Proceedings of the 2020 IEEE International Joint Conference on Neural Networks (IJCNN), Glasgow, UK, 19–24 July 2020; pp. 1–6.

Article

Dynamic Gesture Recognition System with Gesture Spotting Based on Self-Organizing Maps

Hiroomi Hikawa *, Yuta Ichikawa, Hidetaka Ito and Yutaka Maeda

Faculty of Engineering Science, Kansai University, Osaka 564-8680, Japan; k423998@kansai-u.ac.jp (Y.I.); h.ito@kansai-u.ac.jp (H.I.); maedayut@kansai-u.ac.jp (Y.M.)
* Correspondence: hikawa@kansai-u.ac.jp

Featured Application: Human–computer interface.

Abstract: In this paper, a real-time dynamic hand gesture recognition system with gesture spotting function is proposed. In the proposed system, input video frames are converted to feature vectors, and they are used to form a posture sequence vector that represents the input gesture. Then, gesture identification and gesture spotting are carried out in the self-organizing map (SOM)-Hebb classifier. The gesture spotting function detects the end of the gesture by using the vector distance between the posture sequence vector and the winner neuron's weight vector. The proposed gesture recognition method was tested by simulation and real-time gesture recognition experiment. Results revealed that the system could recognize nine types of gesture with an accuracy of 96.6%, and it successfully outputted the recognition result at the end of gesture using the spotting result.

Keywords: dynamic gesture recognition; gesture spotting; self-organizing map

Citation: Hikawa, H.; Ichikawa, Y.; Ito, H.; Maeda, Y. Dynamic Gesture Recognition System with Gesture Spotting Based on Self-Organizing Maps. *Appl. Sci.* **2021**, *11*, 1933. https://doi.org/10.3390/app11041933

Academic Editor: Attila Kovari

Received: 23 December 2020
Accepted: 17 February 2021
Published: 22 February 2021

Publisher's Note: MDPI stays neutral with regard to jurisdictional claims in published maps and institutional affiliations.

Copyright: © 2021 by the authors. Licensee MDPI, Basel, Switzerland. This article is an open access article distributed under the terms and conditions of the Creative Commons Attribution (CC BY) license (https://creativecommons.org/licenses/by/4.0/).

1. Introduction

Hand gestures are one of the most important communication tools frequently used in our daily lives, and they can be used as an attractive means of human–computer interaction (HCI). Hand gestures are generally either static hand signs or dynamic hand gesture. Hand signs are static hand poses without any movements, and the hand sign recognition system identifies the meaning of a hand pose. Meanwhile, in the dynamic gesture recognition, each gesture is defined as the trajectory of the hand movement or a sequence of hand poses.

A number of video-based hand gesture recognition algorithm and systems have been proposed [1]. This approach can use a conventional camera that most laptop PCs are equipped with. Thus, the video-based gesture recognition system can easily be implemented on widely available platforms. Another approach is based on three-dimensional hand image, which has attracted researchers in gesture recognition because the use of 3D image can improve performance [2]. However, the 3D gesture recognition requires a special device such as a Microsoft Kinect and a Leap Motion.

The gesture recognition system should work in real-time for practical use. One of the important function required for the real-time dynamic gesture recognition system is gesture spotting. The gesture spotting segments a meaningful portion from a continuous data stream, and it finds the start and end of gesture. The simplest way to provide the gesture spotting is to define key posture that indicates the start and end of gesture. However, this approach disturbs the natural flow of the intended sequence of gesture. Thus, a new approach that can detect the start and end of gesture naturally in continuous sequence of hand motion, is desired.

In our previous work, a hardware hand sign recognition system was proposed, which was video based system and recognized static hand signs [3]. Its recognition algorithm of the system consisted of feature vector generation and a vector classifier, and the whole system was implemented as a custom hardware on a field programmable gate array (FPGA). Self-organizing map (SOM) and Hebbian learning network were combined to form a SOM-Hebb classifier, which was used as the vector classifier. The SOM [4] is an unsupervised neural network that has been used in pattern recognition, data analysis, and visualization by using its clustering or vector quantization capabilities. The feature vector was computed from video frames and the hand sign recognition was carried out in real time by taking advantage of its high speed computation power of the dedicated hardware.

This paper proposes a new video-based dynamic hand gesture recognition system with the gesture spotting. The SOM-Hebb classifier is enhanced to SOM-SOM-Hebb classifier for the dynamic gesture classification. The proposed system consists of feature vector computation and two SOMs and a Hebbian learning network. The feature vectors computed from video frames are quantized by the first SOM, and a posture sequence vector that represents the current gesture is generated. Then, the SOM-Hebb classifier that contains the second SOM, recognizes the input gesture. During the gesture classification, the end of gesture is detected by the SOM-Hebb classifier, and a recognized gesture class is outputted when the gesture's end is detected. As a result, natural gesture spotting without any key pose is implemented. This paper examines detailed performance of the proposed recognition system by simulation and experiment by using nine types of dynamic gesture.

2. Related Work

In the gesture recognition, a hand segmentation is carried out first, which detects the hand position or hand shape. A popular segmentation method in the vision based system is skin color detection that extracts hand portion from cluttered background [5,6]. Yun et al. [7] proposed a multi-feature fusion method that improved recognition results by extracting angle count, skin color angle, and non-skin color angle in combination with Hu invariant moments features. Some gesture recognition systems simplified hand extraction from the background with the help of inexpensive color-coded gloves for hand segmentation. A glove providing color-coding with six unique colors were used in [8,9]. Wang and Popovi [10] employed an ordinary cloth glove being printed with a custom pattern that was designed to estimate the poses. Our previous work [3] also employed a two-colored glove for hand segmentation. Another option for gesture segmentation is the use of the 3D image that is taken through depth sensors, such as the Microsoft Kinect depth camera and the Leap Motion. The 3D camera views the subject in the front plane and generates a depth image of the subject, and the depth image is used for background removal, followed by the generation of the depth profile of the subject. Gesture recognition systems with the Kinect are found in [11–15]. Molina et al. [16] used another depth camera called Time-Of-Flight range camera that supplied real-time depth information per pixel. In terms of applicability, the vision-based gesture segmentation is desirable since it requires only a conventional camera available on most laptop PCs, and no special depth sensor is needed.

Unlike the Kinect sensor and other depth sensors, the output of the Leap Motion is the depth data which consists of palm direction, fingertips positions, palm center position, and other relevant points. Therefore, no extra computational work is needed to get these information. Due to its unique features, the Leap Motion has been applied to dynamic hand gesture recognition by by researchers. Lu et al. [17] proposed a dynamic gesture recognition system, in which the Leap Motion was used to compute feature vector of the gesture, and a hidden conditional neural field (HCNF) classifier was used to recognize dynamic hand gesture. Another example is the work done by H. Li et al. [18]. Their hand gesture recognition system was based on the Leap Motion and a spatial fuzzy matching (SFM). Hand–eye coordination means the ability to combine seeing and hand movement. Ujbanyi et al. [19,20] examined the correlations between eye motion and the motion of the mouse cursor regarding hand–eye coordination, and they used an hand–eye tracking system which was made of the Leap Motion and Eye Tribe tracker.

Challenge of real-time dynamic gesture recognition is the gesture spotting or temporal segmentation that detects when the gesture starts and ends. In the system proposed by Varshini et al. [13], each dynamic gesture was defined as a sequence of trigger-poses, and the start and end of the gesture were detected by finding the start and end triggers. Chai et al. [21] used hand positions to perform the temporal segmentation by assuming that a user put hands-up pose at the start of gesture and put hands-down pose at end of the gesture. A real-time dynamic hand gesture recognition system proposed by Chen et al. [15] used two hand configurations (open-hand, closed-hand) to achieve gesture spotting and its 3D motion trajectory of the dynamic gesture was captured by the Kinect sensor. These approaches disturb the natural flow of gesture, and thus a new approach that can detect the start and end of gesture naturally, is desirable.

A static hand gesture recognition can be achieved by applying standard pattern recognition techniques such as template matching, whereas dynamic gesture recognition requires time-series pattern recognition algorithm such as a hidden Markov model (HMM) or dynamic time warping (DTW) algorithm. The HMM is a statistical Markov model in which the system being modeled is assumed to be a Markov process. The HMM is a doubly stochastic process with an underlying stochastic process that is not observable, but can be observed through another set of stochastic processes that produce a sequence of observed symbols, and the model is known for their applications to various fields including the gesture recognition such as [22]. Problem of the gesture recognition with the HMM is that its recognition accuracy decreases if the behavior during the gesture transition has not been precisely trained. The DTW is one of the algorithms for measuring similarity between two temporal sequences which may vary in speed. Plouffe et al. [14] and Molina et al. [16] employed the DTW algorithm for their dynamic gesture recognition systems.

Another popular recognition algorithm is a neural network and its derivatives, especially deep learning methodologies [23]. Most modern deep learning models are based on convolutional neural networks (CNNs). The CNNs have been well studied and applied to fields of image recognition. The most crucial challenge in deep learning based gesture recognition is the handling of the temporal dimension. One approach uses 3D filters in the convolutional layer of the CNN. The 3D-CNN captures features of both spatial and temporal dimensions while maintaining a certain temporal structure. Another approach combines a temporal sequence modeling with a 2D (or 3D) CNN. One of the most used networks for the temporal modeling is a recurrent neural network (RNN), which can take into account the temporal data using recurrent connections in hidden layers. The drawback of this network is its short-term memory, and long short-term memory (LSTM) was proposed to solve the problem.

Molchanov et al. [24] proposed a recurrent 3D-CNN that performed simultaneous detection and classification of dynamic hand gesture from multi-modal data. Wu et al. [25] employed a novel method called deep dynamic neural networks (DDNN) for multimodal gesture recognition. The multimodal gesture recognition method based on 3D convolutional LSTM network was proposed by Zhu et al. [26]. Naguri [27] proposed a gesture recognition system based on the LSTM and a convolutional neural network (CNN) that were trained to process input sequences of 3D hand positions and velocity. Chai et al. [21] proposed a continuous gesture recognition method with a two-stream RNN (2S-RNN) for the RGB-depth image recognition. John et al. [28] proposed a vision-based gesture recognition system for automotive user interface, and they employed a long-term recurrent convolution network to classify the video sequence of the dynamic hand gesture.

A recognition system proposed by Chen et al. [15] employed a Support Vector Machine (SVM) as the recognition algorithm. Kim et al. [29] proposed a novel method to measure the video-to-video volume similarity by extending a canonical correlation analysis (CCA). Then, the proposed matching method was demonstrated for action classification by a simple nearest neighbor classifier. Jordan recurrent neural network (JRNN) is a class of recurrent neural networks, which is a three-layer network with addition of a set of context units [30]. The context units are fed from the output layer, and they have a recurrent connection to themselves. This allows the JRNN to exhibit temporal dynamic behavior and can be applied for the gesture recognition. Araga et al. [31] employed the JRNN to implement their dynamic gesture recognition system.

3. Gesture Recognition System

Figure 1 outlines the flow for the gesture recognition algorithm. The proposed system consists of a feature vector generator, a sequence vector generator, and the SOM-Hebb classifier. Input to the system is video frames, and a dynamic hand gestures are assumed to be made of a sequence of F video frames. Since each frame contains different types of posture, the dynamic gesture can be classified by examining change of the posture in the F consecutive video frames.

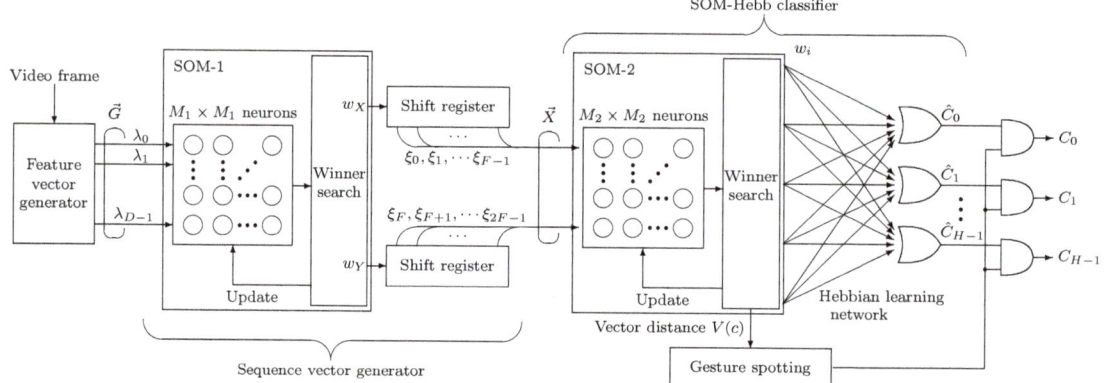

Figure 1. Dynamic gesture recognition system.

3.1. Feature Vector Generation

In the feature vector generator, the image frame that is $P \times Q$ pixels in RGB color format is converted to the feature vector $\vec{G}$. The feature vector proposed in [3] is employed. Computation to obtain the feature vector is shown in Figure 2, which consists of a binary quantization, horizontal and vertical projection histogram calculations, and two discrete Fourier transforms (DFTs). Output is the D dimensional feature vector $\vec{G}$.

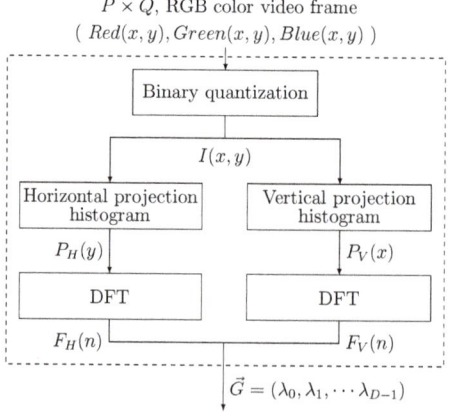

Figure 2. Feature vector generation.

Firstly, the input color frame image is converted to a binary image $I(x,y)$. For the system to remove the background image including the arm as well as to extract the finger segments, the user is required to wear a glove, finger portion of which is colored in red. If color of pixel is red the pixel is treated as 1, otherwise 0. Then horizontal and vertical histograms $P_H(y)$ and $P_V(x)$ of $I(x,y)$ are calculated as follows:

$$P_H(y) = \sum_{x=0}^{P-1} I(x,y) \tag{1}$$

$$P_V(x) = \sum_{y=0}^{Q-1} I(x,y) \tag{2}$$

After the histogram calculations, DFTs are carried out on the histograms.

$$A_H(k) = \sum_{y=0}^{Q-1} P_H(y) \cdot \cos\left(\frac{2\pi y k}{Q}\right) \tag{3}$$

$$B_H(k) = \sum_{y=0}^{Q-1} P_H(y) \cdot \sin\left(\frac{2\pi y k}{Q}\right) \tag{4}$$

$$A_V(k) = \sum_{x=0}^{P-1} P_V(x) \cdot \cos\left(\frac{2\pi x k}{P}\right) \tag{5}$$

$$B_V(k) = \sum_{x=0}^{P-1} P_V(x) \cdot \sin\left(\frac{2\pi x k}{P}\right) \tag{6}$$

Here, $A_H(k)$, $A_V(k)$ and $B_H(k)$, $B_V(k)$ are real and imaginary parts of the frequency components of the histograms. Then, $F_H(n)$ and $F_V(n)$, i.e., the magnitude spectra of $P_H(y)$ and $P_V(x)$ are computed as

$$F_H(k) = \sqrt{A_H^2(k) + B_H^2(k)} \tag{7}$$

$$F_V(k) = \sqrt{A_V^2(k) + B_V^2(k)} \tag{8}$$

The $F_H(k)$ and $F_V(k)$ of the same hand posture images placed in different positions are identical because they are the magnitude spectra lacking the phase information related to the hand posture position. Since most of the image's feature information is concentrated in the lower frequency components, they are used as the feature vector. The D-dimensional feature vector, $\vec{G}$ is formed from $F_H(k)$ and $F_V(k)$ as;

$$\vec{G} = \{\lambda_0, \lambda_1, \cdots, \lambda_{D-1}\} \in \Re^D \tag{9}$$

$$\lambda_i = \begin{cases} F_H(i) & \left(0 \le i < \frac{D}{2}\right) \\ F_V\left(i - \frac{D}{2}\right) & \left(\frac{D}{2} \le i < D\right) \end{cases}$$

This feature vector $\vec{G}$ is fed to the sequence vector generator.

3.2. Sequence Vector Generator

The SOM-1 in the sequence vector generator quantizes the input vectors, and the quantization results are sequentially stored in the shift registers. The contents of the shift registers form the sequence vector, which represents temporal change of the input posture, and is fed to the next SOM-Hebb classifier. The SOM-1 includes $M_1 \times M_1$ neurons, and D-dimensional vector $\vec{m}_j$ that is called a weight vector is included in each neuron.

$$\vec{m}_j = \{\mu_{j0}, \mu_{j1}, \cdots, \mu_{jD-1}\} \in \Re^D, \tag{10}$$

where, j is the neuron number.

Operation of the SOM is divided into learning and recall phases. The weight vectors of the neurons are trained with a set of input vectors in the learning phase. The learning phase is made of a winner search and weight update. During the recall phase, only the winner search is carried out by using the map of the trained weight vectors.

The winner neuron has the weight vector that is the nearest to the input vector. Euclidean distance $V_1(j)$ between the input vector and weight vector of neuron-j, is calculated for the winner search.

$$V_1(j) = \sqrt{\sum_{i=0}^{D-1} (\mu_{ji} - \lambda_i)^2} \qquad (11)$$

The winner neuron-c is then determined.

$$c = \arg\min_j V_1(j) \qquad (12)$$

In the weight update, weight vectors of the winner and its neighborhood neurons are updated to be closer to the input vector as;

$$\vec{m}_j(t+1) = \vec{m}_j(t) + h(c,j,t) \cdot \left[\vec{G}(t) - \vec{m}_j(t)\right], \qquad (13)$$

where t is time index, and $h(c,j,t)$ is a function called neighborhood function, which is defined as;

$$h(c,j,t) = \alpha(t) \cdot \exp\left(-\frac{\|\vec{r}_c - \vec{r}_j\|}{2\sigma(t)^2}\right), \qquad (14)$$

where $\alpha(t)$ is a learning coefficient, $(0 < \alpha(t) < 1)$. The $\vec{r}_c$ and $\vec{r}_j$ are the coordinate vectors of the winner neuron-c, and a neuron-j, respectively. The $\sigma(t)$ represents the neighborhood radius, and the weight vectors within the radius from the winner neuron are updated.

After the learning phase, all weight vectors are kept unchanged and the weight map is used in the recall phase. The winner neuron represents the cluster to which the input vector belongs, and the coordinates (w_X, w_Y) of the winner neuron for the input vector are treated as the quantization result. These coordinates are stored sequentially in the shift registers, so their contents represent the sequence of the input video frames. In this paper, this vector is called the sequence vector, $\vec{X}$, which is a 2F-dimensional vector. Its vector element ζ_m is defined as:

$$\vec{X} = \{\zeta_0, \zeta_1, \cdots, \zeta_{2F-1}\} \qquad (15)$$

$$\zeta_m = \begin{cases} w_X(m) & (0 \leq m < F) \\ w_Y(m-F) & (F \leq m < 2F). \end{cases} \qquad (16)$$

Figure 3 shows examples explaining operation of the system. As shown in Figure 3A, a gesture is made of 10 posture images in different video frames. In the example, SOM-1 is composed of 8×8 neurons and Figure 3B shows the transition of the winner neuron with respect to the input video frames. Posture in the first video frame ($f = 0$) makes a neuron at $(w_X, w_Y) = (0,0)$ the winner. Then the coordinates of the winner neuron are stored in the registers in the sequence vector generator as shown in Figure 3C. Question marks in Figure 3C are the coordinates of the winner neurons of the previous gesture, which are not related to the current gesture. The winner for the second posture ($f = 1$) is a neuron at $(4,2)$. The registers are shifted to the right and the new winner coordinates $(4,2)$ is stored into the registers' most left position. For the third posture at $f = 2$, $w_X = 6$ and $w_Y = 6$ are loaded into the registers. In this way, the information of the previous gesture in the registers are gradually replaced with that of the current gesture. Therefore, the sequence vector $\vec{X}$ representing the current gesture approaches completion as the video frame progresses, and $\vec{X}$ is completed at the 10th frame ($f = 9$). The vector $\vec{X}$ is fed to the SOM-Hebb classifier that is described in the next subsection.

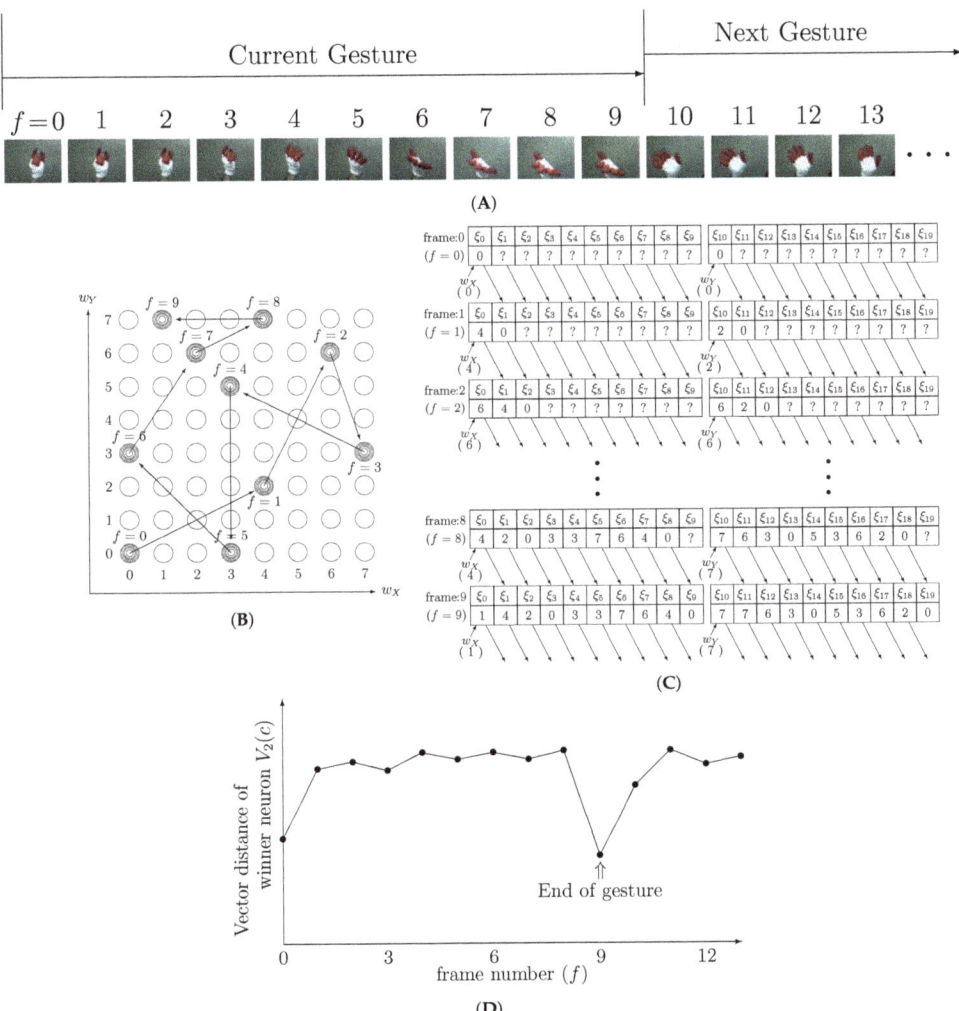

Figure 3. Example of operation of the proposed system, (**A**) Input gesture frames, (**B**) Winner neurons in the SOM-1, (**C**) Development of the shift registers, (**D**) Vector distance $V(c)$ of winner neurons in the SOM-2.

3.3. SOM-Hebb Classifier for Sequence Vector Classification

The SOM-Hebb classifier classifies the sequence vector $\vec{X}$ and identifies the input gesture. This classifier is the same one that was proposed in our previous work [3]. The SOM-2 in this classifier consists of $M_2 \times M_2$ neurons and $2F$-dimensional weight vectors are included in the neurons. The SOM-2 is trained in the same way as was explained in the previous section. Note that $V_2(c)$ is the vector distance of the winner neuron's weight vector to the input vector that is the sequence vector $\vec{X}$, and $V_2(c)$ is used to implement the gesture spotting function.

During the recall phase, the class to which the input vector belongs can be identified from the winner neuron of the SOM-2. Here, H represents the number of classes. The Hebb network generates its output $\hat{C}_h$ from the winner neuron. Each neuron represents a single cluster in the input vector space. Since a single gesture class may consist of combination of multiple clusters, multiple neurons must be associated to the single class in that case.

Selection of the neurons belonging to the same class is done by a single layer feedforward network. This network is trained by the Hebbian training algorithm, which is a supervised training. During the Hebb training, training vectors with their class data are sequentially fed to the network. Every training vector makes one of the neurons the winner. If strong correlation is found between training vectors in class h and neuron j, then the neuron j is assigned to the class h. In practice, the class of the input vectors with which the neuron j won the most, is associated to the neuron.

The SOM-2 must have appropriate number of neurons for the SOM-Hebb network to work properly. It happens that some neurons may have no connection to any gesture class. Obviously, the selection of such neuron as the winner in the recognition phase causes false recognition. To avoid this situation, neurons without connections to class ID are culled. The culling replaces the weight vectors of these neurons with huge vectors so that they never win.

3.4. Gesture Spotting

An important function required for the dynamic gesture recognition system is the gesture spotting which detects when gesture ends so that a meaningful gesture is segmented from the sequence of hand motions. The gesture spotting is implemented in the SOM-Hebb classifier by using $V_2(c)$ that is the vector distance of the winner neuron's weight vector in the SOM-2 to the input vector. The SOM-Hebb classifier performs the recognition for every input frame and generates its recognition results $\hat{C}_h$. However most of the $\hat{C}_i$ are not correct because the contents of the shift registers are not complete vector sequence for the current input gesture until the last gesture frame is input. The recognition result C_h is outputted only when the spotting module detects the end of gesture.

The end of gesture is detected by observing the transition of the vector distance $V_2(c)$. Figure 3C shows the transitions in the shift registers, which is development process of the posture sequence vector, $\vec{X}$. Each gesture consisted of 10 frames in this example, therefore the shift register is filled with appropriate vector's elements at 10th frame ($f = 9$) and posture sequence vector $\vec{X}$ is completed as shown in Figure 3C. The completed vector $\vec{X}$ matches with one of the weight vectors in SOM-2, which decreases the vector distance $V_2(c)$ remarkably as shown in Figure 3D. After that, the distance increases because the next gesture vector elements are loaded into the register. Therefore, the end of gesture can be detected by searching a dip in the transition of the vector distance $V_2(c)$. However, the actual distance transition is not as smooth as that plotted in Figure 3C. The transition in the actual input fluctuates, which makes it difficult to find the dip. In order to solve the problem, a moving average of the vector distance is employed. The moving average $\overline{V_c(f)}$ is computed as;

$$\overline{V_c(f)} = \frac{1}{L}\sum_{l=0}^{L-1} V_c(f-l), \qquad (17)$$

where $V_c(f)$ is the $V_2(c)$ at video frame f, and L is the number of samples to be averaged.

4. Simulation and Experiment

Performance of the proposed system was examined by computer simulation and experiment.

4.1. Simulation

The system was configured as follows.

Frame size: $P \times Q = 128 \times 128$.
Feature vector dimension: $D = 32$.
Sequence vector dimension: $F = 20$.
The numbers of neurons in the SOMs: $M_1 \times M_1 = M_2 \times M_2 = 16 \times 16 = 256$.
Moving average: $L = 4$ in Equation (17).
The number of gesture classes: $H = 9$.

Data set for the simulation was vector sequence taken from video frames of recorded gesture video. Nine types of gesture shown in Figure 4 were used for the test. We defined the gesture by using the Cambridge Hand Gesture Data set [29] for reference. As Figure 4 shows, the data set consisted of nine classes. Each class gesture was defined with 10 frames, therefore the dimension of the sequence vector $\vec{X}$ was 20. Class labels 1 to 9 are assigned to every types of gesture. Note that the labels are used to distinguish class of the gesture types, and the number does not represent numerical character. Gesture motions are combinations of three basic poses (Flat, Spread, V-shape) and three movements (Leftward, Rightward, Contract). Thus, the gesture classes are made of three groups, i.e., 1-2-3, 4-5-6, and 7-8-9. Note that the last posture of gesture 1 is also the first posture of gesture 2, and the last posture of class 2 is the first posture of gesture class 3. The other gesture groups were designed in the same way so that the gesture classes in the same group could be performed seamlessly. The number of training vectors per class was 50, and 100 test vectors were used.

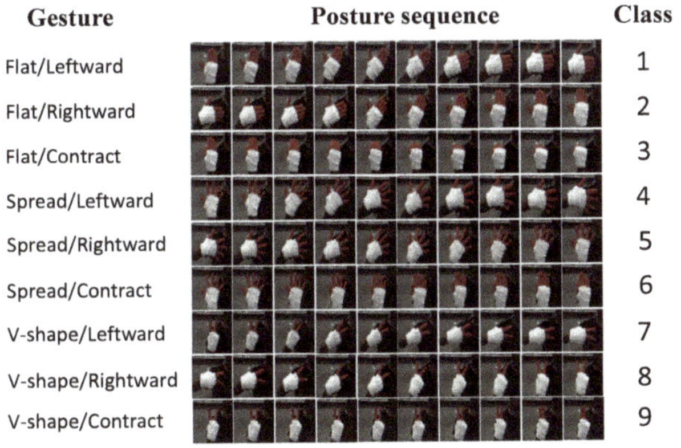

Gesture	Posture sequence	Class
Flat/Leftward		1
Flat/Rightward		2
Flat/Contract		3
Spread/Leftward		4
Spread/Rightward		5
Spread/Contract		6
V-shape/Leftward		7
V-shape/Rightward		8
V-shape/Contract		9

Figure 4. Nine types of gesture.

For comparison purpose, recognition performance of the JRNN [30] was examined using the same feature vectors $\vec{G}$. Table 1 summarizes the recognition accuracies of the two methods.

Table 1. Gesture recognition accuracies of various classifiers.

Method	Gesture Class									Average
	1	2	3	4	5	6	7	8	9	
JRNN	100%	100%	82%	89%	72%	60%	99%	99%	46%	83.0%
This work	82%	66%	100%	89%	98%	100%	100%	100%	100%	92.8%

4.2. Real-Time Gesture Recognition

To conduct the experiment, real-time gesture recognition system was developed in software that ran on a PC. The input gesture was taken by the USB camera, and fed to the system. The recognition result was outputted only when the spotting function detected the end of the gesture. Figure 5 is a screen shot of the implemented real-time gesture recognition system.

The system was tested with the same gesture that was used in the simulation. The system was trained off-line by using pre-captured gesture data set. The number of training data for the off-line training was 50 samples for each gesture class. Each training sample was acquired by simply capturing 10 consecutive frames. Neither setting up the key posture that represented the gesture, nor manual selection of key frame was done in the acquisition of the training data set. The recognition system then classified the dynamic gesture presented in real-time using the weight vectors obtained from the off-line training. In the experiment, recognition test was carried out 100 times for each gesture by the same person who had provided the training data set. Most of the gesture ends were correctly detected even though the gesture groups 1-2-3, 4-5-6, and 7-8-9 were performed in succession (video of the experiment (10 fps) is available on http://www2.kansai-u.ac.jp/hikawa/ichikawa.mp4, 19 February 2021). Since the gesture in our method is defined by the fixed number of posture types, the input gesture must be made of the same number of posture types. Therefore, the speed of gesture to be recognized depends on the speed of gesture that has been captured as the training data. The recognition tests were carried out with two speeds, i.e., 5 frames per second (fps) and 10 fps. Since $F = 20$ (10 frames for one gesture), the gesture speeds were 2 second/gesture in case of 5 fps, and 1 second/gesture in case of 10 fps.

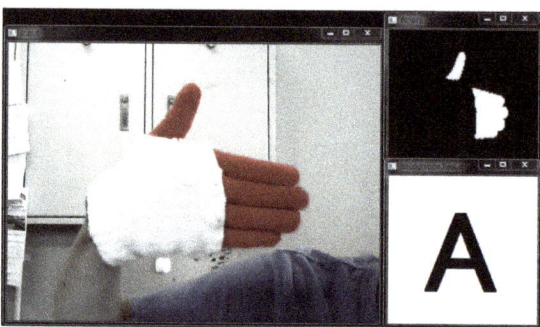

Figure 5. Real-time gesture recognition system.

Table 2 shows the experimental results of 5 fps frame rate. The average accuracy of the recognition was 96.6%. NS in the table is the number of cases where no spotting was detected, and MS is the number of cases where multiple spotting occurred. Both cases were counted as errors. Table 3 summarizes the experimental results of 10 fps frame rate. The average accuracy of the recognition rate was 97.0%, and no significant difference is found due to the speed difference.

Table 2. Confusion table (5 fps).

Input Gesture	Recognition Results									Spotting		Accuracy
	1	2	3	4	5	6	7	8	9	NS	MS	%
1	98	2										98
2		97			3							97
3			91		2					2	5	91
4				89						5	6	89
5					100							100
6						100						100
7							96			2	2	96
8								100				100
9									98		2	98
											Average	96.6

Table 3. Confusion table (10 fps).

Input Gesture	Recognition Results									Spotting		Accuracy %
	1	2	3	4	5	6	7	8	9	NS	MS	
1	93									7		93
2		95			3					2		95
3			98		1	1						98
4	2			94						1	3	94
5					100							100
6					1	99						99
7							95			3	2	95
8								99		1		99
9									100			100
											Average	97.0

5. Discussion

The simulation results show that the proposed method outperformed the JRNN. Difficulty of this gesture data set is that class pairs 1–2, 4–5, 7–8 are reverse gestures. Appearance of poses in the gesture is reverse order and those pairs include the same hand poses. For classes 1 and 2, the proposed system is inferior to the JRNN, but the proposed method recognized class 5 better. Another noticeable point is that the recognition accuracies of the JRNN for classes 6 and 9 are significantly worse than those of the proposed system. This is caused by the another difficulty of the data set. As shown in Figure 4, the class pairs 4–6 and 7–9 have the same poses in their beginning, which confuse the classifiers.

The experimental results shown in Tables 2 and 3 disclose that the recognition and spotting performances of the proposed system are very high. Regarding the spotting, the spotting may be easily implemented by counting the frames because the number of frames of gestures are fixed. To do so, the start of gesture must be detected correctly, and a possible method is the use of key poses to indicate the start of gestures, which was used in [15,21]. However, these approaches disturb the natural flow of gesture. Meanwhile the proposed spotting finds the end of gesture automatically when the sequence of frames matches one of the pre-trained ones, therefore the user can start gesture at any time without the key poses.

The tables also indicate that the most of the recognition errors were caused by the spotting errors. In case of NS, no spotting was detected, and the recognition result was not available. During the experiment, we observed that the proposed spotting detected the end of gesture twice in all MS cases, and two recognition results were outputted. In most of such cases, recognition results at the first spotting were incorrect and the second ones were correct. Therefore, if the detection accuracy of the spotting function is more precise, recognition results would be improved.

Table 4 compares the recognition accuracies of the proposed system with the state-of-the-art in the literature. Since experimental conditions are not the same, the accuracies in the table should not be directly compared. Six of them are real-time recognition systems, and the others were tested with various gesture data sets. Four of them are vision based systems, and the others used 3D gesture images taken from the special sensors. Vision based system is more challenging than the 3D gesture recognition since it uses limited 2D information, but it can be realized with the simple readily available cameras. Additional burdens of the real-time recognition system are high speed computation and the gesture spotting. Note that the proposed system provides the natural spotting function with no special key posture that indicates the start or end of gesture. Considering features of vision-based, real-time accurate gesture recognition and spotting function, the overall contribution of the proposed system in dynamic gesture recognition application is very high. However, even though the proposed system does not require the special sensors, it still requires users to wear the color glove, which may prevent it being used in everyday life. To solve the problem, use of the skin color detection [5,6] is one of the choice for the hand segmentation without the colored glove.

Table 4. Comparison of accuracies.

Method	Input	Data Set	Accuracy (%)
[11]	3D (Kinect)	RT, 8 gestures	84.9
[14]	Kinect	RT, 9 gestures	96.25
[15]	Kinect	RT, 36 gestures	95.42
[16]	time-of-flight	RT, 9 gestures	95.1
[17]	Leap Motion	Handcraft-gesture dataset, 10 gestures	95.0
[18]	Leap Motion	RT, 4 gestures	97.5
[24]	color+depth	SKIG	97.7
	color+depth+optical flow	ChaLearn	98.6
[26]	RGB+depth	SKIG	98.89
[27]	RGB+velocities	RT, 6 gestures	97.0
[28]	Vision (barehand)	Cambridge	91.0
[29]	Vision (barehand)	Cambridge	86.0
[31]	Vision (with colored glove)	RT, 9 gestures	94.3
this work	Vision (with colored glove)	RT, Cambridge	96.6

RT: Real-time recognition; SKIG: Sheffield Kinect Gesture dataset, 10 gestures; ChaLearn: ChaLearn dataset, 20 gestures; Cambridge: Cambridge gesture recognition data set [29], 9 gestures.

6. Conclusions

This paper proposed a vision-based real-time dynamic hand gesture recognition system with a gesture spotting function. In order to recognize the dynamic gesture, the SOM-SOM-Hebb classifier was newly devised. To provide the spotting function, end of gesture was detected from transitions of the vector distance between input and winner neuron's weight vectors. This gesture spotting capability made the system much more practical.

The proposed recognition algorithm was examined by simulation and real-time experiment. The results revealed that the system could recognize the nine types of gesture with an accuracy of 96.6%, which was better than that of other recognition sysems. Other advantages of our system over the compared methods are its real-time operability and the gesture spotting function.

Major drawback of the proposed system is the use of the color glove, and implementation of the hand segmentation without the glove is left for our future work. Another future research objective is to develop a hardware gesture recognition system with faster recognition speed, higher portability, and lower power consumption than those of PC implementation.

Author Contributions: Conceptualization, H.H. and Y.M.; methodology, H.H.; software, Y.I.; validation, Y.I. and H.H.; formal analysis, H.I. and H.H.; writing—original draft preparation, H.H. and Y.I.; writing—review and editing, H.I.; supervision, Y.M. All authors have read and agreed to the published version of the manuscript.

Funding: This work was supported by JSPS KAKENHI Grant Number JP18K11483, JP20K11999, and the Kansai University Fund for the Promotion and Enhancement of Education and Research, "Development of System Analysis and Design Methods Integrating Model and Data Premised on IOT Technology."

Conflicts of Interest: The authors declare no conflict of interest.

Sample Availability: http://www2.kansai-u.ac.jp/hikawa/ichikawa.mp4, accessed on 19 February 2021.

References

1. Pavlovic, V.I.; Sharma, R.; Huang, T.S. Visual Interpretation of Hand Gestures for Human-Computer Interaction: A Review. *IEEE Trans. Pattern Anal. Mach. Intell.* **1997**, *19*, 677–695. [CrossRef]
2. Cheng, H.; Yang, L.; Liu, Z. Survey on 3D Hand Gesture Recognition. *IEEE Trans. Circuits Syst. Video Technol.* **2016**, *26*, 1659–1673. [CrossRef]
3. Hikawa, H.; Kaida, K. Novel FPGA Implementation of Hand Sign Recognition System with SOM-Hebb Classifier. *IEEE Trans. Circuits Syst. Video Technol.* **2015**, *25*, 153–166. [CrossRef]
4. Kohonen, T. *Self-Organizing Maps*, 3rd ed.; Springer Series in Information Sciences 30; Springer: New York, NY, USA, 2001.
5. Alon, J.; Athitsos, V.; Yuan, Q.; Sclaroff, S. A Unified Framework for Gesture Recognition and Spatiotemporal Gesture Segmentation. *IEEE Trans. Pattern Anal. Mach. Intell.* **2009**, *31*, 685–1699. [CrossRef] [PubMed]
6. Dardas, N.H.; Georganas, N.D. Real-Time Hand Gesture Detection and Recognition Using Bag-of-Features and Support Vector Machine Techniques. *IEEE Trans. Instrum. Meas.* **2011**, *60*, 3592–3607. [CrossRef]
7. Yun, L.; Lifeng, Z.; Shujun, Z. A hand gesture recognition method based on multi-feature fusion and template matching. *Procedia Eng.* **2012**, *29*, 1678–1684. [CrossRef]
8. Lamar, M.V.; Bhuiyan, M.S.; Iwata, A. Hand Gesture Recognition Using T-CombNet: A New Neural Network Model. *IEICE Trans. Inf. Syst.* **2000**, *E-38-D*, 1986–1995.
9. Bragatto, T.A.C.; Ruas, G.S.I.; Lamar, M.V. Real-Time Hand Postures Recognition using Low Computational Complexity Artificial Neural Networks and Support Vector Machines. In Proceedings of the ISCCSP 2008, Saint Julian's, Malta, 12–14 March 2008; pp. 1530–1535. [CrossRef]
10. Wang, R.Y.; Popovi, J. Real-time handtracking with a color glove. In Proceedings of the ACM Transactions on Graphics (SIGGRAPH 2009), New Orleans, LA, USA, 3–7 Augst 2009; Volume 28, pp. 1–8. [CrossRef]
11. Biswas, K.K.; Basu, S.K. Gesture recognition using Microsoft Kinect R. In Proceedings of the 5th International Conference on Automation, Robotics and Applications, Wellington, New Zealand, 6–8 December 2011; pp. 100–103. [CrossRef]
12. Li, Y. Hand gesture recognition using Kinect. In Proceedings of the 2012 IEEE International Conference on Computer Science and Automation Engineering, Beijing, China, 22–24 June 2012; pp. 196–199. [CrossRef]
13. Varshini, M.R.L.; Vidhyapathi, C.M. Dynamic fingure gesture recognition using KINECT. In Proceedings of the 2016 International Conference on Advanced Communication Control and Computing Technologies (ICACCCT), Ramanathapuram, India, 25–27 May 2016. [CrossRef]
14. Plouffe, G.; Cretu, A.-M. Static and dynamic hand gesture recognition in depth data using dynamic time warping. *IEEE Trans. Instrum. Meas.* **2015**, *65*, 305–316. [CrossRef]
15. Chen, Y.; Luo, B.; Chen, Y.-L.; Liang, G.; Wu, X. A real-time dynamic hand gesture recognition system using kinect sensor. In Proceedings of the 2015 IEEE International Conference on Robotics and Biomimetics (ROBIO), Kuala Lumpur, Malaysia, 12–15 December 2015; pp. 2026–2030. [CrossRef]
16. Molina, J.; Pajuelo, J.A.; Martinez, J.M. Real-time motion-based hand gestures recognition from time-of-flight video. *J. Signal Process. Syst.* **2017**, *86*, 17–25. [CrossRef]
17. Lu, W.; Tong, Z.; Chu, J. Dynamic Hand Gesture Recognition with Leap Motion Controller. *IEEE Signal Process. Lett.* **2016**, *23*, 1188–1192. [CrossRef]
18. Li, H.; Wu, L.; Wang, H.; Han, C.; Quan, W.; Zhao, J. Hand Gesture Recognition Enhancement Based on Spatial Fuzzy Matching in Leap Motion. *IEEE Trans. Ind. Inform.* **2020**, *16*, 1885–1894. [CrossRef]
19. Ujbanyi, T. Examination of eye-hand coordination using computer mouse and hand tracking cursor control. In Proceedings of the 2018 9th IEEE International Conference on Cognitive Infocommunications (CogInfoCom), Budapest, Hungary, 22–24 August 2018; pp. 000353–000354. [CrossRef]
20. Ujbanyi, T.; Kovari, A.; Sziladi, G.; Katona, J. Examination of the eye-hand coordination related to computer mouse movement. *Infocommun. J.* **2020**, *XII*. [CrossRef]
21. Chai, X.; Liu, Z.; Yin, F.; Liu, Z.; Chen, X. Two streams recurrent neural networks for large-scale continuous gesture recognition. In Proceedings of the IEEE 2016 23rd International Conference on Pattern Recognition (ICPR), Cancun, Mexico, 4–8 December 2016; pp. 31–36. [CrossRef]
22. Deng, J.W.; Tsui, H.T. An HMM-based approach for gesture segmentation and recognition. In Proceedings of the 15th International Conference on Pattern Recognition. ICPR-2000, Barcelona, Spain, 3–7 September 2000; Volume 3, pp. 679–682. [CrossRef]
23. Asadi-Aghbolaghi, M.; Clapes, A.; Bellantonio, M.; Jair Escalante, H.; Ponce-Lopez, V.; Baro, X.; Guyon, I.; Kasaei, S.; Escalera, S. A Survey on Deep Learning Based Approaches for Action and Gesture Recognition in Image Sequences. In Proceedings of the 2017 12th IEEE International Conference on Automatic Face & Gesture Recognition (FG 2017), Washington, DC, USA, 30 May–3 June 2017; pp. 476–483. [CrossRef]
24. Molchanov, P.; Yang, X.; Gupta, S.; Kim, K.; Tyree, S.; Kautz, J. Online Detection and Classification of Dynamic Hand Gestures with Recurrent 3D Convolutional Neural Networks. In Proceedings of the 2016 IEEE Conference on Computer Vision and Pattern Recognition (CVPR), Las Vegas, NV, USA, 27–30 June 2016; pp. 4207–4215. [CrossRef]

25. Wu, D.; Pigou, L.; Kindermans, P.; Le, N.D.; Shao, L.; Dambre, J.; Odobez, J. Deep Dynamic Neural Networks for Multimodal Gesture Segmentation and Recognition. *IEEE Trans. Pattern Anal. Mach. Intell.* **2016**, *38*, 1583–1597. [CrossRef] [PubMed]
26. Zhu, G.; Zhang, L.; Shen, P.; Song, J. Multimodal Gesture Recognition Using 3-D Convolution and Convolutional LSTM. *IEEE Access* **2017**, *5*, 4517–4524. [CrossRef]
27. Naguri, C.R.; Bunescu, R.C. Recognition of Dynamic Hand Gestures from 3D Motion Data Using LSTM and CNN Architectures. In Proceedings of the 2017 16th IEEE International Conference on Machine Learning and Applications (ICMLA), Cancun, Mexico, 18–21 December 2017; pp. 1130–1133. [CrossRef]
28. John, V.; Boyali, A.; Mita, S.; Imanishi, M.; Sanma, N. Deep learning-based fast hand gesture recognition using representative frames. In Proceedings of the 2016 International Conference on Digital Image Computing: Techniques and Applications (DICTA), Gold Coast, QLD, Australia, 30 November–2 Decembere 2016; pp. 1–8. [CrossRef]
29. Kim, T.K.; Cipolla, R. Canonical correlation analysis of video volume tensors for action categorization and detection. *IEEE Trans. Pattern Anal. Mach. Intell.* **2009**, *31*, 1415–1428. [CrossRef]
30. Jordan, M. Serial order: A parallel distributed processing approach. *Adv. Psychol.* **1997**, *211*, 471–495.
31. Araga, Y.; Shirabayashi, M.; Kaida, K.; Hikawa, H. Real time gesture recognition system using posture classifier and Jordan recurrent neural network. In Proceedings of the 2012 International Joint Conference on Neural Networks (IJCNN), Brisbane, QLD, Australia, 10–15 June 2012; pp. 1–8. [CrossRef]

Article

Evaluation of Emotional Satisfaction Using Questionnaires in Voice-Based Human–AI Interaction

Jong-Gyu Shin [1], Ga-Young Choi [2], Han-Jeong Hwang [2] and Sang-Ho Kim [1,*]

[1] Department of Industrial Engineering, Kumoh National Institute of Technology, Gumi 39177, Korea; shinjg@kumoh.ac.kr
[2] Department of Electronics and Information Engineering, Korea University, Sejong 30019, Korea; gychoi@korea.ac.kr (G.-Y.C.); hwanghj@korea.ac.kr (H.-J.H.)
* Correspondence: kimsh@kumoh.ac.kr; Tel.: +82-54-478-7656

Abstract: With the development of artificial intelligence technology, voice-based intelligent systems (VISs), such as AI speakers and virtual assistants, are intervening in human life. VISs are emerging in a new way, called human–AI interaction, which is different from existing human–computer interaction. Using the Kansei engineering approach, we propose a method to evaluate user satisfaction during interaction between a VIS and a user-centered intelligent system. As a user satisfaction evaluation method, a VIS comprising four types of design parameters was developed. A total of 23 subjects were considered for interaction with the VIS, and user satisfaction was measured using Kansei words (KWs). The questionnaire scores collected through KWs were analyzed using exploratory factor analysis. ANOVA was used to analyze differences in emotion. On the "pleasurability" and "reliability" axes, it was confirmed that among the four design parameters, "sentence structure of the answer" and "number of trials to get the right answer for a question" affect the emotional satisfaction of users. Four satisfaction groups were derived according to the level of the design parameters. This study can be used as a reference for conducting an integrated emotional satisfaction assessment using emotional metrics such as biosignals and facial expressions.

Keywords: human–AI interaction; interaction design; Kansei engineering; user satisfaction; voice-based intelligent system

1. Introduction

Recently, artificial intelligence based on various technologies, such as machine learning, natural language processing, machine vision, and big data, has been applied to systems in various fields to be used as user agents or mutual cooperation models [1]. A representative intelligent system is a voice-based intelligent system (VIS) in the form of a chatbot, which is created from developed speech recognition technology, such as natural language processing and text-to-speech. Fierce competition is underway to preempt the market for agent technology using VIS in artificial intelligence speakers and smartphones [2]. Currently, VIS is mainly used to meet customer information requirements through the role of a chatbot helper or virtual assistant, which is used to interact with customers. Interaction with AI-infused systems is defined as human–AI interaction (HAII) [3] and conceived as a modified version of human–computer interaction (HCI), which provides differentiated interaction [4]. Humans have performed system-driven interaction in HCI, whereas intelligent systems are required to provide user-centered interaction that adapts to a specified context of use in HAII. Purington et al. [5] investigated the impact of satisfaction and personalization using user reviews of VIS integrated into real life, and Cárdenas et al. [6] reported that there is a need for personalization and customization in interaction with intelligent systems. Therefore, there is a demand to create an interaction design that can secure the emotional satisfaction of various users as an intelligent system is developed.

According to Walter's "hierarchy of human needs", the final expectation of a user from the system is defined as satisfaction (Figure 1) [7]. This means that the user wants to be satisfied emotionally while interacting with the system, beyond usability. Satisfaction is a feeling that arises from the mind during the process of perceiving and recognizing the information presented by the system, indicating a state or feeling in which a need is fulfilled. Therefore, it can be stated that a person is satisfied when that person is fulfilled emotionally during interaction with the system.

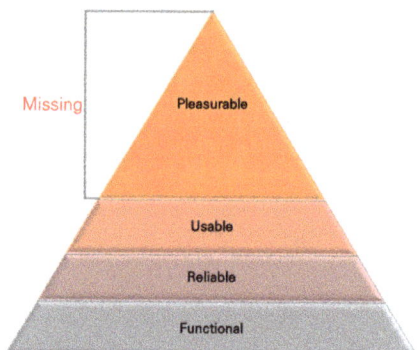

Figure 1. Hierarchy of human needs.

To quantitatively identify personal satisfaction, human emotions should first be addressed accurately. Brain–computer interface (BCI) technology is a way to quantify human emotions that can be used to infer satisfaction [8]. The BCI uses one of the biosignals, brain activity, to decode human thoughts, such as emotion [9,10], attention [11–15], and vigilance [16]. The decoded outputs can be used to identify the type of emotion (e.g., happy, angry, and excited), level of attention, or vigilance directly. Eye movement can also be used to explore the current status (emotion) of a human during a specific task [17]. To utilize these methods to decode human emotion in detail, however, the characteristics of biosignals in each emotional state need to be defined precisely in advance. It is difficult to secure sufficient sensitivity for classification without these precise criteria.

On the contrary, there is another way to quantitatively identify emotional satisfaction, called Kansei engineering or affective engineering [18]. Through this, it is possible to evaluate the user's emotional satisfaction with the product, or to identify the design parameters of the system to enhance the emotional satisfaction of the user. Li et al. [19] presented a method for optimizing user emotion using the Kansei engineering approach for product design, and Kim et al. [20] proposed a framework for evaluating users' emotional satisfaction using Kansei engineering. One advantage of the Kansei engineering approach is its relatively higher resolution in classifying users' emotional states, because they are evaluated in terms of a questionnaire consisting of a variety of adjectives that express feelings and emotions [21]. If it is possible to address the changes in emotional states of the users in detail using the Kansei engineering approach when interacting with intelligent systems, this information can be provided as the criteria for classifying biosignals of different emotions.

In this study, we utilized the Kansei engineering approach to identify users' emotional satisfaction during interaction with a VIS using a variety of conversation styles. The purpose of this study is to confirm that the Kansei engineering approach can be used as a sensitivity measure in the evaluation of emotional satisfaction for HAII in VIS. In addition, design parameters that have a significant influence on emotional satisfaction and the degree of their influence are identified.

2. Materials and Methods

To build human-centered interaction, the VIS needs to modify its style of conversation according to the context of use by changing the design parameters that complete the style of VIS. This section describes the development of the VIS software, evaluation scenarios, and design parameters of the VIS to be evaluated.

2.1. Voice-Based Intelligent System Design

The currently launched VIS (Bixby (Samsung Electronics), Siri (Apple Inc.), etc.) are complete systems, but the design parameters cannot directly be controlled; therefore, we have conducted research to develop an evaluation system using the Wizard of Oz (WoZ) experiment [22]. In the field of HCI, a WoZ experiment is a research experiment in which subjects interact with a computer system that they believe to be autonomous, but the system is actually being operated or partially operated by an unseen human being [23,24]. Large et al. [25] developed and evaluated a WoZ-type voice interface to implement an interaction situation with a voice-based interface during driving conditions. In addition, Howland and Jackson [26] conducted a study to mitigate the problems arising from interaction with users using a WoZ-type VIS. In this study, a WoZ-based adjustable VIS was constructed to evaluate the changes in user satisfaction according to the design parameters. The final four design parameters were selected based on the user requirements revealed in previous studies [27–29] and are listed in Table 1.

Table 1. Scope of voice-based intelligent system design parameters.

Design Parameters	Scope
Response time to get to the next interaction	Between 1 and 5 s
Number of trials to get the right answer for a question	1 trial 2 trials 3 trials
Pace of the answer	4 syllables/s 6 syllables/s 8 syllables/s
Sentence structure of the answer	Answer only Repeat the question and then answer Repeat the question and then answer with a clear reference source

The level of the design parameters was selected based on existing research and Bixby. The "response time to get to the next interaction" is defined as the time taken by the VIS to respond to the user's request after the user requests information. The range of this parameter was selected based on Bixby's average response time of 2 s in order to check if there was a difference in user satisfaction when a certain amount of time was spent on the information to be presented. The experiment was designed such that the information was presented randomly between 1 and 5 s, so that the subject was unaware of the presented time. For the analysis, this parameter was classified as 1, 2, 3, and 4 s.

The "number of trials to get the right answer for a question" is defined as the number of times a user tries to obtain the desired answer. This parameter was set based on a study in which the system is reset if an error occurs up to two times; it was designed to obtain the correct answer after two attempts [30].

The "pace of the answer" is defined as the number of syllables per second of the answer; here, 4, 6, and 8 syllables/s were selected and set in three steps. Four syllables and eight syllables were added to determine the highest level of satisfaction, focusing on the six syllables per second applied to the existing Bixby.

Finally, the "sentence structure of the answer" is defined as the amount of information present in the response based on the user's needs. The "sentence structure of the answer" was divided into three levels—"answer only", "repeat the question and then answer", and "repeat the question and then answer with a clear reference source"—considering the format of the answer of Bixby.

When the structure was designed by combining these four parameters, 108 interactions per subject were determined. However, considering the time, cost, and fatigue of the subject, "response time to get to the next interaction" and "pace of the answer" were designed to appear randomly in all the experiments, and nine interaction scenarios were determined by combining the remaining two parameters ("number of trials to get the right answer for a question" and "sentence structure of the answer").

The user's emotional satisfaction was evaluated based on the set parameters and their levels, and the hypotheses of the final study were as follows:

Hypothesis 1 (H1). *User satisfaction differs depending on the level of "response time to get to the next interaction".*

Hypothesis 2 (H2). *User satisfaction differs depending on the level of "number of trials to get the right answer for a question".*

Hypothesis 3 (H3). *User satisfaction differs depending on the level of "pace of the answer".*

Hypothesis 4 (H4). *User satisfaction differs depending on the level of the "sentence structure of the answer".*

2.2. Experiment Procedure

The experiment is conducted in such a way that the VIS is produced by the WoZ method, and the user confirms the answer to the question presented through the interaction. After confirming the proposed task, the user presses the "microphone" button and inputs the task through voice. In response to this, the VIS presents one of the scenarios that have been prepared by considering the design parameters in advance. If the user finds that the suggested answer is wrong, the user presses the "X" button and enters the command again until the correct answer is presented. If the answer is correct, the user presses the "check" button, conducts a satisfaction evaluation, and waits for the next task to appear. The interaction process can be expressed according to the flowchart shown in Figure 2.

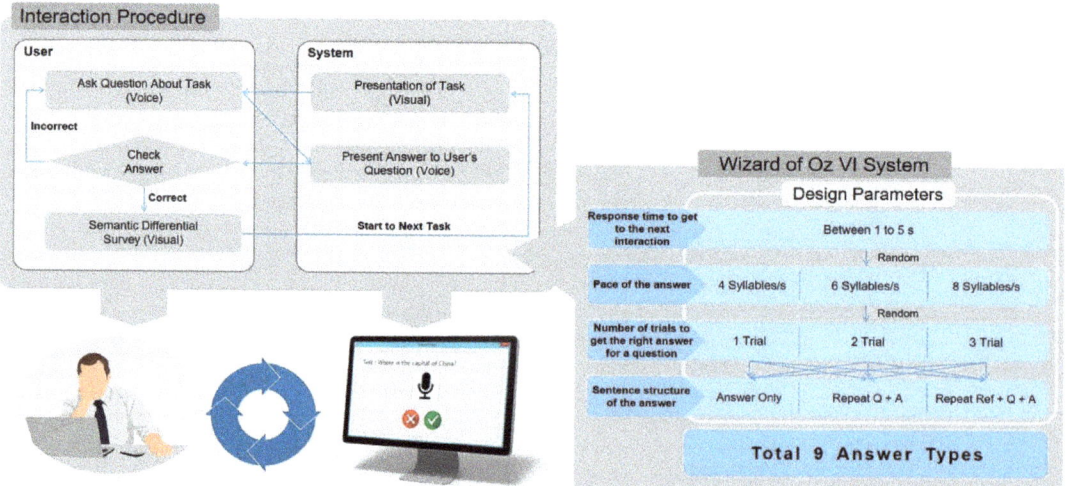

Figure 2. Voice-based intelligent system satisfaction evaluation procedure.

2.3. Participants

Participants who used a VIS were selected for the experiments on satisfaction evaluation. As a basic study to measure emotional satisfaction using a VIS, 23 participants (17 males and 6 females having an average age of 27.4 years) were selected; these were young people with little reluctance to new technologies. Before the experiment was conducted, the VIS, the method of using the system, the experimental procedure, and the purpose of the study were described. After conducting the experiment, it was observed that three datasets had outliers in the data and were, therefore, not used for analysis. Thus, the evaluation results were derived based on the datasets of the final 20 experiments. Figure 3 shows the environment of the interaction experiment.

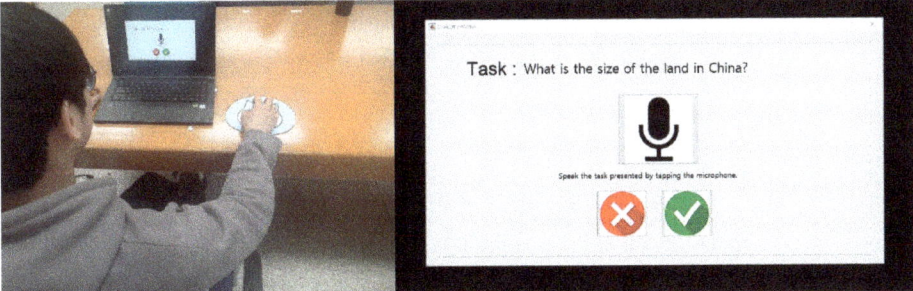

Figure 3. Environment of the experiment.

2.4. Data Collection

In this study, Kansei engineering was used to measure emotional satisfaction with the HAII. Kansei engineering can measure emotions using Kansei words (KWs). A KW is a word that expresses emotion and is mainly used as an adjective. This was derived using the semantic differential method. Osgood et al. [31] developed the semantic differential method as an application of Osgood's more general attempt to measure the semantics or meaning of words, particularly adjectives, and their referent concepts. The KWs derived using the semantic differential method extracted only the significant words that helped in expressing the user's emotion through three preliminary tests (Table 2). For the nine interaction scenarios of the VIS, the final 30 KWs were measured on a 7-point Likert scale to check which user's emotion was closer.

Table 2. Pairs of Kansei words.

No	Positive	Negative	No	Positive	Negative
1	Refreshing	Uncomfortable	16	Clever	Silly
2	Interesting	Indifferent	17	Proper	Ridiculous
3	Anticipated	Broken hearted	18	Spontaneous	Awkward
4	Glad	Perplexed	19	Suitable	Inappropriate
5	Natural	Abrupt	20	Good	Irritated
6	Confident	Anxiety	21	Pleasant	Boring
7	Detailed	Sloppy	22	Concentrated	Dejected
8	Perspicuous	Extensive	23	Organized	Confused
9	Reliable	Unreliable	24	Friendly	Picky
10	Bracing	Unpleasant	25	Lighthearted	Bothersome
11	Simple	Vague	26	Obvious	Suspicious
12	Easy	Difficult	27	Feel Unburdened	Stuffy
13	Fresh	Trite	28	Satisfied	Insufficient
14	Reassured	Concerned	29	Attractive	Banal
15	Stable	Precarious	30	Hopeful	Hopeless

2.5. Statistical Method

In this study, exploratory factor analysis (EFA) and analysis of variance (ANOVA) were used to statistically analyze emotional satisfaction data. All analyses were conducted using Minitab 18. EFA serves to identify potential structures based on the correlations between measured variables [32]. In this study, EFA was used to derive KW clusters with commonality based on the correlation between KW scores, and ANOVA was used to test the four aforementioned hypotheses. Differences in each parameter level were statistically analyzed using the mean and variance values of the KW score. The overall statistical analysis was performed based on a significance level (α) of 0.05.

3. Results

The emotional axes and the user's emotional scores were derived using EFA. ANOVA was performed using the mean and variance values of the emotional scores to analyze whether there was a difference in satisfaction for each design parameter. Finally, the emotional differentiation of the user based on the design parameters was performed on the emotional axes.

3.1. Exploratory Factor Analysis

The emotional scores collected using 30 pairs of KWs were analyzed using EFA. The analysis was conducted using varimax rotation and principal component extraction. From the scree plot, it was confirmed that the KWs were structured by two factors. Factor 1 includes ten KWs, and factor 2 includes nine KWs, as shown in Table 3.

Table 3. Kansei word cluster obtained using factor analysis.

Kansei Word	Factor 1	Kansei Word	Factor 2
Natural–Abrupt	0.804	Lighthearted–Bothersome	0.789
Bracing–Unpleasant	0.783	Interesting–Indifferent	0.768
Reassured–Concerned	0.751	Hopeful–Broken hearted	0.744
Refreshing–Uncomfortable	0.739	Detailed–Sloppy	0.744
Spontaneous–Awkward	0.690	Friendly–Picky	0.734
Reliable–Unreliable	0.679	Feel Unburdened–Stuffy	0.728
Good–Irritated	0.677	Satisfied–Insufficient	0.685
Concentrated–Dejected	0.674	Suitable–Inappropriate	0.680
Glad–Perplexed	0.658	Pleasant–Boring	0.655
Attractive–Banal	0.646		
Variance	10.253	Variance	9.233
% Variance	0.342	% Variance	0.308
Pleasurability		Reliability	

Table 3 summarizes the two factors that reduced using factor analysis and the KWs belonging to those factors. It is evident that 10 and 9 KWs, respectively, were aggregated as two factors, and 65% of the total KWs explained the emotion in the interaction situation. Based on the factor score of the KWs, which belong to the two factors, the names of the factors were determined based on the opinions of three Kansei engineering experts. The name of the suggested factor was set as the emotion that can be expressed on behalf of the feeling that occurs when interacting with the VIS. The first axis was named "pleasurability" because the cluster of emotions that occur when interacting can be referred to as pleasant and lighthearted with no difficulty. The second axis was called "reliability" because the cluster of emotions that arise refers to communication skills, such as the ability to speak clearly and respond to VIS interactions. To check the difference in emotions according to the design parameters of the VIS around the derived emotion axis, the user's emotional scores were mapped, and ANOVA was used to evaluate the difference in emotional satisfaction.

3.2. Analysis of Variance

As shown in Figure 4, the emotional scores of users were analyzed based on the four design parameters and two derived axes (pleasurability and reliability). The ANOVA results obtained for the four design parameters are presented in Appendix A.

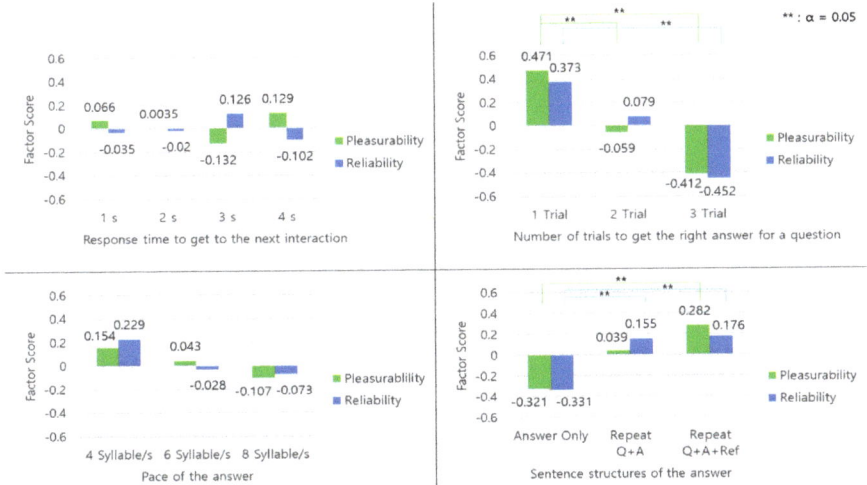

Figure 4. Emotional factor scores obtained using ANOVA for different voice-based intelligent system design parameters.

Of the four design parameters, "response time to get to the next interaction" and "pace of the answer" exhibited no difference between levels in terms of both pleasurability and reliability. Among them, "pace of the answer" exhibited relatively high emotional satisfaction with four syllables/s; however, statistically, there was no significant difference. The "number of trials to get the right answer for a question" shows that both pleasurability and reliability decrease as the number of trials increases. On the pleasurability axis, when the question needs to be asked more than once, the satisfaction decreases significantly, and there is a statistical difference. For reliability, it was observed that up to two trials, the statistically significant difference did not decrease the satisfaction sufficiently, but when three trials were conducted, satisfaction decreased significantly. The "sentence structure of the answer" showed that the satisfaction in the pleasurability axis increased significantly when detailed information was presented rather than a simple answer, and it was confirmed that there exists a statistically significant difference. In the case of reliability, because the emotions of "repeat the question and then answer" and "repeat the question and then answer with a clear reference source" are similar, it is thought that the action that informs the system that the user understands the intention of the user's question is the key to securing trust.

3.3. Classification of Emotional Satisfaction According to the Design Parameters

The ANOVA results proved that among the four hypotheses, H2 and H4 caused a difference in emotional satisfaction. The results of the classification of user emotional satisfaction for the final VIS obtained by integrating the proven "number of trials to get the right answer for a question" (three levels) and "sentence structure of the answer" (three levels) are shown in Figure 5.

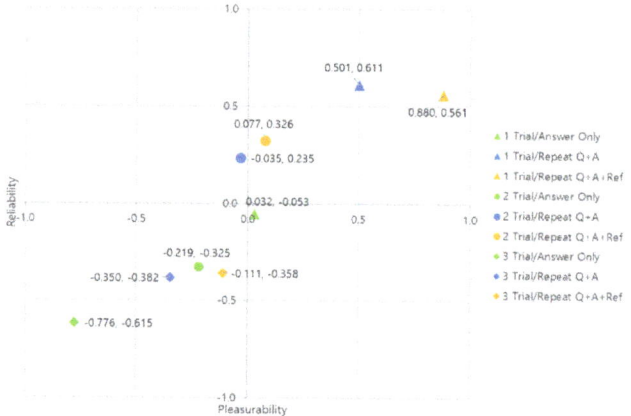

Figure 5. Classification map of emotional satisfaction.

In the graph, "number of trials to get the right answer for a question" can be distinguished by shape, and "sentence structure of the answer" can be identified by color. The influence of these two parameters on each other was confirmed using ANOVA, and it was confirmed that the two parameters independently influenced emotional satisfaction with pleasurability ($p = 0.749$) and reliability ($p = 0.808$) axes. Furthermore, emotional satisfaction was classified using Tukey analysis, and the pleasurability and reliability axes were classified into three and four levels, respectively, as shown in Table 4.

Table 4. Emotional classification using Tukey analysis.

Pleasurability					Reliability						
Number of Trials	Sentence Structure	Avg	Cluster		Number of Trials	Sentence Structure	Avg	Cluster			
1 Trial	Q + A + Ref	0.88	A			1 Trial	Repeat Q + A	0.611	A			
1 Trial	Repeat Q + A	0.501	A	B		1 Trial	Q + A + Ref	0.561	A	B		
2 Trials	Q + A + Ref	0.077	A	B	C	2 Trials	Q + A + Ref	0.326	A	B	C	
1 Trial	Answer only	0.032	A	B	C	2 Trials	Repeat Q + A	0.235	A	B	C	D
2 Trials	Repeat Q + A	−0.035		B	C	1 Trial	Answer only	−0.053	A	B	C	D
3 Trials	Q + A + Ref	−0.111		B	C	2 Trials	Answer only	−0.325		B	C	D
2 Trials	Answer only	−0.219		B	C	3 Trials	Q + A + Ref	−0.358		B	C	D
3 Trials	Repeat Q + A	−0.349		B	C	3 Trials	Repeat Q + A	−0.382			C	D
3 Trials	Answer only	−0.776			C	3 Trials	Answer only	−0.615				D

Clusters that share texts have similar relationships. Both pleasure and confidence are detailed concerning the answer and obtain high emotional satisfaction when the user understands it at once. In terms of pleasure, "2 Trials/Answer Only" has a lower satisfaction than "3 Trials/Q + A + Ref". Therefore, the "sentence structure of the answer" seems to have a greater influence on the satisfaction of the pleasurability axis as compared to that of the "number of trials". In addition, in terms of reliability, "number of trials" is the key to increasing the user's voice recognition rate of the system to obtain reliability because attempting three questions is the least satisfactory, regardless of the structure of the answer.

4. Discussion

4.1. Validity of the Kansei Engineering Approach

Identifying the user's emotional satisfaction that occurs in the process of interacting with an intelligent system is a very important point in designing high-quality HAII. In particular, it is necessary to understand what kind of emotions are generated and develop in a direction that can increase the satisfaction level of emotions. In this study, we confirmed the emotions generated by users in voice-based HAII using the Kansei engineering approach. By quantitatively evaluating the emotions through 30 KWs, two representative emotions were derived—"pleasurability" and "reliability"—and emotion classification was conducted. Human emotions are mostly classified based on Ekman's six basic human emotions (joy, sadness, anger, fear, disgust, and surprise) [33] or valence-arousal (positive/negative and active/passive, VA) [34]. This method is simple for setting emotion standards; however, it may have the disadvantage of being limited in expressing the complex emotions of human beings specifically. The Kansei engineering approach used in this study can measure complex emotions because it is free to select any KW. In the case of pleasurability, it is possible to confirm it as a basic emotion, but reliability is difficult to confirm with basic emotions or the VA model. Several studies have been conducted to evaluate satisfaction with a VIS in advance. Purington et al. [5] conducted a qualitative evaluation of Amazon Echo, focusing on interaction satisfaction and personalization based on user reviews. Pyae and Joelsson [35] qualitatively evaluated Google Home using a survey on usability and user experience. Qualitative evaluation is difficult to express clearly in expressing satisfaction in a way that draws conclusions through subjective evaluation of experts based on individual opinions of users. In contrast, this study designed a VIS using the WoZ method based on user requirements and implemented an environment in which users can interact directly. The data were collected and evaluated using KWs, which express the user's emotions. This study can be seen as a more detailed study in that it goes beyond the verification of the effect of the VIS on the user's emotions, and identifies design parameters of the system that affect the user's emotions. Thus, it is believed that this study is the first to quantitatively confirm the direct emotion of the user in the interaction. In addition, our study can be viewed as more detailed in that it is possible to identify design parameters that influence the user's emotion among design parameters of the VIS and to quantitatively grasp the degree of its influence. Through this, it was confirmed that the Kansei engineering approach can derive the user's detailed emotions in HAII and clearly present the level that is suitable for use in evaluating the user's emotional satisfaction.

4.2. User Emotion Classification According to Design Parameters

In this study, various styles of conversation based on four design parameters were presented in dialog with the VIS, and the emotions felt toward the users were evaluated. As a result of identifying the difference in emotion according to the level of the used design parameters, a difference in emotional satisfaction was observed depending on the number of trials and the sentence structure, and the response time and pace of the answer did not significantly affect the user's "pleasurability" and "reliability." It can be seen that the number of trials and the sentence structure are more sensitive to changes in the user's emotions than the response time and pace of the answer. The response time and pace of the answer did not significantly affect emotion in this study, but since they are important parameters that can be changed in real products, it is judged that additional research needs to be conducted by diversifying the level of the parameters in order to confirm the change in emotion. In contrast, it was confirmed that the two design parameters (number of trials and sentence structure) that showed a statistically significant difference in emotional satisfaction were classified into three levels of pleasurability and four levels of reliability. Table 5 shows the structure of the level of satisfaction classification in an integrated manner for pleasurability and reliability.

Table 5 shows that the user's emotions can be subdivided according to the style of conversation, rather than simply classifying satisfaction as high/low. As a result of this study, the form of dialogue that belongs to the highest level of satisfaction in the four levels of satisfaction classification was found to be one trial, when the question or the reference was mentioned. However, in terms of pleasurability, the number of trials was 1 and the reference was mentioned; in the case of reliability, the number of trials was 1, and the satisfaction was highest in the form of mentioning the question. This shows that there is a need to lead a conversation in a form tailored to a group that values reliability more among users and a group that values pleasurability more. To overcome this, the VIS must understand the characteristics of the user and secure a conversation that considers the user, and finally, establish a customized interaction. In this study, the possibility of classifying design parameters was secured, but an experiment was performed on 20 subjects that did not consider their characteristics. Therefore, among the variety of styles of conversations classified into four levels of satisfaction, there is a limitation in that it was not possible to identify the most suitable conversation type for individual users. Personalization and customization need to be fully implemented in order to optimize user-adaptive interactions, but there are still technical limitations. To overcome these limitations, it is necessary to secure as many user characteristics (human factors) as possible, and design accordingly. Moscato et al. described a novel music recommendation technique based on the identification of personality traits, moods, and emotions [36], and a novel recommender system that provides recommendations based on the interaction between users and multimedia content [37]. Cárdenas et al. [6] conducted a study to derive the human factors of a user that need to be considered when designing information systems for autonomous vehicles. If the user's human factors are structured, the VIS can be designed, and personalized interaction according to the human factors is possible; it is then possible to obtain a considerable level of personalization and customization. The limitations of this study can be supplemented by conducting subsequent evaluation studies.

Table 5. Classification of emotional satisfaction of voice-based intelligent systems.

Level of Satisfaction	Design Parameters
High Satisfaction	1 Trial, Repeat the question and then answer with a clear reference source 1 Trial, Repeat the question and then answer
Mid-High Satisfaction	1 Trial, answer only 2 Trials, Repeat the question and then answer with a clear reference source 2 Trials, Repeat the question and then answer
Mid-Low Satisfaction	2 Trials, answer only 3 Trials, Repeat the question and then answer with a clear reference source 3 Trials, Repeat the question and then answer
Low Satisfaction	3 Trials, answer only

4.3. Integration of Kansei Engineering Approach and Biosignals

In this study, evaluation was conducted through a questionnaire using the Kansei engineering approach. The Kansei engineering approach has the advantage of being able to identify various sensibilities and easy classifications. However, since the questionnaire is conducted based on the perceived sensibility that occurs after the interaction, the truth of the user's emotional expression may be distorted. In addition, the evaluation method in the form of a questionnaire has a major disadvantage in that it cannot be evaluated in real time during HAII, which is an obstacle for constructing adaptive interaction in real time. To overcome these disadvantages, efforts should be made to overcome the limitations of the survey method presented in this study. In recent years, many studies using biosignals or facial expressions have been conducted to evaluate emotions. Electroencephalography is used to check real-time emotional changes [38], and it is possible to evaluate integrated emotions using multiple biosignals, such as photoplethysmography or electrocardiograms [39]. However, because most studies do not have a standard for evaluating emotions using biosignals, they display it using pictures or photos that can induce emotions, and infer the emotion based on the changes in biosignals that occur accordingly [40,41]. It is believed that in the future, it will be extremely important to collect real-time biodata in an interaction situation and analyze the emotional changes in terms of "pleasurability" and "reliability" using biosignals in this research environment. Therefore, in future research, a framework that integrates biosignals and Kansei engineering data should be created to evaluate the user's emotional satisfaction, and the questionnaire data using the Kansei engineering approach are used as supervised learning labels for biosignal data. It is expected that HAII will be able to evaluate real-time user emotional satisfaction. This can lay an important foundation for evaluating a user's emotional satisfaction during interaction with various intelligent systems, which can ensure user satisfaction when designing intelligent systems.

5. Conclusions

In this study, the user satisfaction of HAII, which is different from the existing HCI, was evaluated using a VIS. Among the design parameters constituting the VIS, parameters for user emotional satisfaction were identified, and the degree of emotional classification between the levels of design parameters was evaluated. In this process, it was confirmed that the emotions generated in HAII were pleasure and trust. In addition, the validity of the Kansei engineering approach was established. As a result of classifying emotions around the two design parameters that affect satisfaction among the four design parameters, it was proved that emotions can be classified by subdividing them into four levels of satisfaction. Furthermore, future studies should evaluate the optimal form of dialog for a group of subjects that considers human factors to establish personalization and customization, and more varying design parameters defined in advance should be applied [42]. In addition, we will conduct research integrating the Kansei engineering approach and biosignal data to secure real-time data.

In this study, a method for evaluating emotional satisfaction was presented to implement a user-adaptive interaction. Our findings and future research can be applied to other intelligent HCI applications that can utilize a VIS, such as autonomous cars and smart homes, through changes in design parameters to be considered. Furthermore, our findings can be extended to the evaluation of multimodal interaction using other senses, such as vision and sensation.

Author Contributions: Conceptualization, J.-G.S., G.-Y.C., H.-J.H., and S.-H.K.; Methodology, J.-G.S. and S.-H.K.; Software, J.-G.S. and G.-Y.C.; Validation, J.-G.S. and G.-Y.C.; Formal analysis, J.-G.S. and G.-Y.C.; Investigation, J.-G.S. and G.-Y.C.; Resources, S.-H.K.; Data curation, H.-J.H. and S.-H.K.; Writing—original draft preparation, J.-G.S., G.-Y.C., and H.-J.H.; Writing—review and editing, J.-G.S., G.-Y.C., H.-J.H., and S.-H.K.; Visualization, J.-G.S. and G.-Y.C.; Supervision, S.-H.K.; Project administration, H.-J.H. and S.-H.K.; Funding acquisition, H.-J.H. and S.-H.K. All authors have read and agreed to the published version of the manuscript.

Funding: This research was supported by the Basic Research Program through the National Research Foundation of Korea (NRF) funded by the MSIT (2020R1A4A1017775).

Institutional Review Board Statement: The study was conducted according to the guidelines of the Declaration of Helsinki and approved by the Ethics Committee of Kumoh National Institute of Technology (202007-HR-003-01).

Informed Consent Statement: Informed consent was obtained from all subjects involved in the study.

Data Availability Statement: The data presented in this study are available on request from the corresponding author. The data are not publicly available due to privacy.

Conflicts of Interest: The authors declare no conflict of interest.

Appendix A. ANOVA Tables for Design Parameters

Table A1. One-way ANOVA for "Response time to get to the next interaction".

Pleasurability					
Source	DF	Adj SS	Adj MS	F-Value	p-Value
Response time	3	1.373	0.458	0.45	0.715
Error	176	177.627	1.009		
Total	179	179.000			
Reliability					
Source	DF	Adj SS	Adj MS	F-Value	p-Value
Response time	3	1.068	0.356	0.35	0.788
Error	176	177.932	1.011		
Total	179	179.000			

Table A2. One-way ANOVA for "Number of trials to get the right answer for a question". **: p-value < 0.05.

	Pleasurability				
Source	DF	Adj SS	Adj MS	F-Value	p-Value
Number of trials	2	23.72	11.859	13.52	0.000 **
Error	177	155.28	0.877		
Total	179	179.000			
	Reliability				
Source	DF	Adj SS	Adj MS	F-Value	p-Value
Number of trials	2	20.96	10.482	11.74	0.000 **
Error	177	158.04	0.8929		
Total	179	179.000			

Table A3. One-way ANOVA for "Pace of the answer".

	Pleasurability				
Source	DF	Adj SS	Adj MS	F-Value	p-Value
Pace of the answer	2	1.793	0.897	0.9	0.41
Error	177	177.207	1.0012		
Total	179	179.000			
	Reliability				
Source	DF	Adj SS	Adj MS	F-Value	p-Value
Pace of the answer	2	2.187	1.094	1.09	0.337
Error	177	176.813	0.999		
Total	179	179.000			

Table A4. One-way ANOVA for "Sentence structure of the answer". **: p-value < 0.05.

	Pleasurability				
Source	DF	Adj SS	Adj MS	F-Value	p-Value
Sentence structure	2	11.05	5.527	5.83	0.004 **
Error	177	167.95	0.949		
Total	179	179.000			
	Reliability				
Source	DF	Adj SS	Adj MS	F-Value	p-Value
Sentence structure	2	9.884	4.942	5.17	0.007 **
Error	177	169.116	0.955		
Total	179	179.000			

Table A5. Two-way ANOVA for "Number of Trials and Sentence structure". **: p-value < 0.05.

	Pleasurability				
Source	DF	Adj SS	Adj MS	F-Value	p-Value
Number of trials	2	23.718	11.859	14.22	0.000 **
Sentence structure	2	11.055	5.527	6.63	0.002 **
Trials X Structure	4	1.608	0.402	0.048	0.749
Error	171	142.619	0.834		
Total	179	179.000			
	Reliability				
Source	DF	Adj SS	Adj MS	F-Value	p-Value
Number of trials	2	20.964	10.482	12.21	0.000 **
Sentence structure	2	9.884	4.942	5.76	0.004 **
Trials X Structure	4	1.375	0.345	0.4	0.808
Error	171	146.777	0.858		
Total	179	179.000			

References

1. Mhatre, O.; Barshe, S.; Jadhav, S. Invoice: Intelligent Voice System for Mobile Phones. *Int. J. Innov. Adv. Comput. Sci.* **2018**, *7*, 153–160.
2. Choe, J.H.; Kim, H.T. A Survey Study on the Utilization Status and User Perception of the VUI of Smartphones. *J. Soc. e-Bus. Stud.* **2017**, *21*, 29–40. [CrossRef]

3. Amershi, S.; Weld, D.; Vorvoreanu, M.; Fourney, A.; Nushi, B.; Collisson, P.; Suh, J.; Iqbal, S.; Bennett, P.N.; Inkpen, K.; et al. Guidelines for Human-AI Interaction. In Proceedings of the 2019 CHI Conference on Human Factors in Computing Systems, Glasgow, Scotland, 4–9 May 2019; pp. 1–13.
4. Pandita, R.; Bucuvalas, S.; Bergier, H.; Chakarov, A.; Richards, E. Towards JARVIS for Software Engineering: Lessons Learned in Implementing a Natural Language Chat Interface. In Proceedings of the Workshops at the Thirty-Second AAAI Conference on Artificial Intelligence, New Orleans, LA, USA, 2–7 February 2018; pp. 779–782.
5. Purington, A.; Taft, J.G.; Sannon, S.; Bazarova, N.N.; Taylor, S.H. Alexa is my new BFF: Social Roles, User Satisfaction, and Personification of the Amazon Echo. In Proceedings of the 2017 CHI Conference Extended Abstracts on Human Factors in Computing Systems, Denver, CO, USA, May 2017; pp. 2853–2859.
6. Cárdenas, J.F.S.; Shin, J.G.; Kim, S.H. A Few Critical Human Factors for Developing Sustainable Autonomous Driving Technology. *Sustainability* **2020**, *12*, 3030. [CrossRef]
7. Walter, A. *Designing for Emotion*, 2nd ed.; A Book Apart: New York, NY, USA, 2011; pp. 1–978.
8. Hwang, H.J.; Kim, S.; Choi, S.; Im, C.H. EEG-based brain-computer interfaces: A thorough literature survey. *Int. J. Hum. Comput. Interact.* **2013**, *29*, 814–826. [CrossRef]
9. Pan, J.; Xie, Q.; Huang, H.; He, Y.; Sun, Y.; Yu, R.; Li, Y. Emotion-related consciousness detection in patients with disorders of consciousness through an EEG-based BCI system. *Front. Hum. Neurosci.* **2018**, *12*, 198. [CrossRef]
10. Al-Nafjan, A.; Hosny, M.; Al-Wabil, A.; Al-Ohali, Y. Classification of human emotions from electroencephalogram (EEG) signal using deep neural network. *Int. J. Adv. Comput. Sci. Appl.* **2017**, *8*, 419–425. [CrossRef]
11. Katona, J.; Kovari, A. A Brain–Computer Interface Project Applied in Computer Engineering. *IEEE Trans. Educ.* **2016**, *59*, 319–326. [CrossRef]
12. Katona, J.; Ujbanyi, T.; Sziladi, G.; Kovari, A. Electroencephalogram-based brain-computer interface for internet of robotic things. In *Cognitive Infocommunications, Theory and Applications*; Springer: Berlin/Heidelberg, Germany, 2019; Volume 13, pp. 253–275.
13. Katona, J.; Ujbanyi, T.; Sziladi, G.; Kovari, A. Speed control of Festo Robotino mobile robot using NeuroSky MindWave EEG headset based brain-computer interface. In Proceedings of the 2016 7th IEEE international conference on cognitive infocommunications, Wroclaw, Poland, 16–18 October 2016; pp. 251–256.
14. Katona, J. Examination and comparison of the EEG based Attention Test with CPT and TOVA. In Proceedings of the 2014 IEEE 15th International Symposium on Computational Intelligence and Informatics, Budapest, Hungary, 19–21 November 2014; pp. 117–120.
15. Katona, J.; Ujbanyi, T.; Sziladi, G.; Kovari, A. Examine the effect of different web-based media on human brain waves. In Proceedings of the 2017 8th IEEE International Conference on Cognitive Infocommunications, Debrecen, Hungary, 11–14 September 2017; pp. 407–412.
16. Katona, J.; Kovari, A. Examining the learning efficiency by a brain-computer interface system. *Acta Polytech. Hung.* **2018**, *15*, 251–280.
17. Kovari, A.; Katona, J.; Costescu, C. Evaluation of eye-movement metrics in a software debugging task using gp3 eye tracker. *Acta Polytech. Hung.* **2020**, *17*, 57–76. [CrossRef]
18. Nagamachi, M. Kansei engineering: A new ergonomic consumer-oriented technology for product development. *Int. J. Ind. Ergon.* **1995**, *15*, 3–11. [CrossRef]
19. Li, Y.; Shieh, M.D.; Yang, C.C. A posterior preference articulation approach to Kansei engineering system for product form design. *Res. Eng. Des.* **2019**, *30*, 3–19. [CrossRef]
20. Kim, S.H.; Kim, S.A.; Shin, J.K.; Ahn, J.Y. A Human Sensibility Ergonomic Design for Developing Aesthetically and Emotionally Affecting Glass Panels of Changing Colors. *J. Ergon. Soc. Korea* **2016**, *35*, 535–550. [CrossRef]
21. Daud, N.A.; Aminudin, N.I.; Redzuan, F.; Ashaari, N.S.; Muda, Z. Identification of persuasive elements in Islamic knowledge website using Kansei engineering. *Bull. Electr. Eng. Inform.* **2019**, *8*, 313–319. [CrossRef]
22. Shin, J.G.; Kim, J.B.; Kim, S.H. A Framework to Identify Critical Design Parameters for Enhancing User's Satisfaction in Human-AI Interactions. *J. Phys. Conf. Ser.* **2019**, *1284*, 237–243. [CrossRef]
23. Kelley, J.F. An iterative design methodology for user-friendly natural language office information applications. *ACM Trans. Inf. Syst.* **1984**, *2*, 26–41. [CrossRef]
24. Martin, B.; Hanington, B. *Universal Methods of Design 100 Ways to Research Complex Problems, Develop Innovative Ideas, and Design Effective Solutions*; Rockport Publishers: Beverly, MA, USA, 2012.
25. Large, D.R.; Clark, L.; Quandt, A.; Burnett, G.; Skrypchuk, L. Steering the conversation: A linguistic exploration of natural language interactions with a digital assistant during simulated driving. *Appl. Ergon.* **2017**, *63*, 53–61. [CrossRef]
26. Howland, K.; Jackson, J. Investigating Conversational Programming for End-Users in Smart Environments through Wizard of Oz Interactions. In Proceedings of the Psychology of Programming Interest Group—29th Annual Workshop, London, UK, 5–7 September 2018; pp. 107–110.
27. Dybkjær, L.; Minker, W. *Recent Trends in Discourse and Dialogue*; Springer Science & Business Media: Berlin, Germany, 2008.
28. Farinazzo, V.; Salvador, M.; Kawamoto, A.L.S.; de Oliveira Neto, J.S. An empirical approach for the evaluation of voice user interfaces User Interfaces. *IntechOpen* **2010**, 153–164. [CrossRef]
29. Kouroupetroglou, G.; Spiliotopoulos, D. Usability Methodologies for Real-Life Voice User Interfaces. *Int. J. Inf. Technol. Web Eng.* **2009**, *4*, 78–94. [CrossRef]

30. Lee, M.J.; Hong, K.H. Design and implementation of a usability testing tool for user-oriented design of command and control voice user interfaces. *Phon. Speech Sci.* **2011**, *3*, 79–87.
31. Osgood, C.E.; Suci, G.J.; Tannenbaum, P.H. *The Measurement of Meaning*, 1st ed.; University of Illinois press: Urbana, IL, USA, 1957; pp. 18–30.
32. Fabrigar, L.R.; Wegener, D.T.; MacCallum, R.C.; Strahan, E.J. Evaluating the use of exploratory factor analysis in psychological research. *Psychol. Methods* **1999**, *4*, 272–299. [CrossRef]
33. Eckman, P. Universal and cultural differences in facial expression of emotion. *Neb. Symp. Motiv.* **1972**, *19*, 207–283.
34. Russell, J.A. A circumplex model of affect. *J. Personal. Soc. Psychol.* **1980**, *39*, 1161–1178. [CrossRef]
35. Pyae, A.; Joelsson, T.N. Investigating the usability and user experiences of voice user interface: A case of Google home smart speaker. In Proceedings of the 20th International Conference on Human-Computer Interaction with Mobile Devices and Services Adjunct, Barcelona, Spain, 3–6 September 2018; pp. 127–131.
36. Moscato, V.; Picariello, A.; Sperli, G. An emotional recommender system for music. *IEEE Intell. Syst.* **2020**. [CrossRef]
37. Amato, F.; Moscato, V.; Picariello, A.; Sperlí, G. Recommendation in social media networks. In Proceedings of the 2017 IEEE Third International Conference on Multimedia Big Data, Laguna Hills, CA, USA, 19–21 April 2017; pp. 213–216.
38. Liu, Y.J.; Yu, M.; Zhao, G.; Song, J.; Ge, Y.; Shi, Y. Real-time movie-induced discrete emotion recognition from EEG signals. *IEEE Trans. Affect. Comput.* **2017**, *9*, 550–562. [CrossRef]
39. Ayata, D.; Yaslan, Y.; Kamasak, M.E. Emotion Recognition from Multimodal Physiological Signals for Emotion Aware Healthcare Systems. *J. Med Biol. Eng.* **2020**, *40*, 1–9. [CrossRef]
40. McFarland, D.J.; Parvaz, M.A.; Sarnacki, W.A.; Goldstein, R.Z.; Wolpaw, J.R. Prediction of subjective ratings of emotional pictures by EEG features. *J. Neural Eng.* **2016**, *14*, 016009. [CrossRef] [PubMed]
41. Costa, A.; Rincon, J.A.; Carracsosa, C.; Julian, V.; Novais, P. Emotions detection on an ambient intelligent system using wearable devices. *Future Gener. Comput. Syst.* **2019**, *92*, 479–489. [CrossRef]
42. Shin, J.G.; Jo, I.G.; Lim, W.S.; Kim, S.H. A Few Critical Design Parameters Affecting User's Satisfaction in Interaction with Voice User Interface of AI-Infused Systems. *J. Ergon. Soc. Korea* **2020**, *39*, 73–86. [CrossRef]

Article

Cognitive Patterns and Coping Mechanisms in the Context of Internet Use

Cristina Costescu [1], Iulia Chelba [1], Adrian Roșan [1], Attila Kovari [2,3,*] and Jozsef Katona [4]

1 Special Education Department, Faculty of Psychology and Educational Sciences, Babes-Bolyai University, Sindicatelor Street no 7, 400029 Cluj-Napoca, Romania; christina.costescu@gmail.com (C.C.); chelbaiulia@gmail.com (I.C.); adrian.rosan@ubbcluj.ro (A.R.)
2 Department of Natural Sciences and Environmental Protection, Institute of Engineering, University of Dunaujvaros, Tancsics M 1/A, 2400 Dunaujvaros, Hungary
3 Alba Regia Technical Faculty, Obuda University, Budai u 45., 8000 Szekesfehervar, Hungary
4 Institute of Computer Engineering, University of Dunaujvaros, Tancsics M 1/A, 2400 Dunaujvaros, Hungary; katonaj@uniduna.hu
* Correspondence: kovari@uniduna.hu; Tel.: +36-25-551-635

Abstract: Recent research indicates there are different cognitive patterns and coping mechanisms related to increased levels of Internet use and emotional distress in adolescents. This study aims to investigate the relationship between coping mechanisms, dysfunctional negative emotions, and Internet use. A total of 54 participants aged between 14 and 19 years old completed a questionnaire containing several measures and demographics information. We measured participants' coping strategies, emotional distress, social and emotional loneliness, and their online behavior and Internet addiction using self-report questionnaires. In order to identify the relation between the investigated variables, we used correlation analysis and regression, and we tested one mediation model. The results showed that maladaptive coping strategies and Internet use were significant predictors of dysfunctional negative emotions. Moreover, passive wishful thinking, as a pattern of thinking, was associated with anxious and depressed feelings. The relation between Internet use and dysfunctional negative emotions was mediated by participants' coping mechanisms. Therefore, we can conclude that the level of negative feelings is associated with the coping strategies used while showing an increased level of Internet addiction. Future studies should also consider different and multiple types of measurement other than self-reports, especially related to Internet addiction.

Keywords: Internet addiction; dysfunctional emotions; coping strategies; emotional problems

Citation: Costescu, C.; Chelba, I.; Roșan, A.; Kovari, A.; Katona, J. Cognitive Patterns and Coping Mechanisms in the Context of Internet Use. *Appl. Sci.* **2021**, *11*, 1302. https://doi.org/10.3390/app11031302

Academic Editor: Elisa Quintarelli
Received: 4 December 2020
Accepted: 28 January 2021
Published: 1 February 2021

Publisher's Note: MDPI stays neutral with regard to jurisdictional claims in published maps and institutional affiliations.

Copyright: © 2021 by the authors. Licensee MDPI, Basel, Switzerland. This article is an open access article distributed under the terms and conditions of the Creative Commons Attribution (CC BY) license (https://creativecommons.org/licenses/by/4.0/).

1. Introduction

The Internet and the use of social networks are becoming important influences in the lives of adolescents. More than 70% of adolescents use one or more social networks, and 92% of 13–14-year-old adolescents use the Internet daily, as reported by the Pew Research Center [1]. Considering the growing importance of social networks in adolescents' daily lives, there is an increasing interest in the side effects of the use of these platforms. Several studies have shown that the use of the Internet may affect adolescents' well-being and mental health. [2–4].

Internet addiction refers to the loss of control over Internet use [5] and can be linked to serious emotional distress and socialization problems in everyday life [6,7]. Excessive use of the Internet means, among other things, investing a lot of time and effort in being present on social networks and neglecting basic needs, such as sleep, physical activity, food, or hygiene [8]. Regarding the emotional problems that are associated with Internet addiction, studies have found that anxiety and depression are among the most frequent issues among adolescents [9–11]. There is a bidirectional relationship between internalizing problems and Internet use; in other words, adolescents with negative dysfunctional emotions are more likely to find refuge in the virtual world, and on the other hand, increased use of the Internet exacerbates negative dysfunctional emotions and loneliness over time. Internet addiction functions similarly to substance addictions, with characteristics that include withdrawal and loss of control [12,13].

When it comes to negative emotions that adolescents may experience, the binary model of distress may explain how the qualitative differences between emotions can dissociate between functional and dysfunctional emotions [14,15]. Contrary to the unitary model of emotions [16], which states that dysfunctional negative emotions (e.g., anxiety) differ from functional negative emotions (e.g., concern) only based on quantitative reasons, the binary model assumes that dysfunctional emotions appear when irrational beliefs are present, regardless of their intensity [17,18]. Moreover, dysfunctional negative emotions are linked to action tendencies, such as acting in a way that leads to experiencing pain, preventing their goals from being reached, and acting in the opposite direction of their goals [19]. According to Lazarus's transactional stress and coping model [20], the way in which individuals may act in a stressful situation depends on their appraisal of the stressor and their coping strategies. Therefore, in our study, we chose to look at both the adolescents' coping strategies and their dysfunctional emotions. The coping strategies involve cognitive and behavioral resources directed at eliminating or minimizing demands [20]. Usually, there are two major ways of coping with stress: problem-focused coping, which is centered on the management of the sources of stress, and emotion-focused coping, which deals more with the regulation of stressful emotions [20].

Studies have shown that there are adaptive and less-adaptive coping strategies: the first one can facilitate well-being and strong mental health, while the second one can impair mental and physical health [21]. Task-oriented coping is associated with lower psychological distress and less disruptive behaviors, whereas emotion-focused coping was found to be linked with more emotional and behavioral problems. [22]. A recent study [23] that investigated 508 undergraduate students aged 18–24 years showed that less-adaptive coping strategies were the main predictor of emotional problems, such as anxiety depression, and stress. Another study that used a cross-sectional survey design recruited 326 participants to examine coping styles and depressive mood in teenagers and found similar results, namely, that symptoms such as sadness or unhappiness for an unspecific period of time can be predicted by avoidant coping strategies [24]. The relation between less-adaptive (maladaptive) coping strategies and anxiety and depression symptoms has also been investigated in large samples. For example, Meng et al. [25] enrolled 13,512 adolescents aged between 12 and 19 years and investigated how psychological mechanisms were linked to mental health problems. Their results showed that coping mediated the relationship between life events and mental health. Considering all of the above-mentioned evidence, we hypothesized that coping styles could be strongly related to negative dysfunctional emotions, and we decided to investigate their impact on adolescents' mental health. We defined adaptive coping strategies according to a revised version of Lazarus theory [24] as problem-focused strategies, such as rational problem solving and seeking support and ventilation, and maladaptive coping strategies based on avoidance, such as resigned distancing and passive wishful thinking.

Adolescence is considered to be a period of great changes at many levels: social, physical, psychological, moral, and spiritual [26]. This period is characterized by a struggle to define their own selves in relation to the social world [27]. Considering the fact that several studies have shown that social relations are associated with psychological well-being [28,29] and that social isolation is associated with loneliness [30], there is a need for a better understanding of how adolescents perceive their online social life.

Several studies have shown that there are some personality traits and temperamental characteristics that make some individuals more prone to become Internet addicts [31,32]. Among the most investigated features are high impulsivity, low self-esteem, and increased levels of shyness [33,34]. Additionally, recent literature suggests that we should also take into consideration the type of device that adolescents are using when analyzing Internet addiction. Some researchers believe that the omnipresence and ubiquitous characteristics of smartphones may be important components of addictions [35]. Moreover, given the fact that Internet addiction is also associated with poor self-control [36] and that smartphones have become an essential part of our daily activities, the risks of overusing the technology are even higher. Therefore, we propose a theoretical model that researchers should consider when investigating the impact of Internet use on the lives of adolescents. As shown in Figure 1, the following components should be considered when investigating the impact of Internet use on adolescents: Internet addiction with the two components dependency and compulsive behavior, the type of coping mechanism (either adaptive or maladaptive), and emotional distress involving both social and emotional loneliness and functional positive or negative emotions vs. dysfunctional positive or negative emotions. All of the above-mentioned components may impact an adolescent's academic performance, well-being, and social adaptation.

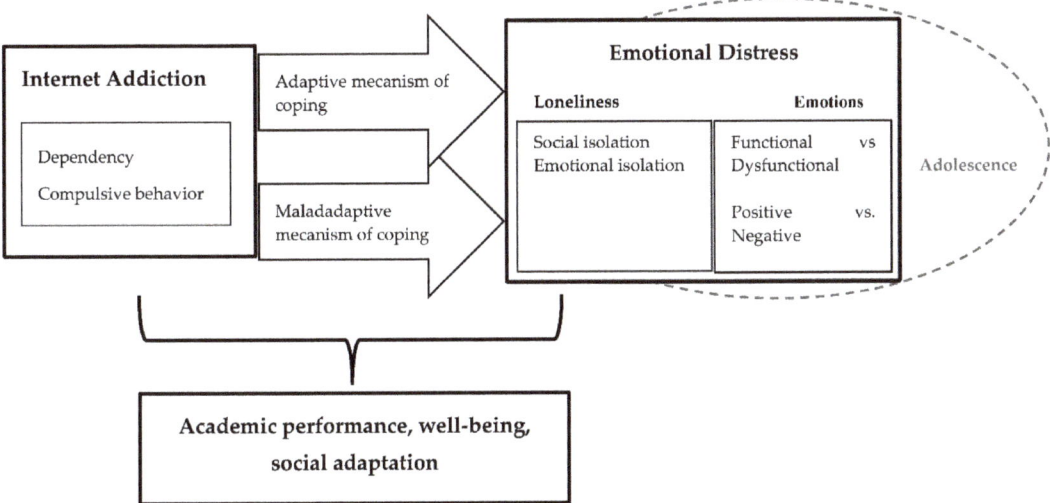

Figure 1. Proposed theoretical model.

In conclusion, adolescents are facing developmental challenges and issues regarding their college lives and social lives. Because of these challenges, they sometimes seek social support in social networks, and other times, they feel that they are ignored or neglected in their online virtual social lives and feel anxious or depressed. Either way, the strategies that they choose to use to cope with these challenges or social loneliness may affect their mental health and well-being. Understanding these bidirectional relations, along with the major factors that are related to adolescents' emotional problems and negative dysfunctional emotions, may lead us to relevant information that can be used by teachers, parents, and mental health professionals to determine the most suitable mental health intervention and prevention strategies. Therefore, this study has six major hypotheses: 1. There is a positive relation between Internet addiction and dysfunctional negative emotions; 2. There is a positive relation between negative dysfunctional emotions and maladaptive coping strategies; 3. There is a positive relation between and social and emotional isolation and coping strategies; 4. Internet addiction predicts increased levels of dysfunctional negative emotions; 5. Maladaptive coping strategies predict dysfunctional negative emotions; and 6. Coping strategies mediate the relation between Internet addiction and dysfunctional negative emotions. Although there are several studies that investigated the link between psychological, social, and family factors and Internet Addiction, to our knowledge, there are no studies that looked at this particular connection between dysfunctional negative emotions, coping strategies, and Internet addiction. The added value of this research work lies in the fact that we analyze this relationship from a different point of view that considers not only the valence of emotions but also the functionality of the emotions (function vs. dysfunctional emotions) that adolescents experience in relation to the use of the Internet. Our study also provides a complex theoretical model that is partly tested in this research and that can represent a starting point for future studies in understanding the dynamics between Internet addiction behaviors and psychological factors.

2. Materials and Methods

2.1. Participants

Fifty-four adolescents aged 14–19 years participated in this study. All of our participants were attending high schools in Romania, mainly in Transylvania, and 68.5% were from urban areas and 31.5% were from rural areas. The mean age was 16.9 with a standard deviation of 1.13. All participants completed the questionnaires online. Before answering the questions from the standardized survey, they filled in some information regarding their demographics, social network preferences, and devices.

2.2. Design and Procedure

In order to test our hypothesis, we used a correlational design and completed a regression and mediation analysis. After gaining informed consent, the questionnaires were completed using Google Forms and sent to one of the researchers of the study. The subjects had the option to receive personalized psychological reports based on what they had completed in the questionnaires regarding their online behavior, coping strategies, functional and dysfunctional emotions, and emotional and social loneliness. The study was carried out according to the law concerning the conduct of psychological studies, including abidance by international ethical standards foreseen in the updated Helsinki Declaration of Human Rights. We obtained the approval of the faculty ethics committee to conduct the study under the code Research Ethical Approval/10 October 2019. Data were used while respecting regulations regarding the subject's privacy and identity protection, and informed consent was obtained from each participant.

2.3. Instruments

The Internet Addiction Test [37] is a 20-item questionnaire that measures the characteristics and the severity of online behaviors, such as dependency and compulsive behavior. The questions also assess difficulties related to personal and social functioning that may occur due to Internet use. The scale was created by adapting the DSM-IV criteria for gambling pathology and uses a 6-point Likert scale, which ranges between "5 = always" and "0 = not applicable" [31]. The instrument was translated into Romanian and adapted for this particular research work by two independent psychologists. *The Internet Addiction Test* has an acceptable psychometric indicator (Cronbach alpha = 0.72).

The Profile of Affective Distress [38] is a 39-item scale that measures functional and dysfunctional negative emotions, such as anxiety, sadness, depression, or fear, and likewise for positive emotions. The items are grouped into 7 subscales and are rated directly from "1 = not at all" to "5 = very much". The overall distress score is obtained by summing the scores for the 26 directly rated negative items and the 13 reversely rated positive items. The higher the score, the more distress the participant experiences. Aside from the global score of distress, scores for all of the subscales can be calculated.

The Emotional/Social Loneliness Inventory [39] (Cronbach alpha = 0.86) scale was developed to measure both loneliness and isolation from a social and emotional perspective. The instrument has 15 items distributed on 4 subscales: social loneliness (items 1–8, first set of questions), emotional loneliness (items 1–8, the second set of questions), social isolation (items 9–15, first set of questions), and emotional isolation (items 9–15, the second set of questions). ESLI uses a 4-point Likert scale, which includes "3 = almost always true", "2 = often true", "1 = sometimes true", and "0 = almost never true", and its items are grouped in pairs to highlight a person's perception of a situation and how they feel about it.

The Ways of Coping Questionnaire [40] is the original form of the instrument used in our study and is a 68-item checklist covering a wide range of cognitive and behavioral coping activities, but after a series of studies developed by distinct research groups produced different results, the WCQ scale was reduced to 16 items. In our study, we used a revised version of the instrument, which has 4 subscales that include adaptive and maladaptive coping strategies such as rational problem solving, resigned distancing, seeking support and ventilation, and passive wishful thinking [35]. The above-mentioned short version of the questionnaire has good psychometric indicators (Cronbach Alpha = 0.62 for resigned distancing and 0.70–0.74 for the other subscales), and it was used to measure regulation strategies among both students and teachers in secondary education.

3. Results
3.1. Descriptive Analysis

When analyzing adolescents' responses to the instruments applied, we found that the majority of them had positive emotions (42%), and only 21% showed negative dysfunctional emotions. However, we found that the mean of affective distress in our sample was 85.64, which represents a high level of distress. Regarding the score of Internet Addiction, with mean = 33.62, our sample indicated the presence of a mild level of Internet addiction. Among the most frequently used coping strategies by our participants, we found that rational problem solving was the most used, followed by resigned distancing and passive wishful thinking. Table 1 presents the mean, standard deviation, and minimum and maximum values of our sample.

Table 1. Means and standard deviations for the measured variables.

	M	SD	Minimum	Maximum
Internet Addiction Test (total score)	33.629	15.313	4.00	70.00
Emotional/Social Loneliness Inventory	23.463	18.282	0.00	70.00
Social Loneliness	6.78	5.05	0.00	21.00
Emotional Loneliness	6.50	6.04	0.00	22.00
Social Isolation	4.79	4.04	0.00	15.00
Emotional Isolation	5.39	5.12	0.00	18.00
Ways of Coping Questionnaire	30.388	8.605	0.00	46.00
Rational Problem Solving	8.42	2.71	0.00	12.00
Resigned Distancing	5.81	2.49	0.00	10.00
Seeking Support and Ventilation	8.07	3.29	0.00	12.00
Passive Wishful Thinking	8.07	3.12	0.00	12.00
The Profile of Affective Distress	85.648	26.53	46.00	163.00
Positive Emotions	41.98	13.372	15.00	64.00
Negative Dysfunctional Emotions	21.83	10.58	14.00	60.00

Correlation analysis (* $p < 0.05$; ** $p < 0.01$).

Considering the high score that our participants had on the Internet Addiction Test, we analyzed the way in which they spent their time on the Internet. Our findings showed that 90% of the participants spent their time on social networks and listening to music. The teenagers' favorite devices included smartphones, laptops, and tablets, and when it comes to favorite social networking sites, the applications that obtained the highest scores among the subjects were Messenger, YouTube, WhatsApp, Instagram, and Facebook (see Figure 2). Generally, the subjects reported less than an hour of activity on each social network; nevertheless, some of them spent more than 6 h on the Internet (see Figure 3).

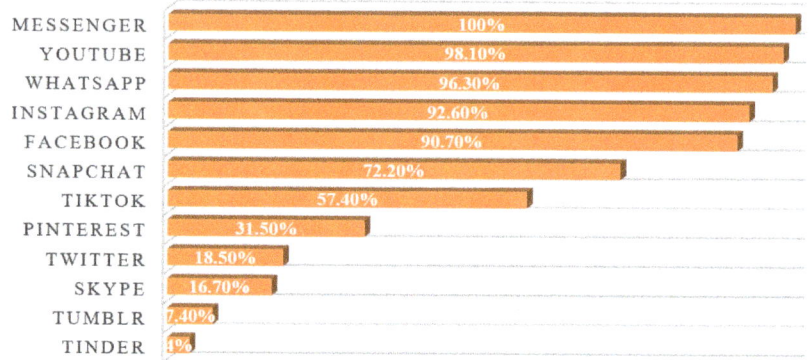

Figure 2. Favorite apps and social networks.

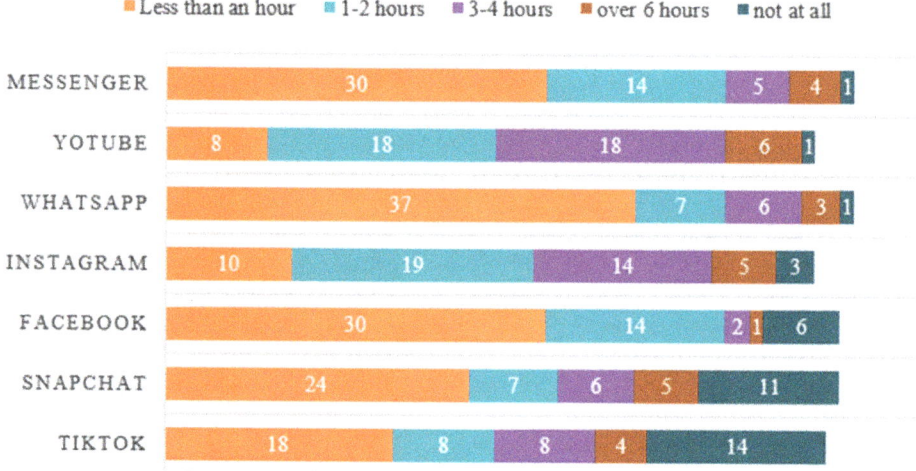

Figure 3. Time spent on each social network.

3.2. Correlation Analysis

When analyzing the relationship between Internet addiction and dysfunctional negative emotions, we found a significant positive correlation between the two variables ($r(54) = 0.351$, $p < 0.001$), meaning that if Internet dependence increased, dysfunctional negative emotions also increased. We were interested to see which negative emotions were associated with Internet dependence. Firstly, we looked at emotions that are linked to depressed mood, and we found that Internet addiction was strongly correlated with feeling grief ($r(54) = 0.360$, $p < 0.001$), hopeless ($r(54) = 0.342$, $p < 0.05$), useless ($r(54) = 0.344$, $p < 0.001$), and desperate ($r(54) = 0.332$, $p < 0.001$) (see Table 2). Afterwards, we found that there were emotions linked to anxious mood, which were also correlated with Internet addiction, such as terrified ($r(54) = 0.357$, $p < 0.001$), frightened ($r(54) = 0.287$, $p < 0.05$), and nervous ($r(54) = 0.317$, $p < 0.05$) (see Table 3).

Table 2. Association between Internet dependency and the dysfunctional negative emotions desperate, useless, hopeless, and grief.

	1	2	3	4	5
Internet Addiction Test	1				
Desperate	0.332 *	1			
Useless	0.344 *	0.493 **	1		
Hopeless	0.342 *	0.638 **	0.688 **	1	
Grief	0.360 **	0.777 **	0.599 **	0.614 **	1

Correlation analysis (* $p < 0.05$; ** $p < 0.01$).

Table 3. Association between Internet dependency and the dysfunctional negative emotions terrified, frightened, and nervous.

	1	2	3	4	5
Internet Addiction Test	1				
Terrified	0.357 **	1			
Frightened	0.278 *	0.899 **	1		
Fearful	0.136	0.709 **	0.707 **	1	
Nervous	0.317 *	0.350 **	0.425 **	0.317 *	1

Correlation analysis (* $p < 0.05$; ** $p < 0.01$).

3.2.1. The Association between Emotional Problems and Coping Strategies

When it comes to the association between social and emotional isolation and the coping strategies that the participants used, we found that passive wishful thinking was positively associated with social loneliness (r(54) = 0.526, $p < 0.001$), emotional loneliness, (r(54) = 0.580, $p < 0.001$), social isolation (r(54) = 0.515, $p < 0.001$), and emotional isolation (r(54) = 0.552, $p < 0.001$). This means that maladaptive coping strategies are associated with emotional problems. Meanwhile, adaptive coping strategies are associated with reduced emotional loneliness. For example, seeking support and ventilation negatively correlates with emotional isolation (r(54) = −0.325, $p < 0.05$). Additionally, we found a strong association between seeking support and ventilation and adaptive coping (r(54) = 0.357, $p < 0.001$) (see Table 4).

Table 4. Association between coping strategies and social and emotional loneliness.

	1	2	3	4	5	6	7	8
Social Loneliness	1							
Emotional Loneliness	0.801 **	1						
Social Isolation	0.662 **	0.798 **	1					
Emotional Isolation	0.525 **	0.868 **	0.825 **	1				
Rational Problem Solving	0.195	0.168	0.109	0.160	1			
Resigned Distancing	0.052	−0.039	−0.160	−0.129	0.478 **	1		
Seeking Support and Ventilation	−0.173	−0.325 *	−0.225	−0.227	0.478 **	0.546 **	1	
Passive Wishful Thinking	0.526 **	0.580 **	0.515 **	0.552 **	0.438 **	0.315 **	0.176	1

Correlation analysis (* $p < 0.05$; ** $p < 0.01$).

3.2.2. Regression Analysis

In order to identify possible predictors of dysfunctional negative emotions, we conducted a multiple regression. As a predictive analysis, multiple linear regression is used to explain the relationship between one continuous dependent variable and two or more independent variables. We included the following variables in the regression model: Internet addiction and coping strategies (according to Hypotheses 4 and 6). Among the four types of coping strategies investigated, i.e., rational problem solving, resigned distancing, seeking support and ventilation, and passive wishful thinking, only the last one represented a significant predictor of negative dysfunctional thinking. The results of the regression indicated that the model explained 45% of the variance and was a significant predictor of dysfunctional negative emotions $F(2,51) = 6.732$, $p < 0.05$. Both the use of the Internet in an addictive way (B = 2.734, SE = 1.162,003 β = 0.302, $p < 0.05$) and using an avoidant coping strategy (B = 0.192, SE = 0.089, β = 0.277, $p < 0.05$) predicted negative dysfunctional emotions (see Table 5).

Table 5. Multiple regression to predict dysfunctional negative emotions from Internet addiction and passive wishful thinking.

	Model 1			Model 2		
	B	SE B	β	B	SE B	β
Constant	16.386	2.332		10.932	3.385	
Passive wishful thinking	3.343	1.166	0.370 *	2.734	1.162	0.302 *
Internet addiction				0.192	0.089	0.277 *

$p < 0.05$; * $p < 0.01$; $R^2 = 0.457$; $\Delta R^2 = 0.209$ (model 2).

3.3. Mediation Analysis

In order to identify possible mediators of the relation between Internet addiction and dysfunctional negative emotions, we used a bootstrapping procedure for assessing indirect effects; this approach is based on a methodology proposed by Preacher and Hayes [41]. This method was proven to be more reliable when compared either to Baron and Kenny's mediation procedure [42] or to Sobel test approach [43] because it is not dependent on the sample size and it does not assume a normal sampling distribution of the indirect effect [44]. In our study, we used the Preacher and Hayes mediation script for SPSS. We used bootstrapping tests with 500 re-samples and calculated the bias-corrected and accelerated confidence interval [45]. Mediation is considered to be present when the confidence interval for the estimation of the indirect effect does not contain 0. Theoretically speaking, while a mediation effect would imply a significant correlation between the independent variable and the outcome (i.e., a significant total effect), an indirect effect is not based on this assumption [46]. Our correlations suggest that a maladaptive coping strategy (passive wishful thinking) is linked to both Internet addiction and dysfunctional negative emotions; therefore, we conducted a mediation analysis with Internet addiction as an independent variable, negative dysfunctional emotions as a dependent variable, and passive wishful thinking as a mediator. The results revealed that passive wishful thinking (indirect effect = 0.1069, 95% CI = 0.0321–1819) mediated the relation between Internet addiction and dysfunctional negative emotions. The indirect effect diagrams are presented in Figure 4.

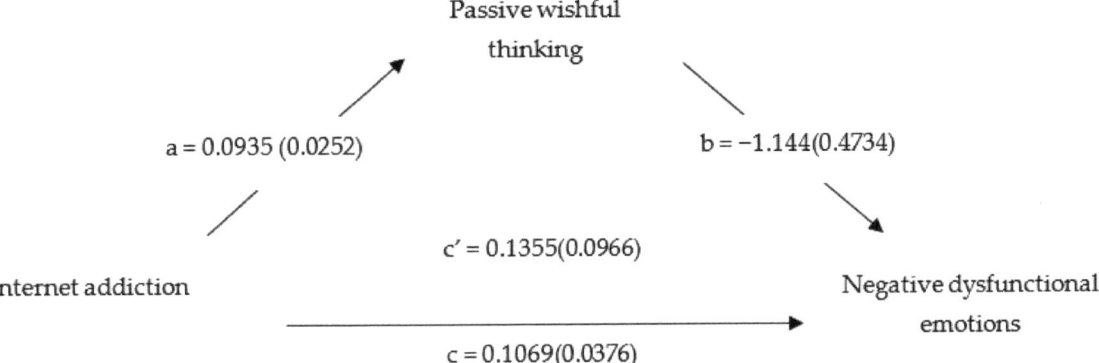

Figure 4. Values are path coefficients representing unstandardized regression weights and standard errors. The paths are represented as follows: a = independent variable to mediator; b = mediator of dependent variable; c = total effects of independent variable on dependent variable; c' = direct effect of independent variable on dependent variable.

4. Discussion and Conclusions

Considering the fact that Internet addiction is so high among adolescents, a fact that was also proven in our sample, meaning that our results show that the adolescents who participated in the study have a mild level of Internet addiction, there is a need for more studies to highlight the consequences of this phenomenon. In our research, we looked at not only the amount of time spent online but also how adolescents from Romania spent their time online. Our findings suggest that the most preferred activity online among youngsters (90.7%) is surfing on social networks, followed by listening to music online and watching movies. The most used social networks are Messenger, WhatsApp, Instagram, and Facebook. We also analyzed the amount of time that youngsters spend on their favorite network, and we found that 55% of them spend less than 1 hour on Messenger and 25% of them spend between 1 and 2 hours on Messenger. YouTube seems to be the online activity that is most engaging: 33% of the participants answered that they spend between 3 and 4 h per day on YouTube. After analyzing the online behaviors of our participants, we focused on their emotions and coping mechanisms.

Regarding their emotional pattern, almost half of them (43%) declared that they have positive emotions, such as happy, content, and joyful; this illustrates that the majority of the participants can manage their emotional state and express functional negative emotions. Only 21% of them showed dysfunctional negative emotions, of which the most prevalent are anxious, depressed, hopeless, and desperate. Even if only a small number of adolescents from our sample present this type of emotion, from our results, we found that the affective distress level is pretty high compared to the general population. These results are relevant considering the fact that there are studies showing that there is an association between Internet addiction and mental disorders, such as depression or social anxiety [47]. Therefore, we can conclude that almost a quarter of our sample is at risk of developing serious mental health problems, such as depression or anxiety.

Our findings suggest that there is an association between Internet addiction and dysfunctional negative emotions, especially those emotions related to sadness and anxiety. This was proven by the Pearson correlations developed in order to test our first hypothesis. Considering the fact that the main activity online is being on social networks, we believe that there are some particularities of online activity that need to be further investigated in order to prevent the appearance of mental health problems. For example, 40% of the participants in our study responded that their main activity on social networks is to share their experience with their friends and chat about it. The contents of online stories need to be further investigated in order to better reveal the source of the teenagers' distress.

When testing our second hypothesis, we found that the way in which adolescents interpret reality or a social situation is associated with the way that they feel. Specifically, we found that among the investigated coping strategies, passive wishful thinking is associated with dysfunctional negative emotions. This type of coping refers to the fact that individuals imagine different positive future scenarios and dream about them, but they do not act in order to fulfill their dreams. Compared to rational problem solving, which is a more adaptive means of coping whereby individuals take actions and prepare themselves for the worst scenarios, previous research has shown that passive wishful thinking is more associated with emotional distress [23–25]. Our fourth and fifth hypotheses investigated this relation in terms of prediction. Therefore, we used a multiple regression model to investigate whether coping strategies and Internet addiction predict dysfunctional negative emotions. Our results show that both Internet addiction and one specific maladaptive coping strategy (i.e., passive wishful thinking) predict emotional distress. This relation was previously investigated in both typical and atypical populations, and similar results to ours were found [48]. Internet addiction predicts negative emotions, but when we tested our sixth hypothesis, we found that the coping style mediates the relation between Internet addiction and emotional distress. Thus, even when adolescents from our group show a mild level of Internet addiction, the appearance of dysfunctional emotions is mediated by the way that they interpret the online behaviors of others and their own.

There are several important implications of our research; firstly, it is important to know which are the most common activities of adolescents online and from which type of applications and sites they get their information. Considering the fact that they use a lot of applications that allow them to communicate in writing with their friends (Messenger, WhatsApp), we suggest that mental health specialists involved in school programs focus more on written assertive communication and social skills when using online platforms. On the other hand, children should also benefit from classes regarding how to select and interpret the information that is posted on social networks. Secondly, our findings show that there is a connection between Internet use and emotional distress; even though there are also other relevant factors that may influence the level of distress in adolescents, parents, teachers, and mental health specialists should pay more attention to their online behavior. Depression and anxiety symptoms may influence their academic performances and quality of life, and therefore, we should be aware and consider all the factors that may contribute to an increased level of emotional distress. Thirdly, even though depressed mood and anxiety symptoms may be linked to Internet use, we found in our research that the way that adolescents interpret the situation mediates the relation between the two. Therefore, task-oriented coping strategies seem to be more effective in reducing emotional distress, whereas emotion-focused coping strategies are less effective. Passive wishful thinking—as a coping strategy—may also be linked to the content that adolescents access online; for example, they watch other friends' posts, activities, and achievements, without having real information about the necessary steps in order to achieve certain goals. Proactivity and reappraisal strategies could be part of their educational programs and also be considered as alternatives to annual evaluation systems.

Although our results are in line with the findings of other researchers [23–25] who showed that there is a link between coping strategies and negative emotions, the use of the Internet and its consequences add an interesting view of this relation. More specifically, when analyzing the types of maladaptive coping strategies, we found that passive wishful thinking is highly correlated with dysfunctional negative emotions, meaning that adolescents imagine different positive future scenarios and dream about them, but they do not act in order to fulfill their dreams, and this can increase their anxiety and depressive mood. However, there are also studies that claim [35] that the possible negative impact of technology overuse on our daily activities could be less significant than our current level of concern regarding these issues.

There are a few limitations that need to be considered when interpreting the results of our research. Our measurements do not cover all of the possible factors that may influence the level of emotional distress, such as irrational beliefs, family background, personality traits, etc. Moreover, even the collected information regarding Internet use and emotional distress must be interpreted with caution because it reflects the opinions of the participants (self-report). More objective measures and information from different sources should be included in future studies, such as direct observation, daily motorizing, and interviews with parents, teachers, and friends.

Through our findings, we only partially validated the proposed theoretical model, mainly because we did not include instruments that measure well-being, academic performance, or social adaptation in our study. These variables represent important features that can be linked to Internet addiction and emotional distress. Future studies should involve measurements such as well-being, social adaptation, and academic performance for a better understanding of the phenomena of Internet cognitive factors and Internet use and should also use a larger sample of adolescents. Furthermore, it may be interesting to analyze the way that their online behavior and coping strategies change during development, i.e., at different ages in adolescence.

Author Contributions: Conceptualization, C.C., I.C., A.R., A.K., and J.K.; methodology, C.C. and I.C.; formal and data analysis, A.K., J.K.; data collection, I.C.; writing—original draft preparation, C.C., I.C., A.R., and A.K.; writing—review and editing, A.K. and C.C.; funding acquisition, A.K. All authors have read and agreed to the published version of the manuscript.

Funding: This research and APC were funded by European Social Fund, EFOP (Human Resources Development Operational Programme), grant number EFOP-3.6.2-16-2017-00018, "Produce together with the nature" and EFOP-3.6.1-16-2016-00003, "Consolidate long-term R and D and I processes at the University of Dunaujvaros".

Institutional Review Board Statement: The study was conducted according to the guidelines of the Declaration of Helsinki and approved by the Ethics Committee of Babes-Bolyai University, Special Education Department (10 October 2019).

Informed Consent Statement: Informed consent was obtained from all subjects involved in the study.

Conflicts of Interest: The authors declare no conflict of interest.

References

1. Lenhart, A.; Smith, A.; Anderson, M.; Duggan, M.; Perrin, A. *Teens, Technology Friendship*; Pew Research Center: Washington, DC, USA, 2015; Volume 15.
2. Twenge, J.M.; Campbell, W.K. Associations between screen time and lower psychological well-being among children and adolescents: Evidence from a population-based study. *Prev. Med. Rep.* **2018**, *12*, 271–283. [CrossRef] [PubMed]
3. Shapira, N.A.; Goldsmith, T.D.; Keck, P.E., Jr.; Khosla, U.M.; McElroy, S.L. Psychiatric features of individuals with problematic internet use. *J. Affect. Disord.* **2000**, *57*, 267–272. [CrossRef]
4. Morahan-Martin, J.; Schumacher, P. Incidence and correlates of pathological Internet use among college students. *Comput. Hum. Behav.* **2000**, *16*, 13–29. [CrossRef]
5. Young, K.S. Internet addiction: Evaluation and treatment. *BMJ* **1999**, *319*, 9910351. [CrossRef]
6. Nalwa, K.; Anand, A. Internet addiction in students: A cause of concern. *Cyberpsychol. Behav.* **2003**, *6*, 653–656. [CrossRef]
7. Stavropoulos, V.; Gentile, D.; Motti-Stefanidi, F. A multilevel longitudinal study of adolescent Internet addiction: The role of obsessive–compulsive symptoms and classroom openness to experience. *Eur. J. Dev. Psychol.* **2016**, *13*, 99–114. [CrossRef]
8. Hussain, Z.; Griffiths, M.D. Gender swapping and socializing in cyberspace: An exploratory study. *Cyber Psychol. Behav.* **2008**, *11*, 47–53. [CrossRef]
9. Casale, S.; Fioravanti, G. Satisfying needs through Social Networking Sites: A pathway towards problematic Internet use for socially anxious people? *Addict. Behav. Rep.* **2015**, *1*, 34–39. [CrossRef]
10. Kaess, M.; Durkee, T.; Brunner, R.; Carli, V.; Parzer, P.; Wasserman, C.; Sarchiapone, M.; Hoven, C.; Apter, A.; Balazs, J.; et al. Pathological Internet use among European adolescents: Psychopathology and self-destructive behaviours. *Eur. Child Adolesc. Psychiatry* **2014**, *23*, 1093–1102. [CrossRef]
11. Liang, L.; Zhou, D.; Yuan, C.; Shao, A.; Bian, Y. Gender differences in the relationship between internet addiction and depression: A cross-lagged study in Chinese adolescents. *Comput. Hum. Behav.* **2016**, *63*, 463–470. [CrossRef]
12. Billieux, J. Problematic Use of the Mobile Phone: A Literature Review and a Pathways Model. *Curr. Psychiatry Rev.* **2012**, *8*, 299–306. [CrossRef]
13. Billieux, J.; Gay, P.; Rochat, L.; Van der Linden, M. The role of urgency and its underlying psychological mechanisms in problematic behaviours. *Behav. Res. Ther.* **2010**, *48*, 1085–1096. [CrossRef] [PubMed]

14. Ellis, A. *Reason and Emotion in Psychotherapy*; Birch Lane: Secaucus, NJ, USA, 1994.
15. David, D.; Schnur, J.; Belloiu, A. Another search for the "hot" cognitions: Appraisal, irrational beliefs, attributions, and their relation to emotion. *J. Ration.-Emot. Cogn.-Behav. Ther.* **2002**, *20*, 93–131. [CrossRef]
16. Russell, J.; Barrett, L. Core affect, prototypical emotional episodes, and other things called emotions:dissecting the elephant. *J. Pers. Soc. Psychol.* **1999**, *76*, 805–819. [CrossRef] [PubMed]
17. David, D.; Montgomery, G.H.; Macavei, B.; Bovbjerg, D. An empirical investigation of Albert Ellis' binary model of distress. *J. Clin. Psychol.* **2005**, *61*, 499–516. [CrossRef]
18. Opriș, D.; Macavei, B. The distinction between functional and dysfunctional negative emotions; An empirical analysis. *J. Cogn. Behav. Psychother.* **2005**, *5*, 181–195.
19. Dryden, W.; Di Giuseppe, R. *Ghid de Terapie Rațional-Emotivă și Comportamentală*; Editura ASCR: Cluj-Napoca, Romania, 2003.
20. Folkman, S.; Lazarus, R.S. *Stress, Appraisal, and Coping*; Springer Publishing Company: New York, NY, USA, 1984; pp. 150–153.
21. Endler, N.S.; Parker, J.D. Assessment of multidimensional coping: Task, emotion, and avoidance strategies. *Psychol. Assess.* **1994**, *6*, 50. [CrossRef]
22. McWilliams, L.A.; Cox, B.J.; Enns, M.W. Use of the Coping Inventory for Stressful Situations in a clinically depressed sample: Factor structure, personality correlates, and prediction of distress 1. *J. Clin. Psychol.* **2003**, *59*, 1371–1385. [CrossRef]
23. Mahmoud, J.S.; Staten, R.T.; Lennie, T.A.; Hall, L.A. The relationships of coping, negative thinking, life satisfaction, social support, and selected demographics with anxiety of young adult college students. *J. Child Adolesc. Psychiatr. Nurs.* **2015**, *28*, 97–108. [CrossRef]
24. Chan, S.M. Early adolescent depressive mood: Direct and indirect effects of attributional styles and coping. *Child Psychiatry Hum. Dev.* **2012**, *43*, 455–470. [CrossRef]
25. Meng, X.H.; Tao, F.B.; Wan, Y.H.; Yan, H.U.; Wang, R.X. Coping as a mechanism linking stressful life events and mental health problems in adolescents. *Biomed. Environ. Sci.* **2011**, *24*, 649–655. [CrossRef] [PubMed]
26. Rew, L. *Adolescent Health: A Multidisciplinary Approach to Theory, Research, and Intervention*; Sage: Thousand Oaks, CA, USA, 2005.
27. Nakkula, M.J.; Toshalis, E. *Understanding Youth: Adolescent Development for Educators*; Harvard Education Press: Cambridge, MA, USA, 2006; p. 02138.
28. Borge, L.; Martinsen, E.W.; Ruud, T.; Watne, Ø.; Friis, S. Quality of life, loneliness, and social contact among long-term psychiatric patients. *Psychiatr. Serv.* **1999**, *50*, 81–84. [CrossRef] [PubMed]
29. Lauder, W.; Sharkey, S.; Mummery, K. A community survey of loneliness. *J. Adv. Nurs.* **2004**, *46*, 88–94. [CrossRef] [PubMed]
30. Wang, J.; Lloyd-Evans, B.; Giacco, D.; Forsyth, R.; Nebo, C.; Mann, F.; Johnson, S. Social isolation in mental health: A conceptual and methodological review. *Soc. Psychiatry Psychiatr. Epidemiol.* **2017**, *52*, 1451–1461. [CrossRef] [PubMed]
31. Tian, Y.; Qin, N.; Cao, S.; Gao, F. Reciprocal associations between shyness, self-esteem, loneliness, depression and Internet addiction in Chinese adolescents. *Addict. Res. Theory* **2020**, *28*, 1–13. [CrossRef]
32. Munno, D.; Cappellin, F.; Saroldi, M.; Bechon, E.; Guglielmucci, F.; Passera, R.; Zullo, G. Internet Addiction Disorder: Personality characteristics and risk of pathological overuse in adolescents. *Psychiatry Res.* **2017**, *248*, 1–5. [CrossRef]
33. Wang, D. A study of the relationship between narcissism, extraversion, body-esteem, social comparison orientation and selfie-editing behavior on social networking sites. *Pers. Individ. Differ.* **2019**, *146*, 127–129. [CrossRef]
34. Wartberg, L.; Kriston, L.; Thomasius, R. Internet gaming disorder and problematic social media use in a representative sample of German adolescents: Prevalence estimates, comorbid depressive symptoms and related psychosocial aspects. *Comput. Hum. Behav.* **2020**, *103*, 31–36. [CrossRef]
35. Jeong, Y.; Suh, B.; Gweon, G. Is smartphone addiction different from Internet addiction? comparison of addiction-risk factors among adolescents. *Behav. Inf. Technol.* **2020**, *39*, 578–593. [CrossRef]
36. Peris, M.; de la Barrera, U.; Schoeps, K.; Montoya-Castilla, I. Psychological risk factors that predict social networking and internet addiction in adolescents. *Int. J. Environ. Res. Public Health* **2020**, *17*, 4598. [CrossRef]
37. Young, K.S. *The Internet Addiction Test*; Stoeling Co.: Wood Dale, IL, USA, 2016.
38. Opriș, D.; Macavei, B. *Profilul Distresului Afectiv. Sistem de Evaluare Clinică*; Editura RTS: Cluj-Napoca, Romania, 2007.
39. Vincenzi, H.; Grabosky, F. Measuring the emotional/social aspects of loneliness and isolation. *J. Soc. Behav. Personal.* **1987**, *2*, 257–270.
40. Folkman, S.; Lazarus, R.S. *Manual for the Ways of Coping Questionnaire*; Consulting Psycholo-Gists Press: Palo Alto, CA, USA, 1988.
41. Chan, D.W. Stress, coping strategies, and psychological distress among secondary school techers in Hong Kong. *Am. Educ. Res. J.* **1998**, *35*, 145–163. [CrossRef]
42. Preacher, K.J.; Hayes, A.F. Asymptotic and resampling strategies for assessing and comparing indirect effects in multiple mediator models. *Behav. Res. Methods* **2008**, *40*, 879–891. [CrossRef] [PubMed]
43. Baron, R.; Kenny, D. The moderator–mediator variable distinction in social psychological research: Conceptual, strategic, and statistical considerations. *J. Pers. Soc. Psychol.* **1986**, *51*, 1173–1182. [CrossRef]
44. Sobel, M.E. Asymptotic confidence intervals for indirect effects in structural equation models. *Sociol. Methodol.* **1982**, *13*, 290–312. [CrossRef]
45. Hayes, A.F. Beyond Baron and Kenny: Statistical mediation analysis in the new millennium. *Commun. Monogr.* **2009**, *76*, 408–420. [CrossRef]

46. Preacher, K.J.; Hayes, A.F. SPSS and SAS procedures for estimating indirect effects in simple mediation models. *Behav. Res. Methods Instrum. Comput.* **2004**, *36*, 717–731. [CrossRef]
47. Ko, C.-H.M.; Yen, J.-Y.M.; Chen, C.; Yeh, Y.-C.; Yen, C.-F. Predictive Values of Psychiatric Symptoms for Internet Addiction in Adolescents. *Arch. Pediatrics Adolesc. Med.* **2009**, *163*, 937–943. [CrossRef]
48. Predescu, E.; Sipos, R.; Costescu, C.A.; Ciocan, A.; Rus, D.I. Executive Functions and Emotion Regulation in Attention-Deficit/Hyperactivity Disorder and Borderline Intellectual Disability. *J. Clin. Med.* **2020**, *9*, 986. [CrossRef]

Article

EEG-Based Emotion Recognition Using Deep Learning and M3GP

Adrian Rodriguez Aguiñaga [1,†,‡], Luis Muñoz Delgado [1,‡], Víctor Raul López-López [1,‡] and Andrés Calvillo Téllez [2,*,‡]

1. Computation, Aeronautical and Mechanical Departments of the Tecnológico Nacional de México Campus, Tijuana 22414, Mexico; adrian.rodriguez@tectijuana.edu.mx (A.R.A.); lmunoz@tectijuana.edu.mx (L.M.D.); vlopez@tectijuana.edu.mx (V.R.L.-L.)
2. Telecommunications Department of the Instituto Politécnico Nacional, Tijuana 22435, Mexico
* Correspondence: calvillo@citedi.mx; Tel.: +52-664-529-9998
† Current address: Av Castillo de Chapultepec 562, Tomas Aquino, Tijuana 22414, México.
‡ These authors contributed equally to this work.

Abstract: This paper presents the proposal of a method to recognize emotional states through EEG analysis. The novelty of this work lies in its feature improvement strategy, based on multiclass genetic programming with multidimensional populations (M3GP), which builds features by implementing an evolutionary technique that selects, combines, deletes, and constructs the most suitable features to ease the classification process of the learning method. In this way, the problem data can be mapped into a more favorable search space that best defines each class. After implementing the M3GP, the results showed an increment of 14.76% in the recognition rate without changing any settings in the learning method. The tests were performed on a biometric EEG dataset (BED), designed to evoke emotions and record the cerebral cortex's electrical response; this dataset implements a low cost device to collect the EEG signals, allowing greater viability for the application of the results. The proposed methodology achieves a mean classification rate of 92.1%, and simplifies the feature management process by increasing the separability of the spectral features.

Keywords: EEG; emotion; neural networks; M3GP; BED; Emotiv; multiclass; deep learning

1. Introduction

Machine Learning and EEG

The development of techniques that facilitate human behavior analysis has been progressed notably in the last decade, which has driven the development of machine learning tools and signal processing technology that can perform complex analyses of our environments. This has led to a diversification of analysis techniques and even the development of hardware to analyze our environmental signals, such as intelligent voice assistants and wearable devices that can give a detailed follow-up of our vital signs. Similarly, this development has allowed certain fields, such as the analysis of electroencephalography (EEG) signals, to also notably develop in areas that have previously been restricted by costs and technological limitations.

EEG signal analysis is a non-invasive technique that is widely used to analyze brain activity. Traditionally, the analysis of these signals has focused on the diagnosis and monitoring of medical conditions [1]. However, the interest in this signal has diversified, and machine learning (ML) has played a significant role in EEG signal processing and analysis because it has allowed the development of analysis techniques that successfully perform recognition and classification tasks. Promising results have been achieved in fields as varied as the analysis of mental illness, object manipulation, and cognitive processes [2–6].

A central aspect for the study of EEG signals is the premise of obtaining information directly from the cognitive processes of a person and implementing ML to process traits that allow us to identify, analyze, associate, and even control these processes through cognitive responses [3,7,8]. When analyzing the EEG signals, it is expected that the data are closely related to the phenomenon and cannot be manipulated. However, it is also a technique that is highly susceptible to experimental noise, so signal processing methodologies need to be carefully established.

The novelty of our proposal lies in applying the ML tool M3GP to the conversion of the search space through selecting, combining, deleting, and constructing the most suitable features to ease the classification process of the learning method. This process allows us to improve the classification stage without changing the architectures, which are often obtained through extensive processes.

The literature shows that the community has proposed diverse EEG data processing architectures; some use large and complex neural networks since their features require it [9] and others argue that low complexity architectures could handle the problem but use distinct spectral feature extractions [10,11]. The proposed methodology enables the improvement of the recognition rate for this type of analysis without changing the topology.

Several techniques have been developed for signal processing and feature selection, which have been traditionally based on spectral components [12], since they reduce the computational load for ML processes [11]. Recent literature suggests that automatic feature selection can be achieved using ML algorithms, thereby reducing the signal processing stage and allowing an ML strategy to choose the best-fit features [13,14]. This proposal is based on implementing MP3G to transform and select the most suitable features by considering a spectral stage instead of performing a greedy search.

This work focuses on the BED dataset [15], which was developed to evoke and register emotional stimuli using low cost devices that can help to generalize the results of this type of research and also proposes precise experimentation strategies and performs a rigorous signal treatment process. State-of-the-art (SOA) works, as reported in [16], have demonstrated the viability and importance of implementing new and accessible methodologies that seek the greater applicability of this type of research.

2. Materials and Methods
2.1. Emotions in Machine Learning

Russell's arousal and valence levels [17] are used to measure how positive and intense an emotion is. This model subdivides the emotional responses into high arousal–high valence (HAHV), high arousal–low valence (HALV), low arousal–low valence (LALV), and low arousal–high valence (LAHV), as can be seen in Figure 1. We used this four subdivision model to create the classes our the ML process.

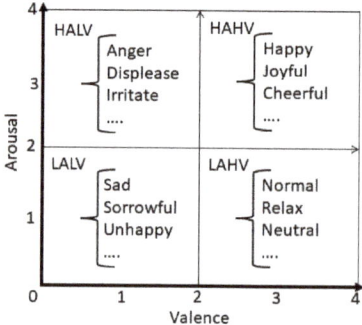

Figure 1. Emotion distribution and classes distribution based on the Russell circumplex model. This structure allowed us to carry out a multiclass classification scheme.

2.2. Dataset

This work used the BED, which is a dataset that is specialized for the analysis of emotional states. Although there are other datasets available, such as DEAP or SEED [18,19], BED was conceived and designed with the specific purpose of studying the brain's response to an emotional stimulus; it is also a novel dataset that demonstrates an extensive and well-designed experimental process. In addition, BED was developed using low cost signal acquisition equipment, specifically the Emotiv EPOC+ [20] headband, which allows the experimental process to have a greater applicability.

The BED implements a methodology of evoking emotional responses through audio-visual stimuli followed by an arousal and valence analysis. The dataset contains metadata that allow for the association of the expected responses from each stimulus. Even more important is that it reports the experience of the user: each experiment is associated with a survey that records the actual level of arousal and valence experienced by the users (scored on the scale 0–4). We relied on the responses provided by the users to associate each experiment to one of the Russell model classes, as shown in Figure 2. In this way, since we were working according to the user responses, we were able to ensure that we were working with the appropriate information.

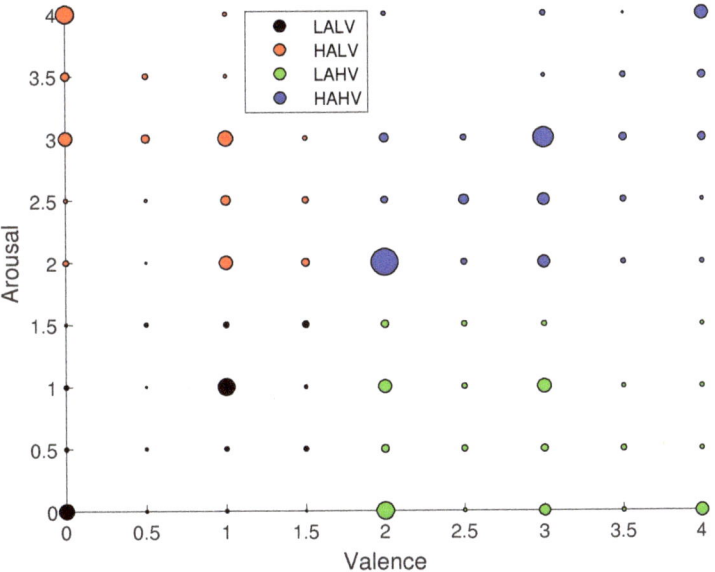

Figure 2. The responses from the research subjects were distributed according to an arousal and valence scale. Higher diameter circles indicate the higher sample frequency of a class.

The dataset contains 21 healthy participants (18 male and 3 female), aged between 23 and 47 years old. The dataset emphasizes that it was created with the specific purpose of evoking emotions using Visual Evoked Potentials (VEP). The experiment implements a prolonged rest period to mitigate against the fatigue or satiety of the participants by dividing the experimental sessions into two days.

Emotiv EPOC+ is a low cost wireless headset that records EEG signals using 16 contact sensors at 256 Hz. BED considers the 14 channels that are closely aligned with the modified combinatorial nomenclature (MCN) of the 10–20 system (AF3, F7, F3, FC5, T7, P7, O1, O2, P8, T8, FC6, F4, F8, and AF4) and the M1 and M2 mastoid locations are used as the references. Using this device was one of the main aspects of this analysis exercise since it is a portable device and a viable alternative to much more expensive equipment. The computational resources and software implemented in the development of this project are: Python Scikit Learn libraries, Python 3.7.6, NumPy 1.18.5, Pillow 8.3.1, imutils 0.5.4, TensorFlow GPU 1.15.0, keras 2.2.4, matplotlib 3.1.1, SciPy 1.2.0., Intel i7 (4th gen), 3.6 GHz, 16 RAM, and Nvidia GeForce 1080.

2.3. Feature Extraction

This work implemented two feature selection strategies. First, we used a configuration similar to that presented in [15] to obtain our reference by using the features already contained in the dataset. Second, we implemented a feature selection and transformation through a genetic programming (GP) [21] algorithm called M3GP [22], which helped us to improve the recognition rate.

Spectral Features Arrangement

This work uses 518,154 spectral features contained in the dataset (mel-frequency cepstral (MFCC), autoregression reflection coefficients (ARRC), and spectral features (SPEC)) distributed into four categories, as presented in Figure 2. Recent advances have proven that artificial intelligence (AI) and ML have numerous applications in all engineering fields for processing and analyzing signals [23–25].

2.4. Feature Transformation by M3GP

M3GP is an ML tool based on genetic programming that improves the search space in terms of locally maximizing the distance between classes and thus, increasing the performance of the classification and recognition algorithms. Our method used a variant of tree-based GP called multidimensional multiclass GP with multidimensional populations (M3GP) [22], which is a wrapper approach for supervised classification and regression [26]. M3GP enables a transformation of the form k from $\mathbb{R}^p \to \mathbb{R}^d$ into $p, d \in \mathbb{N}$ using a special tree representation (see Figure 3), in essence mapping the p input features of the problem to a new feature space of size d. Afterward, M3GP applies the Mahalanobis distance classifier [27] to measure the quality of each transformation based on classification accuracy. In other words, M3GP is a wrapper-based GP classifier for multiclass problems that performs a feature treatment of the input data, thereby obtaining new data features with which the classification method can achieve better results.

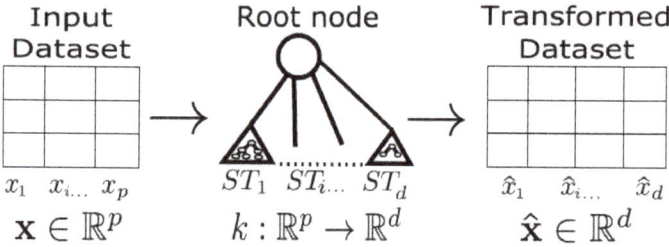

Figure 3. The special tree with d dimensions stemming from the root node.

The evolutionary process of M3GP enables the transformation of the input feature space, such that the transformed data of each class can be grouped into a single cluster, as in Figure 4. To achieve this, M3GP uses a tree representation that allows it to perform the mapping $k : \mathbb{R}^p \to \mathbb{R}^d$. The representation is the same as in regular tree-based GP, except that the root node of the tree exists only to define the number of dimensions d of the new space. Each branch stemming directly from the root performs the mapping in one of the d dimensions (see Figure 3). The initial population can start with a single feature; to search for simpler and smaller transformations, the increase in features occurs through the genetic operators.

As with any evolutionary algorithm, M3GP involves a population that may contain individual features of several different dimensions, as shown in Figure 5. M3GP includes special genetic operators that can add or remove dimensions. It is assumed that the selection process is sufficient to discard the individuals with the least useful dimensions (also referred to as "features") whilst maintaining the best dimensions within the population.

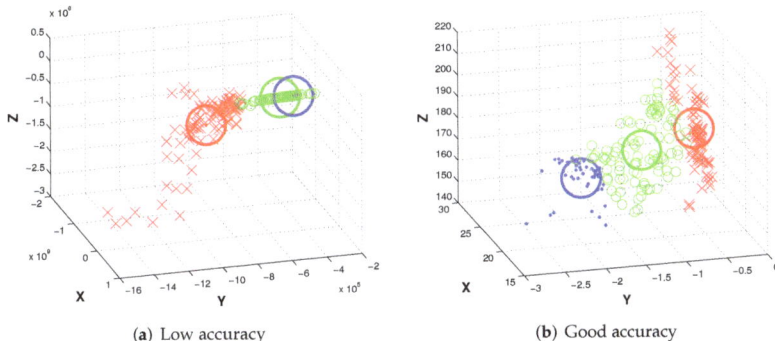

(a) Low accuracy

(b) Good accuracy

Figure 4. An example of a transformation produced by M3GP for a problem with three classes. On the left are the original data with a low classification accuracy Figure 4a. On the right are the same data transformed into a space where the resulting classification achieves a very good accuracy Figure 4b. The large circles represent the centroids of each class and each data point is marked by a different symbol depending on the class (i.e., dot, cross, small circle).

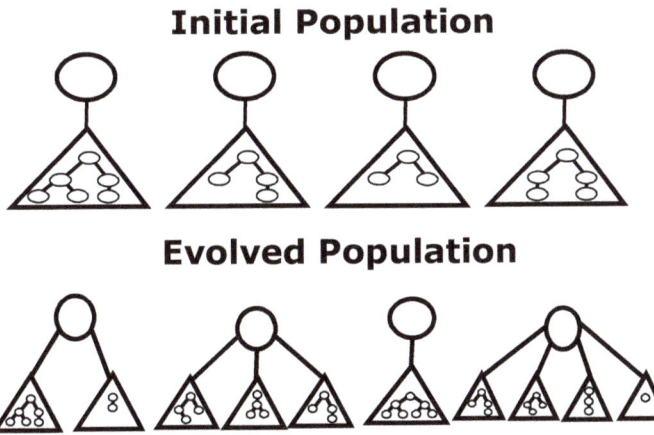

Figure 5. M3GP starts with (but is not limited to) a population of one-dimensional transformations. Through mutation (Figure 6b,c), the search can evolve into a multi-dimensional population.

2.4.1. Initial Population

M3GP starts the search with a random population where all features have only one dimension, see Figure 5. This ensures that the evolutionary search begins by looking for simple one-dimensional solutions before moving on to higher dimension solutions, which could also be more complex. If the problem has a high number of features, it is recommended to start with more features. For the experiments presented in this work, the number of starting dimensions was 10.

2.4.2. Mutation

During the breeding phase, whenever mutation is the chosen genetic operator, one of three actions is performed with equal probability: (1) standard sub-tree mutation, where a new randomly created tree replaces a randomly chosen branch of the parent (excluding the root node); (2) adding a new randomly created tree as a new branch of the root node, effectively adding one dimension to the parent tree; and (3) randomly removing a complete branch of the root node, effectively removing one dimension from the parent tree (see Figure 6).

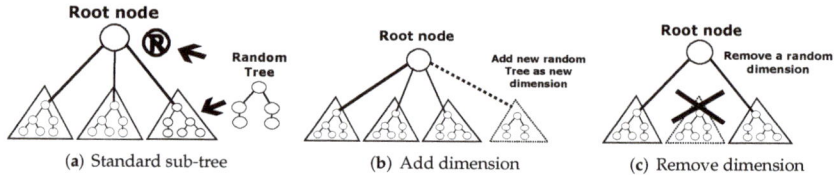

Figure 6. The three possible mutation types in M3GP standard sub-tree, add dimensions and remove dimensions.

In M3GP, mutation is the only way of adding dimensions. Therefore, it has a probability of 0.5 of guaranteeing a proper search for the best number of dimensions. Previous results have shown that this high mutation probability (compared to most GP algorithms) works best with M3GP [22].

2.4.3. Crossover

Whenever crossover is chosen, one of two actions is performed with equal probability: (1) standard sub-tree crossover, where a random node is chosen from each of the parents and the respective branches are swapped (excluding the root node); (2) the swapping of dimensions, where a randomly selected complete branch of the root node is chosen from each parent and swapped, effectively swapping how the input data are transformed into a new feature dimension. The second event is just a variant of the first, where the crossing nodes are guaranteed to belong to the root node (see Figure 7).

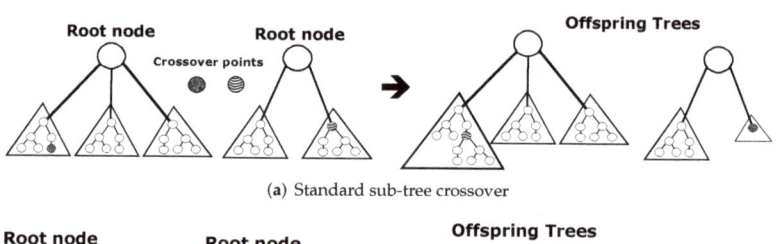

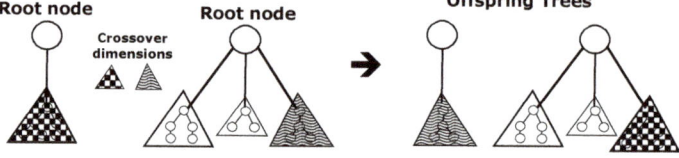

Figure 7. The two possible crossover types in M3GP, standard sub-tree and dimensional.

2.4.4. Pruning

Mutation, as described above, allows M3GP to easily add dimensions to the evolved solutions. However, some dimensions may degrade the individual's fitness and so, they need to be removed. Mutation can also remove dimensions but it does so randomly and is blind to fitness, as described above. We removed the detrimental dimensions by pruning the best individual after the breeding phase to maintain the simplicity and stochasticity of the genetic operators.

The pruning procedure removes the first dimension and re-evaluates the tree. If the fitness improves, the pruned tree replaces the original and moves on to the pruning of the next dimension. Otherwise, the pruned tree is discarded and the original tree moves on to the pruning of the next dimension. The procedure stops after pruning the last dimension. Pruning applied to the whole population has a higher computational cost but controls the growth of dimensions; nevertheless, to offset the computational cost, we only pruned the best solution found in the population for each generation.

2.4.5. Elitism

This criterion ensures the survival of the individuals from one generation to the next. M3GP does not allow the best individual of any generation to be lost and always copies it to the next generation. Let us recall that this individual is highly optimized because it went through the pruning process. Another elitist criterion occurs when two solutions have the same accuracy: only the smallest is conserved, thereby simplifying the transformation tree.

3. Neural Network Configuration

The implemented deep learning (DL) topology and configuration was extracted from the experimental process in [11], which determined that this kind of low-complexity neural network achieves an excellent performance without increasing the computational burden. The topology can be observed in Figure 8 and Table 1. The network uses the one-dimensional convolutional structure to perform a depthwise convolution that acts separately on the channels, followed by a pointwise convolution that mixes the channels [28]. The four emotional states define the four classes, as in the emotions model (see Figure 1). Also, this work considers ReLU, Tanh, and logistics functions to observe performance variations and demonstrate how they are affected by the change in traits. The proposed methodology uses the same topology but distinct features composition in the training process as can be observed in Figure 9. This allowed us to observe whether we could improve the classification rate by using the same topology with distinct feature configurations.

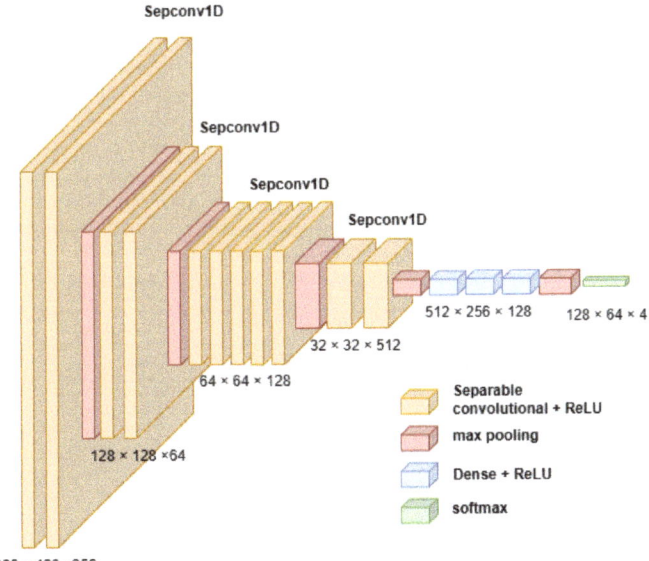

Figure 8. The EEG neural network architecture. The basis for this configuration is presented in [11,28].

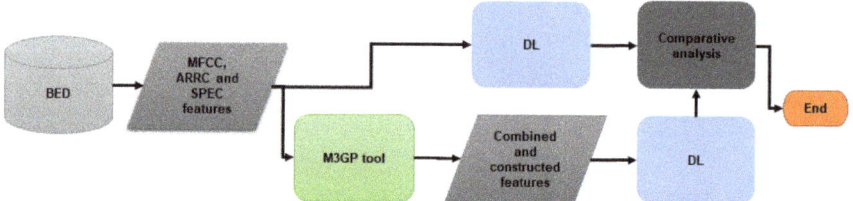

Figure 9. The integration of DL into the recognition process. The DL architecture was trained with spectral coefficients and features created by M3GP.

Table 1. The experimentation parameters considered in our DL architecture.

Parameters	Configuration
Train/Test	70/30
Validation	10-Fold
Alpha value	0.0002
Maximum iterations	200
Functions	ReLU, Tanh, and Logistic
Output layer	Softmax

4. Results

We performed tests with various neural network configurations, using ReLU, Tanh, and logistic functions, to demonstrate the performance improvement when implementing the M3GP. The performance of the network before M3GP is shown in Table 2, where it can be observed that the classification accuracy was around 65.9% for the cases of the ReLU and Tanh functions. In contrast, the logistic function achieved an inferior performance, which was expected since it does not usually perform well on non-binary systems. However, it was in our interest to show the impact of M3GP on the grouping of the elements.

The improvements achieved in the classification rates can be seen in Table 3. It can be seen that the ReLU and Tanh functions had an improvement of 26% and 20.8%, respectively, and each class classification improvement is represented in Figures 10 and 11. The logistic function showed a decrease in the recognition rate since it considers a binary scenario, which M3GP makes less feasible; this can be better appreciated in Figure 12. The best training and testing cross-validations for the ReLU function can be observed in Figure 13 and the training confusion matrix values can be observed in Figure 14.

Table 2. The best results before the M3GP transformation of the dataset.

Function	AUC	CA	F1	Precision
ReLU	0.880	0.650	0.651	0.661 ± 0.021
Tanh	0.880	0.655	0.655	0.657 ± 0.027
Logistic	0.748	0.478	0.477	0.482 ± 0.011

Table 3. the best results after the M3GP transformation of the dataset.

Function	AUC	CA	F1	Precision
ReLU	0.991	0.920	0.920	0.921 ± 0.008
Tanh	0.972	0.859	0.859	0.865 ± 0.012
Logistic	0.692	0.457	0.425	0.457 ± 0.180

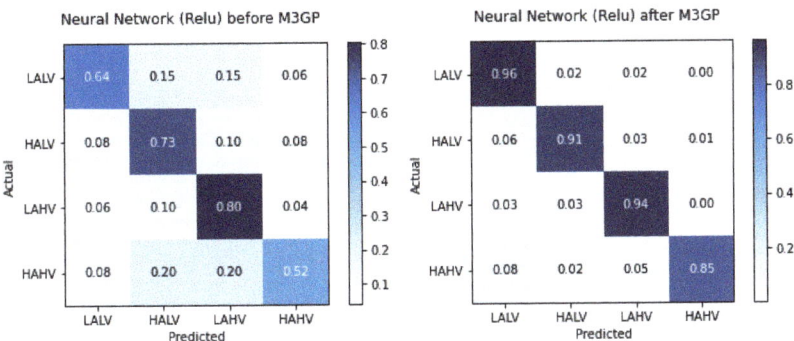

Figure 10. Classification results with ReLU: on the left-hand side is the classification without treatment; on the right-hand side is the classification with the M3GP transformation.

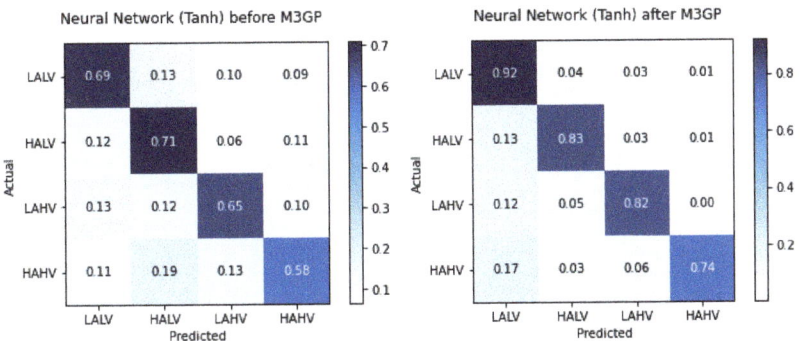

Figure 11. Classification results with Tanh: on the left-hand side is the classification without treatment; on the right-hand side is the classification with the M3GP transformation.

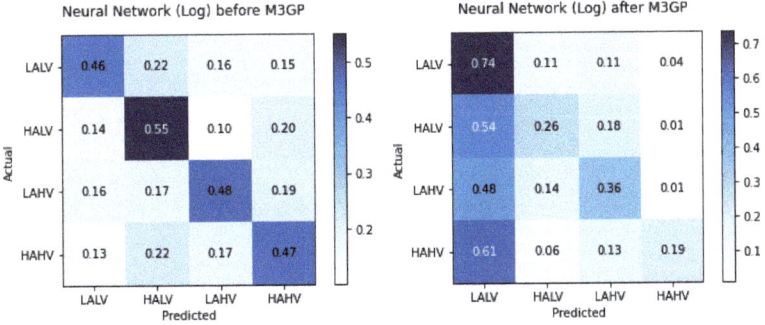

Figure 12. Classification results with Logistic (Log): on the left-hand side is the classification without treatment; on the right-hand side is the classification with the M3GP transformation.

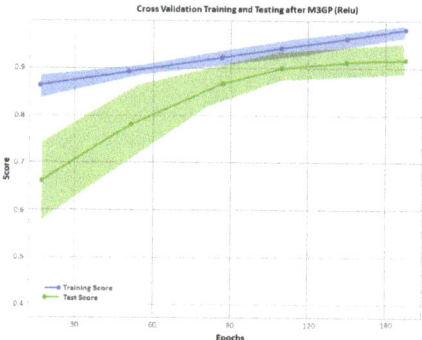

Figure 13. Training and testing validations scores for the DL after M3GP (ReLU function).

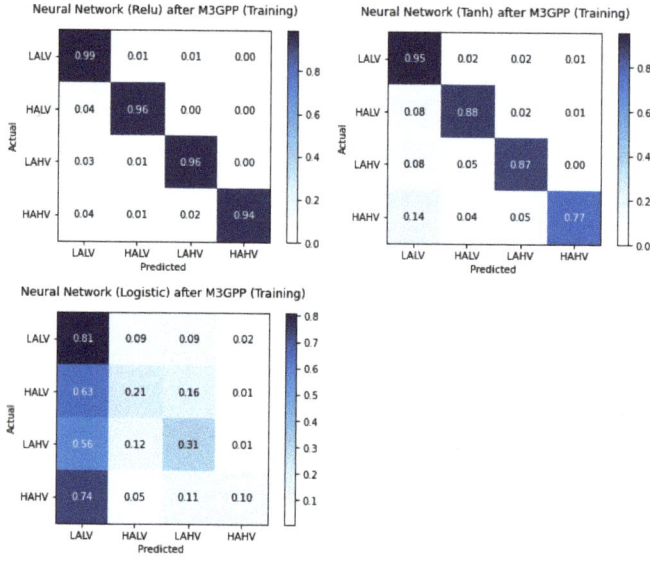

Figure 14. The confusion matrix of the training stages of the reported results obtained after the M3GP transformation for different neural networks (Relu, Tanh and Logistic).

4.1. Standalone M3GP

The main reason for using M3GP was not to obtain a classifier with superior results to the SOA algorithms, but to obtain a feature transformation that could improve the results obtained by the SOA algorithms.

M3GP was run with the settings shown in Table 4 with the spectral feature (MFCC, ARRC, and SPEC) arrangement of the BED dataset. The best results, presented in Table 5, were not significantly different from the SOA methods shown in Table 6. There are many reasons for the low performance of well-known algorithms (e.g., the number of features is too high, there are not enough samples, there is an unbalanced dataset) and when considering all possibilities, users may be tempted to apply some form of data treatment, but not letting the evolutionary process use all available data can affect the performance of the evolutionary algorithms by removing the possibility of letting the natural selection process find the ideal treatment.

Table 4. M3GP setup settings.

Runs	10
Population Size	500 individuals
Generations	100 generations
Initial Dimension	10 dimensions
Initialization	6-depth full initialization [21]
Data	70% Training −30% Tests
Operator Probabilities	Crossover $p_c = 0.5$, Mutation $p_\mu = 0.5$
Function Set	$+(plus), -(minus), \times (times), \div$ protected as in [21], sine, cosine, exponential, logarithm (abs), power^2, power^3, square root (positive), tangent (tan), hyperbolic tangent (tanh)
Terminal Set	Ephemeral random constants $[0, 1]$
Bloat Control	17-depth limit [21]
Selection	Lexicographic tournament [29] of size 5
Elitism	Keep best individual

Table 5. The accuracy obtained by M3GP, the number of root node features of the best solution found in each dataset, and the original features of each set.

M3GP	MFCC	ARRC	SPEC
Train	94.4	59.9	46.3
Test	39.1	39.6	40
Root Node Features	63	23	13
Original Features	168	168	224

Table 6. Performance comparisons between SOA techniques reported in [30]. Similar EEG-based works have reported an overall recognition rate of 82.9%.

Classifier	Average Performance
Neural Network [31–33]	85.80%
Support Vector Machine [34–36]	77.80%
K-Nearest Neighbor [33,37,38]	88.94%
Multi-layer Perceptron [38–40]	78.16%
Bayes [41–43]	69.62%
Extreme Learning Machine [41]	87.10%
K-Means [43]	78.06%
Linear Discriminant Analysis [42]	71.30%
Gaussian Process [44]	71.30%

Table 7. The percentage of classification improvement.

Fuction	AUC %	CA %	F1 %	Precision %
ReLU	11.1	27	26.9	26
Tanh	9.2	20.4	20.4	20.8
Logistic	−5.6	−2.1	−5.2	−2.5

To improve on the performance of the SOA methods, we used the features generated by M3GP in Figure 15. This kind of transfer learning was used in [45] with success and the best training transformation found in M3GP was used to transform the dataset into a new one (M3GP tree in Section 4.2), considering that these new features contain more information to simplify the learning process of the SOA methods. The chosen dataset was MFCC, with a matrix of 1481 × 168 becoming a matrix of 1481 × 63 through the M3GP transformation. The improved result is shown in Table 3.

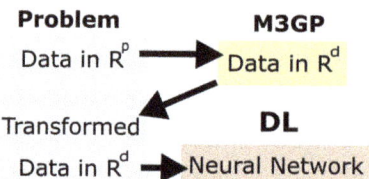

Figure 15. The setup flow: M3GP produces a data transformation tree and the neural network runs with the new data in R^d.

4.2. Transformation Tree

Root node(X23, X24, times(X139,X85), X147, X116, X6, X80, X118, X42, X67, X2, X167, times(X161,tanh(tanh(X130))), X56, X122, X158, X74, X111, X109, X121, X154, X75, X84, X19, X43, X149, X12, X139, X102, X96, X36, X112, X28, X45, X33, X29, plus(X152,X21), X166, X27, X31, X15, X159, X38, squareroot(X86), X34, X89, X126, X65, X161, X82, X123, X151, X136, X156, X103, X168, X162, X51, X57, X87, X155, X132, sine(X30)). Note that each X is an original feature of the BED MFCC dataset.

5. Discussion

The implementation of M3GP in the feature generation stage and the pattern recognition stage through deep learning allowed us to obtain competitive results. We carried out a similar test to that reported in [15] but implementing an architecture that we had previously tested in other benchmarks [12,28]. These results were still not significant compared to the SOA algorithms, although we obtained similar results to the original study (see Table 6). However, the results of maintaining the same architecture and changing the feature selection stage via a transformation of features using M3GP improved to the results obtained in [15] by 20.9% and even exceeded some of the SOA methods.

M3GP has shown competitive results in different benchmarks [22,26,46], especially regarding multiclass problems; but in this case, it was not able to obtain competitive results with EEG from the BED dataset. It also showed a high overfitting problem in the best solution found for the MFCC data. However, this experiment was not wasteful since, based on the results obtained in [45], M3GP feature transformations can still be used as feature optimizers by transferring the knowledge obtained from the evolutionary process to the transformation of features. The effect of the feature can be seen in Table 7.

An interesting result was the overfitting problem when using the Mahalanobis classifier being so high, as observed in Table 5. No other classifiers show any signs of overfitting when working with the transformed dataset, which makes the overfitting problem data independent.

6. Conclusions

A central aspect of the study of mental states through EEG signals is that it does not leave a clear marker on which to focus, unlike certain diseases, such as epilepsy. This means that we often cannot clearly define whether a stimulus exists or not when it begins or ends and which part of the signal is liked to the stimuli. By implementing M3GP, we presupposed that we could accentuate the differences between the behavior of the signals. As shown, once the spectral coefficients of the signals have been extracted, they are processed to increase the distance between them and facilitate the classification process.

This proposal reduced the complexity of the recognition process by reducing features through an evolutionary process that selects those features that best define each of the classes. Traditionally, this analysis process extracts the spectral features of the signals to carry out the pattern search process. However, these features could be linked to diverse phenomena that are inherent to the signals, such as artefact noise, or various physiological phenomena, such as closing eyes, or cognitive processes that are simply not associated with the experimental process. M3GP allowed us, through a selection process, to obtain only those individuals that best fit the description of each of the classes and thus, allowed us to improve the classification rate.

Another essential aspect of this work is the implementation of the BED dataset, which was conceived and built to study users' emotional reactions. BED explores the use of low cost devices in a formal study, thereby promoting and establishing some basis for applicability in less controlled environments.

Our future work will based on integrating the feature selection process into the network architecture configuration, i.e., implementing the neat GP concept in the generalization and optimization of network operation [9,47]. To our knowledge, DL has shown outstanding performances when working with EEG signals; however, the community continues to propose architectures that can efficiently solve this type of problem. We hypothesize that we can establish the criteria for the network features and topology when implementing an evolutionary strategy. Another proposal for future work that is linked to this model lies in implementing GP as an AutoML methodology, with the premise that GP could be used to autonomously select features and DL topology (currently, this process is performed through digital signal processing by extracting spectral coefficients).

Author Contributions: Conceptualization, A.R.A.; methodology, A.R.A. and L.M.D.; software, A.R.A. and L.M.D.; validation, A.R.A., L.M.D., V.R.L.-L. and A.C.T.; investigation, A.R.A. and L.M.D.; resources, V.R.L.-L. and A.C.T.; data curation, A.R.A. and L.M.D..; writing—original draft preparation, A.R.A.; writing—review and editing, L.M.D.; visualization, V.R.L.-L. and A.C.T.; supervision, A.R.A., L.M.D., V.R.L.-L. and A.C.T.; project administration, A.R.A.; funding acquisition, V.R.L.-L. and A.C.T. All authors have read and agreed to the published version of the manuscript.

Funding: This research received funding from the National Technological of Mexico, project "Apply machine learning techniques to correlate EEG and facial expressions produced by emotional states".

Institutional Review Board Statement: Not applicable.

Informed Consent Statement: Not applicable.

Data Availability Statement: BED dataset: https://zenodo.org/record/4309472 (accessed on 21 October 2021).

Acknowledgments: The authors would like to thank the Tecnológico Nacional de México for making this work possible.

Conflicts of Interest: The authors declare no conflict of interest.

References

1. Holzinger, A. Explainable AI and Multi-Modal Causability in Medicine. *I-com* **2021**, *19*, 171–179. [CrossRef]
2. Akbarian, B.; Erfanian, A. Automatic Seizure Detection Based on Nonlinear Dynamical Analysis of EEG Signals and Mutual Information. *Basic Clin. Neurosci.* **2018**, *9*, 227–240. [CrossRef] [PubMed]
3. Vahid, A.; Mückschel, M.; Stober, S.; Stock, A.K.; Beste, C. Applying deep learning to single-trial EEG data provides evidence for complementary theories on action control. *Commun. Biol.* **2020**, *3*, 112. [CrossRef]
4. Chen, Y.; Chang, R.; Guo, J. Effects of Data Augmentation Method Borderline-SMOTE on Emotion Recognition of EEG Signals Based on Convolutional Neural Network. *IEEE Access* **2021**, *9*, 47491–47502. [CrossRef]
5. Hussain, I.; Park, S.J. Quantitative Evaluation of Task-Induced Neurological Outcome after Stroke. *Brain Sci.* **2021**, *11*, 900. [CrossRef]
6. Hussain, I.; Young, S.; Park, S.J. Driving-Induced Neurological Biomarkers in an Advanced Driver-Assistance System. *Sensors* **2021**, *21*, 6985. [CrossRef]
7. Jadhav, N.K.; Momin, B.F. Eye blink pattern controlled system using wearable EEG headband. In Proceedings of the 2018 3rd IEEE International Conference on Recent Trends in Electronics, Information Communication Technology (RTEICT), Bengaluru, India, 18–19 May 2018; pp. 2382–2386. [CrossRef]
8. Guo, Y.; Wang, M.; Zheng, T.; Li, Y.; Wang, P.; Qin, X. NAO robot limb control method based on motor imagery EEG. In Proceedings of the 2020 International Symposium on Computer, Consumer and Control (IS3C), Newcastle upon Tyne, UK, 17–19 November 2020; pp. 521–524. [CrossRef]
9. Stanley, K.O.; Miikkulainen, R. Evolving Neural Networks Through Augmenting Topologies. *Evol. Comput.* **2002**, *10*, 99–127. [CrossRef]

10. Mustaqeem.; Kwon, S. Att-Net: Enhanced Emotion Recognition System Using Lightweight Self-Attention Module. *Appl. Soft Comput.* **2021**, *102*, 107101. [CrossRef]
11. Aguinaga, A.R.; Ramirez, M.A.L.; del Rosario Baltazar Flores, M. Classification model of arousal and valence mental states by EEG signals analysis and Brodmann correlations. *Int. J. Adv. Comput. Sci. Appl.* **2015**, *6*, 230–238. [CrossRef]
12. Aguiñaga, A.R.; Hernández, D.E.; Quezada, Á.; Téllez, A.C. Emotion Recognition by Correlating Facial Expressions and EEG Analysis. *Appl. Sci.* **2021**, *11*, 6987. [CrossRef]
13. Aquino-Brítez, D.; Ortiz, A.; Ortega, J.; León, J.; Formoso, M.; Gan, J.Q.; Escobar, J.J. Optimization of Deep Architectures for EEG Signal Classification: An AutoML Approach Using Evolutionary Algorithms. *Sensors* **2021**, *21*, 2096. [CrossRef] [PubMed]
14. Hernández, D.E.; Trujillo, L.; Rodriguez, A. AutoML for emotion recognition in EEG signals. In Proceedings of the 4th Conference on Computer Science and Computer Engineering, LatinIndex, Tijuana, BC, MX, 13 May 2021; pp. 139–146.
15. Arnau-González, P.; Katsigiannis, S.; Arevalillo-Herráez, M.; Ramzan, N. BED: A New Data Set for EEG-Based Biometrics. *IEEE Internet Things J.* **2021**, *8*, 12219–12230. [CrossRef]
16. Hussain, I.; Park, S.J. HealthSOS: Real-Time Health Monitoring System for Stroke Prognostics. *IEEE Access* **2020**, *8*, 213574–213586. [CrossRef]
17. Posner, J.; Russell, J.A.; Peterson, B.S. The circumplex model of affect: an integrative approach to affective neuroscience, cognitive development, and psychopathology. *Dev. Psychopathol.* **2005**, *17*, 715–734. [CrossRef]
18. Zheng, W.L.; Lu, B.L. Investigating Critical Frequency Bands and Channels for EEG-based Emotion Recognition with Deep Neural Networks. *IEEE Trans. Auton. Ment. Dev.* **2015**, *7*, 162–175. [CrossRef]
19. Koelstra, S.; Muhl, C.; Soleymani, M.; Lee, J.S.; Yazdani, A.; Ebrahimi, T.; Pun, T.; Nijholt, A.; Patras, I. DEAP: A Database for Emotion Analysis; Using Physiological Signals. *IEEE Trans. Affect. Comput.* **2012**, *3*, 18–31. [CrossRef]
20. Emotiv. *Emotiv EPOC+ Headset*; Emotiv: San Francisco, CA, USA, 2022.
21. Koza, J.R. *Genetic Programming: On the Programming of Computers by Means of Natural Selection*; MIT Press: Cambridge, MA, USA, 1992; Volume 1.
22. Muñoz, L.; Silva, S.; Trujillo, L. M3GP—Multiclass Classification with GP. In Proceedings of the European Conference on Genetic Programming, Copenhagen, Denmark, 8–10 April 2015.
23. Sattari, M.A.; Roshani, G.H.; Hanus, R.; Nazemi, E. Applicability of time-domain feature extraction methods and artificial intelligence in two-phase flow meters based on gamma-ray absorption technique. *Measurement* **2021**, *168*, 108474. doi:10.1016/j.measurement.2020.108474. [CrossRef]
24. Reddy, A.P.; Vijayarajan, V. Audio Compression with Multi-Algorithm Fusion and Its Impact in Speech Emotion Recognition. *Int. J. Speech Technol.* **2020**, *23*, 277–285. [CrossRef]
25. Roshani, M.; Sattari, M.A.; Ali, P.J.M.; Roshani, G.H.; Nazemi, B.; Corniani, E.; Nazemi, E. Application of GMDH neural network technique to improve measuring precision of a simplified photon attenuation based two-phase flowmeter. *Flow Meas. Instrum.* **2020**, *75*, 101804. doi:10.1016/j.flowmeasinst.2020.101804. [CrossRef]
26. Muñoz, L.; Trujillo, L.; Silva, S.; Castelli, M.; Vanneschi, L. Evolving multidimensional transformations for symbolic regression with M3GP. *Memetic Comput.* **2018**, *11*, 111–126. [CrossRef]
27. Xiang, S.; Nie, F.; Zhang, C. Learning a Mahalanobis Distance Metric for Data Clustering and Classification. *Pattern Recogn.* **2008**, *41*, 3600–3612. [CrossRef]
28. Rodriguez Aguiñaga, A.; Realyvásquez-Vargas, A.; López R., M.; Quezada, A. Cognitive Ergonomics Evaluation Assisted by an Intelligent Emotion Recognition Technique. *Appl. Sci.* **2020**, *10*, 1736. [CrossRef]
29. Luke, S.; Panait, L. Lexicographic parsimony pressure. In *Proceedings of the GECCO-2002*; Morgan Kaufmann Publishers: San Francisco, CA, USA, 2002; pp. 829–836.
30. Suhaimi, A.S.; Mountstephens, J.T.J. EEG-Based Emotion Recognition: A SOA Review of Current Trends and Opportunities. *Comput. Intell. Neurosci.* **2020**, *19*, 171–179. doi:10.1155/2020/8875426. [CrossRef]
31. Song, T.; Zheng, W.; Song, P.; Cui, Z. EEG Emotion Recognition Using Dynamical Graph Convolutional Neural Networks. *IEEE Trans. Affect. Comput.* **2020**, *11*, 532–541. [CrossRef]
32. Shi, W.; Feng, S. Research on music emotion classification based on lyrics and audio. In Proceedings of the 2018 IEEE 3rd Advanced Information Technology, Electronic and Automation Control Conference (IAEAC), Chongqing, China, 12–14 October 2018; pp. 1154–1159. [CrossRef]
33. Ullah, H.; Uzair, M.; Mahmood, A.; Ullah, M.; Khan, S.D.; Cheikh, F.A. Internal Emotion Classification Using EEG Signal with Sparse Discriminative Ensemble. *IEEE Access* **2019**, *7*, 40144–40153. [CrossRef]
34. Li, Y.; Zheng, W.; Zong, Y.; Cui, Z.; Zhang, T.; Zhou, X. A Bi-Hemisphere Domain Adversarial Neural Network Model for EEG Emotion Recognition. *IEEE Trans. Affect. Comput.* **2021**, *12*, 494–504. [CrossRef]
35. Hidaka, K.; Qin, H.; Kobayashi, J. Preliminary test of affective virtual reality scenes with head mount display for emotion elicitation experiment. In Proceedings of the 2017 17th International Conference on Control, Automation and Systems (ICCAS), Jeju, Korea, 18–21 October 2017; pp. 325–329. [CrossRef]
36. Marín-Morales, J.; Higuera-Trujillo, J.L.; Greco, A.; Guixeres, J.; Llinares, C.; Scilingo, E.P.; Alcañiz, M.; Valenza, G. Affective computing in virtual reality: emotion recognition from brain and heartbeat dynamics using wearable sensors. *Sci. Rep.* **2018**, *8*, 13657. [CrossRef] [PubMed]

37. Fan, J.; Wade, J.W.; Key, A.P.; Warren, Z.E.; Sarkar, N. EEG-Based Affect and Workload Recognition in a Virtual Driving Environment for ASD Intervention. *IEEE Trans. Biomed. Eng.* **2018**, *65*, 43–51. [CrossRef]
38. Soroush, M.Z.; Maghooli, K.; Setarehdan, S.K.; Nasrabadi, A.M. Emotion Classification through Nonlinear EEG Analysis Using Machine Learning Methods. *Int. Clin. Neurosci. J.* **2018**, *5*, 135–149. [CrossRef]
39. Al-Galal, S.A.Y.; Alshaikhli, I.F.T.; bin Abdul Rahman, A.W.; Dzulkifli, M.A. EEG-based emotion recognition while listening to quran recitation compared with relaxing music using valence-arousal model. In Proceedings of the 2015 4th International Conference on Advanced Computer Science Applications and Technologies (ACSAT), Kuala Lumpur, Malaysia, 8–10 December 2015; pp. 245–250. [CrossRef]
40. Saeed, S.M.U.; Anwar, S.M.; Majid, M.; Bhatti, A.M. Psychological stress measurement using low cost single channel EEG headset. In Proceedings of the 2015 IEEE International Symposium on Signal Processing and Information Technology (ISSPIT), Abu Dhabi, UAE, 7–10 December 2015; pp. 581–585. [CrossRef]
41. Subramanian, R.; Wache, J.; Abadi, M.K.; Vieriu, R.L.; Winkler, S.; Sebe, N. ASCERTAIN: Emotion and Personality Recognition Using Commercial Sensors. *IEEE Trans. Affect. Comput.* **2018**, *9*, 147–160. [CrossRef]
42. Bai, J.; Luo, K.; Peng, J.; Shi, J.; Wu, Y.; Feng, L.; Li, J.; Wang, Y. Music emotions recognition by cognitive classification methodologies. In Proceedings of the 2017 IEEE 16th International Conference on Cognitive Informatics & Cognitive Computing, (ICCI*CC), Oxford, UK, 26–28 July 2017; pp. 121–129.
43. Dabas, H.; Sethi, C.; Dua, C.; Dalawat, M.; Sethia, D. Emotion classification using EEG signals. In Proceedings of the 2018 2nd International Conference on Computer Science and Artificial Intelligence; Association for Computing Machinery, New York, NY, USA, 24–26 February 2018; pp. 380–384. [CrossRef]
44. An, Y.; Sun, S.; Wang, S. Naive bayes classifiers for music emotion classification based on lyrics. In Proceedings of the 2017 IEEE/ACIS 16th International Conference on Computer and Information Science (ICIS), Wuhan, China, 24–26 May 2017; pp. 635–638. [CrossRef]
45. Muñoz, L.; Trujillo, L.; Silva, S. Transfer learning in constructive induction with Genetic Programming. *Genet. Program. Evolvable Mach.* **2019**, *21*, 529–569. [CrossRef]
46. Silva, S.; Luis Muñoz, L.T.V.I.M.C.; Vanneschi, L. Multiclass classification through multidimensional clustering. In *Genetic Programming Theory and Practice XIII*; Springer: Berlin/Heidelberg, Germany, 2016.
47. Trujillo, L.; Muñoz, L.; Galván-López, E.; Silva, S. neat Genetic Programming: Controlling bloat naturally. *Inf. Sci.* **2016**, *333*, 21–43. doi:10.1016/j.ins.2015.11.010. [CrossRef]

Article

Privacy Preserving Classification of EEG Data Using Machine Learning and Homomorphic Encryption

Andreea Bianca Popescu [1,2], Ioana Antonia Taca [1,3], Cosmin Ioan Nita [1,2,*], Anamaria Vizitiu [1,2], Robert Demeter [1,2], Constantin Suciu [1,2] and Lucian Mihai Itu [1,2]

1. Advanta, Siemens SRL, 500097 Brașov, Romania; Andreea.Popescu@ctbav.ro (A.B.P.); ioana_antonia29@yahoo.com (I.A.T.); anamaria.vizitiu@siemens.com (A.V.); robert.demeter@siemens.com (R.D.); suciu.constantin@siemens.com (C.S.); lucian.itu@siemens.com (L.M.I.)
2. Department of Automation and Information Technology, Transilvania University of Brașov, 500174 Brașov, Romania
3. Department of Mathematics and Computer Science, Transilvania University of Brașov, 500091 Brașov, Romania
* Correspondence: nita.cosmin.ioan@unitbv.ro

Abstract: Data privacy is a major concern when accessing and processing sensitive medical data. A promising approach among privacy-preserving techniques is homomorphic encryption (HE), which allows for computations to be performed on encrypted data. Currently, HE still faces practical limitations related to high computational complexity, noise accumulation, and sole applicability the at bit or small integer values level. We propose herein an encoding method that enables typical HE schemes to operate on real-valued numbers of arbitrary precision and size. The approach is evaluated on two real-world scenarios relying on EEG signals: seizure detection and prediction of predisposition to alcoholism. A supervised machine learning-based approach is formulated, and training is performed using a direct (non-iterative) fitting method that requires a fixed and deterministic number of steps. Experiments on synthetic data of varying size and complexity are performed to determine the impact on runtime and error accumulation. The computational time for training the models increases but remains manageable, while the inference time remains in the order of milliseconds. The prediction performance of the models operating on encoded and encrypted data is comparable to that of standard models operating on plaintext data.

Keywords: privacy-preserving computations; homomorphic encryption; machine learning; EEG signals

1. Introduction

In recent years, artificial intelligence (AI) algorithms have shown great potential in several fields, including healthcare. Allowing for customized diagnosis, treatment planning and disease prevention, AI has demonstrated its ability to deliver personalized medicine [1]. However, to obtain a satisfactory performance, a significant amount of data needs to be collected, stored and processed. While achieving promising results in patient-specific medical applications, accessing sensitive data for training AI models requires proper anonymization [2]. Even at the inference phase, when an already trained AI model is employed, privacy should not be compromised. Moreover, regulations regarding personal data confidentiality (e.g., GDPR in the EU, HIPAA in the U.S.A.) emphasize the need for developing more effective privacy-preserving techniques.

Therefore, over the last few years, the development of privacy-preserving techniques that allow for machine learning-based analysis to be performed on encrypted data has received increasing interest. Techniques such as secure multiparty computation (SMPC), differential privacy and homomorphic encryption have shown great potential to be adopted in machine-learning applications, but each of them comes with limitations. Mohassel and

Zhang [3] were the first who attempted to train a neural network using SMPC for data privacy, but the usability of this method is limited due to the high computational cost. Although differential privacy is less complex from the computational point of view, it implies a trade-off between precision and privacy that can be managed through a proper noise injection mechanism. This technique was used in machine-learning applications [4–6] and has achieved promising results. Homomorphic encryption, allowing for mathematical operations on encrypted data without revealing its underlying information, represents a promising privacy-preserving method to be combined with machine-learning techniques. An early work that uses an HE cryptosystem together with neural networks was introduced in [7], requiring communication between the server and the data owner. This interaction is eliminated in CryptoNets [8], but the YASHE [9] encryption scheme does not support real numbers. To address limitations regarding model complexity, CryptoDL [10] uses low-degree polynomials to approximate nonlinear functions. On the other hand, using approximated activation functions lowers the prediction accuracy of the models. Moreover, as a result of a higher number of operations and a longer runtime, none of the above-described solutions covers the training phase of the models on encrypted data.

All these approaches prove that the integration of HE schemes in real-world AI applications is a difficult task, as there are still important limitations. For example, the BFV scheme [11], also employed in the SEAL homomorphic encryption library [12], introduces noise over plaintext values, reducing thus the number of possible consecutive operations, and does not allow for the encryption of rational numbers. To overcome this last limitation, SEAL uses a particularity of the CKKS [13] scheme that affects the computational precision. Although noise-free HE schemes also exist, most are only partially homomorphic, supporting only certain operations, e.g., Paillier [14], ElGamal [15] or AHEE [15], allowing for noise-free addition and multiplication on small integers.

While for most HE schemes, the plaintext domain is restricted to integers, medical data typically rely on rational numbers. Lacking the possibility of encrypting rational numbers hinders the utility of an HE scheme. Thus, a major challenge is the extension of HE schemes to enable computations with real-valued numbers. To achieve this, multiple encoding methods that extend existing encryption schemes have been proposed. An adaptation of the BFV scheme is introduced in [16], which allows for real number encryption, and prevents the underlying message to rapidly increase during computations. In this case, it is not the message that is encrypted, but a polynomial encoding of it, as presented in [17]. Thus, polynomial coefficients are the digits of the message m in a certain base b. For real values, the integer part is encoded using positive powers of b, and the rational part uses negative powers of the base. After adding the degree of the polynomial modulus, all polynomial exponents become positive. Another approach for modifying the BFV scheme to render it usable for real values is presented in [18]. Similar to the previous approach, in [19] an encoding method based on a binary system is proposed. An encoding technique that enables the use of the Paillier cryptosystem and its homomorphic properties with rational numbers is introduced in [20], where a real value is written as a fraction of two integers. The restrictions of this method stem from certain conditions imposed on the numerator and denominator: they are bounded by a predefined domain and require an additional step of decryption and approximation with smaller values, to ensure that values remain between the imposed bounds. In [21], real numbers are expressed in a 32-bit binary representation, with sign (1 bit), exponent (8 bit) and mantissa (23 bit), which are then encrypted with HE schemes allowing for bit-wise operations. Similar work was performed by Jaschke and Armknecht [22,23] where an in-depth evaluation of multiple encoding methods was conducted, and a method for choosing the appropriate encoding scheme and parameters for specific use-cases is also described.

Another significant difficulty in performing machine learning using homomorphically encrypted data is that such models require an evaluation of non-linear activation functions and iterative gradient-based optimization methods. An option that was also considered in other similar scenarios (e.g., CryptoNets library [8]) is to approximate the non-linear activation functions with low-degree polynomials. If the approximated activation functions are used with a classical neural network, the function of the entire model can be reduced to a polynomial function containing different combinations of input features with different exponents. Under this assumption, the model-fitting process can be performed using direct analytical methods, rather than iterative gradient-based methods, which allows for model training in a fixed and deterministic number of steps. This provides control over the number of arithmetic operations performed on encrypted data, and over the choice of encryption parameters (e.g., encryption key size). Additionally, because there is no need for performing a comparison between a loss value and a threshold, this approach is suitable for being integrated with existing HE schemes.

The main purpose of this study is to enable machine-learning algorithms to operate directly on encrypted data, employing homomorphic encryption. This way, original data are never distributed, as they are initially encrypted and all subsequent operations are directly performed on the encrypted data, without revealing any information about the underlying data. Thus, we are proposing a method that overcomes certain limitations of the HE schemes. Specifically, we are proposing an encoding scheme that enables HE computations to be performed on rational numbers (considering that, typically, only the addition and multiplication of integers are supported), along with a numerical optimization approach that allows for model training in a fixed and deterministic number of steps, an important property in the HE context. Once the model is trained, the goal is to perform real-time inference on new encrypted data.

To evaluate the feasibility of the proposed method for healthcare applications, we consider two use cases: epileptic seizure recognition and alcoholism predisposition detection, based on EEG recordings. Both use cases are based on publicly available data and are formulated as binary classification problems. Experiments were performed with the proposed method (on encrypted data) but also with classic machine learning methods (on plaintext data) to demonstrate that the prediction performance is similar. Since collecting EEG information for training a model or sending the EEG data to the model owner for the inference phase raises concerns about patient privacy, several studies were conducted regarding EEG signal encryption. Lin et al. [24] and Ahmad et al. [25] proposed a chaos-based EEG encryption system. Since the purpose is to secure EEG information while being stored or transmitted over an insecure channel, the encryption method does not allow for arithmetic operations. Liu et al. [26] used a feed-forward neural network where the activation function is approximated with a linear function to solve an EEG-based classification problem. The training step was performed using plaintext data, while the encryption was employed only during the inference phase. The EEG signals were encrypted using the Paillier algorithm, which only supports addition operations. To allow for the usage of decimals, an encoding method based on scaling was used. The input data was preprocessed by selecting 44 out of the 64 electrode channels, which have the highest Pearson correlation coefficient. Through a performance comparison with other studies in which other classifiers are used, the authors demonstrated that their approach represents the state of the art (90.17% accuracy). However, some differences between the non-privacy-preserving and the privacy-preserving versions of the method were reported.

The paper is structured as follows. Section 2 presents the encoding method, the algorithms, the data pre-processing, along with a description of the use cases. Section 3 details the evaluation of the method performed on synthetic data, and the results obtained for the two real use cases. The last section includes a discussion, further development ideas and conclusions.

2. Methods

2.1. Encoding Scheme

Although integrating homomorphic encryption schemes with encoding methods addresses some of the impediments that hinder their real-world utility, there are still difficulties to be considered for obtaining an efficient encoding technique. As HE schemes are already complex, and encrypted operations require a higher computational cost than their plaintext equivalent, adding a new layer of encoding increases the complexity even more. Furthermore, expressing a value as a sequence of numbers leads to a ciphertext of growing size, as every number from the sequence is translated to its encrypted form. For example, when considering a polynomial representation, the multiplication of two encoded values results in an increased number of terms and polynomials with a higher order. As the number of operations increases, the increasing dimension is more difficult to handle, and performing computations becomes more time-consuming.

As discussed in the introduction, encoding based on a polynomial representation is a common technique used to allow for real number encryption. Our approach is to use a polynomial representation of a number, based on negative powers of 10. Since the focus lies on enabling the encryption of real numbers, we deal with the integer and the rational part separately. Let us consider $A \in R$, a real number represented as $A = a_0, a_1 a_2 \ldots a_n$. All digits of the integer part, denoted by a_0, are encoded in the first coefficient of the polynomial, corresponding to 10 to the power of 0. The rational part is handled differently, by multiplying each digit with a negative power of 10, according to their position in the initial number. Adding these partial products, as in Equation (1), we obtain the encoding of a real value that consists of a sequence of small integers.

$$A = a_0 10^0 + a_1 10^{-1} + a_2 10^{-2} + \ldots + a_n 10^{-n} \tag{1}$$

The n value denotes the number of decimals that are taken into account in this encoded representation and can be set, depending on the required precision for a certain application. To allow for the encryption of negative numbers, our proposed approach consists of replacing each coefficient of the polynomial with a difference, as in Equation (2).

$$A = (a_0 - b_0)10^0 + (a_1 - b_1)10^{-1} + (a_2 - b_2)10^{-2} + \ldots + (a_n - b_n)10^{-n} \tag{2}$$

If A is a negative number, then terms denoted by $a_0, a_1, \ldots, a_n$ are equal to zero and the underlying information is stored in the $b_0, b_1, \ldots, b_n$ terms, while preserving the sign of the number. From this point onward, referring to a polynomial coefficient designates a difference, such as $(a_i - b_i)$.

For instance, let us consider $A = -13.8701$, which can be encoded as follows:

$$-13.8701 = (0 - 13)10^0 + (0 - 8)10^{-1} + (0 - 7)10^{-2} + (0 - 1)10^{-4} \tag{3}$$

We notice in this example that null decimals are ignored, simplifying the representation and allowing for an increase in precision. If, for a certain application, we have $n = 5$, the encoding takes into account only non-zero decimals, storing in the encoded representation more information from the original number. Using this method, it is possible to encode both positive and negative real numbers, with as many decimals as required, and then encrypt each coefficient by applying an HE scheme.

The operations that can be performed on numbers encoded as previously explained are addition, subtraction and multiplication. As shown next, homomorphic properties with respect to these operations are preserved by this encoding technique since they are based on polynomial calculus.

We consider two real numbers A_1 and A_2, with their corresponding generic encoded representations as follows:

$$\begin{aligned} A_1 &= (a_0 - b_0)10^0 + (a_1 - b_1)10^{-1} + \ldots + (a_n - b_n)10^{-n} \\ A_2 &= (c_0 - d_0)10^0 + (c_1 - d_1)10^{-1} + \ldots + (c_n - d_n)10^{-n} \end{aligned} \tag{4}$$

Addition is performed by adding the corresponding coefficients as in (5), while subtraction consists of adding the opposite of the subtrahend that is computed, as in (6).

$$A_1 + A_2 = (a_0 + c_0 - (b_0 + d_0))10^0 + \ldots + (a_n + c_n - (b_n + d_n))10^{-n} \tag{5}$$

$$-A_2 = (d_0 - c_0)10^0 + (d_1 {-}_1)10^{-1} + \ldots + (d_n - c_n)10^{-n} \qquad (6)$$

To compute $A_1 A_2$, we multiply each coefficient of the first factor with all coefficients of the second factor and group them after the corresponding powers. This may lead to an increase in the number of coefficients. Hence, the implementation of each operation allows for selecting a different number of decimals (n) to be stored in the result.

As can be noticed in (1), the entire integer part of a rational number is stored in a single coefficient. Of course, the same value can be represented symmetrically by distributing individual digits between different positive powers of 10, but since most of the machine learning-based techniques make use of data normalization, all input numbers are sub-unitary. Hence, adding more coefficients to the polynomial representation only to encode a zero value leads to an avoidable computational overhead.

In combination with this encoding method, it is possible to use any HE scheme that is homomorphic for addition and multiplication, as these are the only two operations performed on ciphertexts during the encoded computations. Moreover, if a specific scheme enables scalar multiplication, the encoding method can also preserve this property, where a scalar could be, in this context, any integer that is neither encoded nor encrypted, or any encoded (but not encrypted) real value.

As explained above, the encoding step transforms the input rational number in a sequence of integers representing coefficients of a polynomial function. In the following experiments, each of these coefficients is encrypted using the HE scheme proposed in [27]. The resulting ciphertext is also a sequence of integers. This scheme allows for homomorphic addition and multiplication, and it also supports scalar multiplication. The encoding method does not affect the homomorphic properties of the ciphertext. Specifically, in [27], the encryption algorithm is described by Equation (7):

$$E_p(x) = smod((x + sign(x) \cdot rand() \cdot p), n) \qquad (7)$$

where x is the message (the coefficient) to be encrypted, p is a large prime number, and n is the product between p and another large prime number. The $smod$ operation is described in Equation (8), mod is the common modulo operation, $rand$ yields a random positive integer, and the $sign$ function provides the sign of x.

$$smod(m, p) = \begin{cases} mod(m, p), & m \geqslant 0. \\ -(mod(|m|, p)), & m < 0. \end{cases} \qquad (8)$$

An example of a ciphertext obtained by encrypting the coefficients from the Equation (3) with this HE scheme, using a 32-bit key (for a simpler visualization), is represented in Equation (9).

$$\begin{aligned}-13.8701 = &(19214182057 - 24703948372)10^0 + (8234649453 - 19214182065)10^{-1} \\ &+ (13724415755 - 8234649460)10^{-2} + (5489766302 - 10979532605)10^{-4}\end{aligned} \qquad (9)$$

For the experiments using encoded and encrypted data from the two real use-cases, the encoding precision parameter is set equal to 35, and a 256-bit encryption key is used.

2.2. Optimization Formulation

Numerical optimization is typically performed using gradient-based iterative methods that converge in an unknown number of steps. Unfortunately, such formulations are difficult to apply on homomorphically encrypted data because they require an unknown (and typically large) number of operations, there are often nonlinear functions involved, and they require a comparison operation for evaluating the convergence. As most HE schemes are homomorphic only for addition and multiplication, they do not allow for division and do not support nonlinear function evaluation. The comparison between encrypted values is another operation that cannot be performed [28].

To bypass these requirements, we are proposing a solution inspired by previous work, where non-linear functions are approximated with low degree polynomials. Specifically, when performing such approximations, the resulting model turns into a polynomial function. If the loss function to be minimized is also analytically differentiable, the entire optimization problem is convex and can be solved analytically.

Let X be a N by M matrix containing the input data where each row $\mathbf{x}_i$ is a data sample and $\mathbf{y}$ is a column vector of size N containing the output values.

Following the idea of encoding or encrypting the numbers into a polynomial form as suggested in [9,12,18], we assume that the model function to be optimized can be reduced to a multivariate polynomial function as in Equation (10):

$$P(\mathbf{x}) = \sum_{i=0}^{N_t} a_i p_i(\mathbf{x}), \tag{10}$$

where $p_i(\mathbf{x})$ is a monomial term combining components of $\mathbf{x}$ at different powers, e.g., $x_1 x_2, x_1^2 x_2, x_1^2 x_2^3$ etc., $\mathbf{a} = [a_1, a_2, \ldots, a_{N_t}]$ are the model parameters, and N_t is the number of terms. As discussed in [3], to determine the model parameters, a cost function must be defined and optimized. Thus, we assume that the objective is to minimize the sum of squares, described by Equation (11):

$$C(\mathbf{a}) = \sum_{i=0}^{N} (P(\mathbf{x}_i) - y_i)^2, \tag{11}$$

With respect to the inputs x_i, $C(\mathbf{a})$ is a highly nonlinear function. However, with respect to the parameters, $\mathbf{a}$ is a quadratic function where the minimum can be computed directly by solving the normal equation of the over-determined linear system $P\mathbf{a} = \mathbf{y}$. The already derived, widely known and used Equation (12) is employed for this purpose:

$$\mathbf{a} = \left(P^T P\right)^{-1} P^T \mathbf{y}, \tag{12}$$

where P is a N by the N_t matrix containing the numerical values obtained when evaluating each polynomial feature $p_i(\mathbf{x}), i \in [1, N_t]$ for each data sample $\mathbf{x}_i, i \in [1, N]$.

Although all the steps above require only algebraic operations, since the proposed encoding scheme does not allow for the division of two numbers, we cannot completely compute $\left(P^T P\right)^{-1}$ in the encoded–encrypted form. Therefore, we only compute the dot products $P^T P$ and $P^T \mathbf{y}$ in the encrypted form and then perform the remaining operations after decryption. This is a notable limitation that prevents the model fitting process to be performed completely on encrypted data. However, typically, the number of data samples N is much larger than the number of model parameters N_t; therefore, the dot product $P^T P$ results in a small matrix N_t by N_t, while $P^T \mathbf{y}$ results in a vector of size N_t. Since the input data are significantly regressed by these operations, decrypting does not expose the original data.

Referring to overfitting issues, they typically appear when the model is too complex. In our case, if the number of polynomial terms is not too large, the model does not attempt to learn or explain the random error or noise present in the data. Furthermore, overfitting can be addressed by using larger training datasets: the more samples are used for training, the more difficult it is for the model to learn the error or noise. Thus, according to [29], the number of samples should be at least 10–15 for each term in the model to avoid overfitting. In our use cases, this condition is met in all experiments, as the number of samples exceeds the required threshold (e.g., 9200 training samples for 80 features in the seizure detection use case).

For those experiments where EEG signals are used in the unencrypted form, we also applied SVM methods, as they previously delivered state-of-the-art results in studies employing EEG information [30,31]. Specifically, we used nuSVR from the scikit-learn library [32] on plaintext data and compared the results with the performance obtained by applying polynomial regression on both plaintext and encrypted data. We performed a grid search for the parameters of the SVM classifier (nu, which controls the number of support vectors, and the penalty parameter C) to determine their optimal values. The parameters obtained are reported in the Results section.

The matrix multiplication operation was parallelized to reduced the computational time (using the Python multiprocessing package [33]). The elements of the resulting matrix were computed by distributing the corresponding rows and columns of the input matrices across multiple processes. All experiments were run on a machine equipped with an Intel Core i7 CPU running at 4.2 GHz and 32 GB RAM.

2.3. Use Case 1—Seizure Detection

The objective in this use case is to determine if an epileptic seizure activity occurred during an EEG recording. An input sample contains a sequence of real numbers representing an EEG signal, which can stem from surface recordings from healthy individuals (with their eyes closed or open) or intracranial recordings from epilepsy patients (during the seizure-free interval or epileptic seizures) from different brain areas. A 128-channel amplifier system was used to record all EEG signals [34].

The original dataset [34] contains 500 files, each of them corresponding to 23.6 s of EEG recording of one individual, at a sampling rate of 173.61 Hz. Each file represents a single-channel EEG segment that was selected from continuous multichannel EEG recordings. Each time sequence contains 4097 EEG values equally sampled in time. We used a restructured version of this dataset from [35], where every recording was split into 23 chunks corresponding to 1 s of recording, consisting of 178 data points each. The resulting dataset contains 11,500 samples. The models were evaluated using five-fold cross-validation: the dataset was divided into five folds, each with 2300 samples, which led to 9200 training samples and 2300 testing samples for each experiment. A ground truth label is associated with each EEG recording: 1—epileptic seizure activity, 2—tumor area, 3—healthy brain area of a patient with an identified tumor, 4—the patient had their eyes closed during recording, and 5—the patient had their eyes open during recording.

We have formulated a binary classification problem by considering that all recordings labeled as 2, 3, 4 or 5 indicate the absence of an epileptic seizure (ground truth label 0), while ground truth label 1 indicates the presence of an epileptic seizure. The resulting dataset is imbalanced since only 20% of the recordings have a class label of 1 (epileptic seizure recorded). As our experiments employ a linear model, the class imbalance is handled by setting the threshold used to discretize the output, closer to 0. Hence, all samples with an output value greater than 0.1 are classified as class 1. Input data are normalized in the range $[-1, 1]$.

Because in the use case described above all the features of an input sample represent a time sequence, the original signal of each entry was downsampled through interpolation to reduce the complexity of the problem. We used one-dimensional linear interpolation (numpy.interp [36]). Normalized cross-correlation (NCC) was computed between the initial and the resampled signal to evaluate the impact of the interpolation operation. Thus, a trade-off was performed between the model complexity and loss of information, and we chose to use the resampled signal with the minimum number of data points and with an NCC value greater than 0.95. This condition led to 40 data points for the downsampled signal (Figure 1).

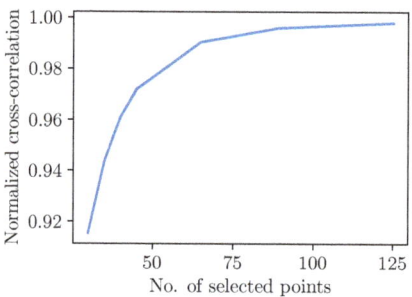

Figure 1. NCC values for different versions of downsampled EEG signals.

Considering the restructured version of the dataset, where 178 points correspond to 1 s of the recording, and the downsampled signal (40 points for 1 s), the frequency was reduced from 178 Hz to 40 Hz. A comparison between the original EEG signal and some downsampled versions is displayed in Figure 2.

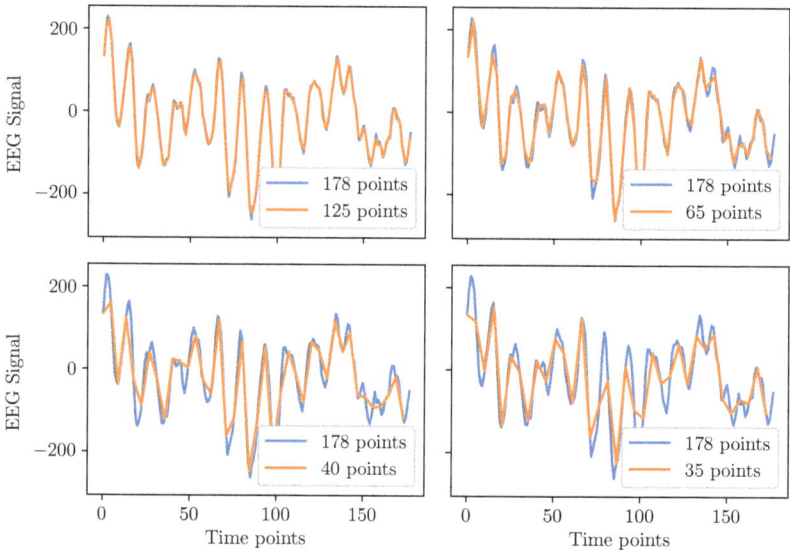

Figure 2. Comparison between the downsampled and the original EEG signals.

2.4. Use Case 2—Predisposition to Alcoholism

Another use case where EEG signals provide valuable information is the analysis of the predisposition to alcoholism, based on the differences between the frequency bands of alcoholic and non-alcoholic individuals [37]. We used a dataset created by Henri Begleiter (Neurodynamics Laboratory, State University of New York Health Center, Brooklyn), made available at [35]. The data were collected during a study evaluating the correlation between EEG signals and genetic predisposition to alcoholism. The signals were measured by 64 electrodes scanned at 256 Hz (3.9 ms epoch) for 1 s. The sensors were placed on the subjects' scalps according to the International 10–20 standard. The subjects were exposed to one stimulus (S_1) or two stimuli (S_1 and S_2) represented by pictures from the Snodgrass and Vanderwart picture set [38]. The two stimuli were shown in a matched condition (S_1 was identical to S_2) or in a non-matched condition (S_1 differed from S_2).

The original dataset is split into two folders—train and test—each of them containing 10 runs for 10 alcoholic and 10 nonalcoholic subjects. The folders consist of trails files with the following columns: trial number, sensor position, sample number, sensor value, subject identifier, matching, channel number, name, and time. In a single trial run, a patient provides 256 samples for each of the 64 electrodes. The pre-processed training dataset contains 153,600 samples of 64 features corresponding to the 64 electrodes; the corresponding ground truth labels are 0 if the subject is non-alcoholic and 1 if the subject is alcoholic. The pre-processed testing dataset also contains 153,600 samples of 64 features and the ground truth label. The input data are normalized in the range $[-1, 1]$.

3. Results

3.1. Synthetic Data

To test and evaluate the applicability and limitations of the proposed method, we performed multiple experiments on synthetic, randomly generated data, which were then encrypted with the scheme proposed in [27], as described in Section 2.1. All the experiments were simultaneously performed on plaintext (unencrypted) data and on the encrypted equivalent to ensure that the same result was obtained. The experiments consisted of performing a multivariate linear regression with different data sizes. Specifically, we generated a matrix X containing random data samples in the $[0, 1]$ interval, and a corresponding random output vector y. We then computed the regression coefficients as described in Section 2.2.

We analyzed the following metrics as a function of data size: run time, the magnitude of the resulting values, and the number of arithmetic operations performed. The magnitudes of the values and the number of operations are of crucial importance in the HE context. More specifically, the chosen HE scheme (and the vast majority of other HE schemes) only operates on a fixed range of integer values. Performing operations on encrypted (and therefore unknown) values can result in values becoming too large, thus exiting the valid range. Therefore, it is imperative that the magnitude of values encountered throughout the computations can be determined beforehand. Clearly, this cannot be generally determined. However, in this synthetic setup, where the magnitude range of the input values and the total number of operations is known, such an analysis can be performed.

To study the impact of the number of features, we fixed the number of samples (500 samples) and varied the number of features between 2 and 102. To study the influence of the number of samples, we varied it from 15 to 19,915 while maintaining the number of features fixed (two features). For consistency reasons, during this analysis, the precision parameter (i.e., number of terms in the polynomial representation) was set to 20, and a 256-bit encryption key was considered.

Figure 3 displays the values of the chosen metrics as a function of the data size. An increasing number of features has an exponential impact on both the number of operations and on the ciphertext size, while an increasing number of samples appears to affect these metrics linearly. The same applies when analyzing the computational time. A similar conclusion can also be drawn for the maximum magnitude of the polynomial coefficients obtained as a result. Although the computational time is significantly increased, the proposed approach is able to handle data sizes of real-world proportions.

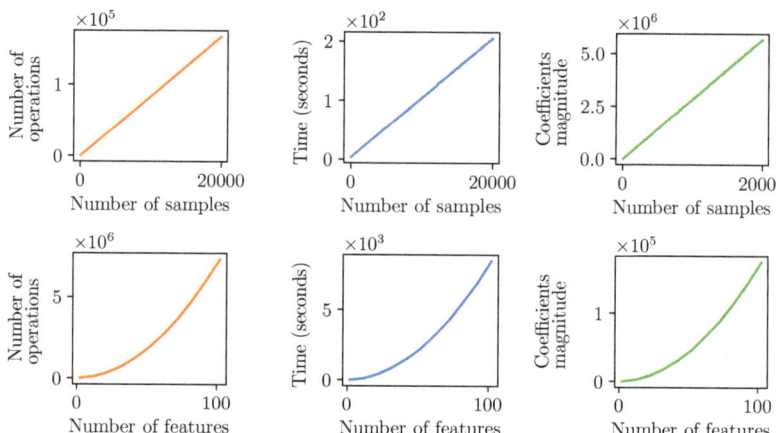

Figure 3. Impact of input data size on the total number of arithmetic operations (**left**), total runtime (**mid**) and maximum value (**right**). The total number of data samples was varied while keeping the number of features fixed (**top**) and the number of input features was varied while keeping the number of samples fixed (**bottom**).

To analyze the prediction error, we computed the mean absolute error and the root-mean-square error for all experiments, where the error is defined as the difference between the outputs obtained, using plaintext and ciphertext values. The error was close to zero (in double precision). On the other hand, as this encoding method allows for a choice of the precision parameter, a large value of the error might indicate that this parameter is too small for a certain application. A randomly generated dataset with 1000 samples and 5 features was used for all experiments synthesized in Figure 4. To determine the dependence between the number of terms in the polynomial encoding (the precision parameter), the numerical error and the execution time, respectively, the same experiment was run multiple times, varying the precision parameter from 1 to 30. Figure 4 displays the trade-off between the accuracy and computational time: an increase in the number of terms used for the encoding leads to a significant decrease in error, but with a cost in terms of the computational time.

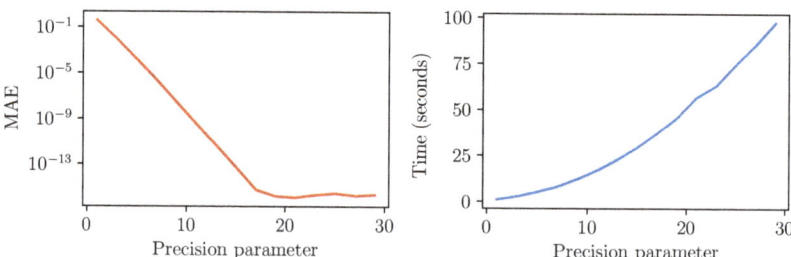

Figure 4. Dependence between the number of terms in the polynomial encoding, the numerical error and the execution time, respectively.

3.2. Use Case 1—Seizure Detection

We ran multiple experiments for the seizure detection use case, relying on different input data configurations:

- Using all 178 initial data points in each sample as input.
- Using all 178 initial data points in each sample and the 178 corresponding quadratic terms (356 features overall).

- Using a downsampled version of each sample. The number of data points was gradually reduced from 178, and the mean NCC over the entire dataset was computed for each downsampled version. We selected the downsampled version with the smallest number of data points for which the average NCC was still larger than 0.95. The selected downsampled version had 40 features (instead of the initial 178) and the corresponding quadratic terms were added, leading to a total of 80 features per sample. Figure 1 displays the variation of the mean NCC as a function of the number of samples. Figure 2 displays, for one signal, the original sample and the resampled sample for different numbers of data points.

For all three input data configurations, the experiments were first conducted using the plaintext data: both a multivariate polynomial regression and an SVM model were evaluated, using five-fold cross-validation. The optimal values of the SVM parameters were determined by performing a grid search. For this use case, the best results were obtained for $\nu = 0.3$ and $C = 1$. Thus, by setting $\nu = 0.3$, we considered that, at most, 30% of the training samples are allowed to be wrongly classified and at least 30% of them act as support vectors.

Table 1 displays the classification metrics aggregated over all five folds: accuracy, sensitivity, specificity, positive predictive value, negative predicted value. For each of the five folds, the metrics were computed only on the testing subset. While the SVM model is largely independent of the input data configuration, the performance of the polynomial regression model increases significantly both when adding the quadratic terms and when reducing the number of data points per sample (increase in performance stems mainly from an increase in sensitivity). After analyzing the performance obtained using SVM, before and after downsampling the signal, we conclude that this pre-processing step does not influence the classification process. For the last input data configuration (80 features) the polynomial regression model was also trained on ciphertext values, considering the encoding precision parameter equal to 35 and a 256-bit encryption key. The results were identical down to double precision when compared to the plaintext experiment. When choosing the precision parameter for this use case, the synthetic date results were taken into account. As the quadratic resampled dataset has a larger number of features (80 compared to 5) and training samples (9200 compared to 1000), to ensure no loss in performance, the precision parameter was chosen to be equal to 35, which is larger than the values used in the synthetic data experiments displayed in Figure 4.

Table 1. Performance comparison when using different input data configurations in the EEG-based seizure detection use case.

	Initial Data 178 Features		Quadratic Data 356 Features		Resampled Data 80 Features		
	PR	SVM	PR	SVM	PR Plaintext	PR Ciphertext	SVM
Accuracy	80.59	95.78	89.22	95.86	92.26	92.26	95.85
Sensitivity	34.25	98.82	84.40	98.60	86.56	86.56	98.70
Specificity	92.17	95.02	90.43	95.17	93.68	93.68	95.14
PPV	52.46	83.23	68.81	83.62	77.44	77.44	83.54
NPV	84.86	99.69	95.86	99.63	95.54	95.54	99.65

In terms of computational time, a comparison between the plaintext and ciphertext polynomial regression experiments is displayed in Table 2. For both training and inference, the time increases for the encrypted version. However, the inference time remains in the order of milliseconds. For the parallelized version, four processes were employed.

Table 2. Training and inference runtime comparison between plaintext and ciphertext implementation versions for the seizure detection use case.

	Plaintext	Ciphertext	Ciphertext Parallelized
Training runtime/fold	0.0119 s	23.60 h	7.05 h
Inference runtime/sample	1.73×10^{-6} s	0.012 s	

Considering the scenario where the model is used only on ciphertext data, without any previous analysis regarding its performance, we chose the precision parameter to be large enough (35 powers) to ensure zero loss in accuracy. This led indeed to the same results as those obtained on plaintext data, but the runtime was significantly larger. To determine the influence of choosing a lower precision parameter, we ran multiple experiments and analyzed the computational time and performance loss. As displayed in Table 3, the runtime can be drastically improved by choosing a smaller number of terms for the encoding step, but after a certain threshold is exceeded (21 powers for this use case), the performance decreases.

Table 3. Influence of the precision parameter on the loss of accuracy and on the computational time for the EEG-based seizure detection use case.

	Precision Parameter (Number of Terms)						
	35	32	29	26	23	21	20
Loss in accuracy	0	0	0	0	0	0	1.66
Runtime/fold (hours)	7.05	6.02	4.69	3.89	2.97	2.41	2.16

3.3. Use Case 2—Predisposition to Alcoholism

The evaluation of the second use case was performed on the pre-processed testing dataset. When designing the experiment, we started from the fact that the dataset consists of 153,600 samples of 64 features corresponding to the 64 electrodes, and the ground truth labels. Moreover, we also took into account that the computational time required for an experiment on the ciphertext data increases considerably once the number of samples becomes larger than 10,000. Thus, as the purpose of this paper is not to find a state-of-the-art model, a balanced subset of 10,000 samples from the training dataset was considered for our experiments.

We have designed the following plaintext experiments:

- E_1— The machine learning models were trained on the original dataset (153,600 samples and 64 features).
- E_2— The machine learning models were trained on 10,000 balanced samples. The number of features was unchanged: 64. This is the baseline experiment to be conducted also on the ciphertext data (E_1 cannot be run on ciphertext since the number of samples is too large).
- E_3— The machine learning models were trained on the entire dataset (153,600 samples) and all polynomial combinations of features with degrees less than or equal to 2 were considered, leading to a total of 2145 features. The idea is to determine whether increasing the model complexity leads to better performance. Due to the large number of features, it is not feasible to run this experiment on ciphertext data.
- E_4— The machine learning models were trained on 10,000 balanced samples with 128 features (the original 64 and their quadratic terms). Considering the computational overhead introduced by the encoding and encryption steps, E_4 represents a suitable trade-off solution.

For the plaintext experiments, we also used the SVM approach. The parameters of the SVM classifier that lead to the best performance were obtained through a grid search ($\nu = 0.5$ and $C = 1$). In Table 4, we display the classification results for samples representing individual time points, recorded by the 64 electrodes at a given moment. All reported metrics were computed using only the testing dataset. The columns represent the results of the experiments described above, where the input was normalized in the range $[-1, 1]$. Because in E_4, only some of the polynomial combinations of features are used (the original and quadratic features) and the training is performed on only the 10,000 samples subset from E_2, the accuracy is increased when compared to E_1 (original dataset), and the computational cost is significantly reduced compared to E_3. While adding the quadratic terms improves the performance of the polynomial regression algorithm, the SVM algorithm leads to similar performance for E_2 and E_4.

For the results in Table 5, a voting scheme is applied to determine a single prediction per patient (alcoholic or non-alcoholic): the outputs of all samples corresponding to a patient (which form a time series) are used as input to the voting scheme, where each sample has the same weight.

In Table 6, we display the runtime for both inference and training. As in the first use case, the runtime increases when processing encrypted data, but it still remains reasonably small for the inference step. For the same reasons stated in Section 3.2, a similar analysis regarding the influence of the precision parameter was performed also in this use case. The same decreasing tendency in the runtime can be noticed in Table 7, but the minimum number of terms required for zero loss in performance is smaller than in the previous use case. Although the number of operations is larger, the initial precision of the training data is also reflected in the encoding precision.

Table 4. Performance comparison when using different input data configurations in the EEG-based predisposition to alcoholism detection use case.

	E_1 (153,600, 64)		E_2 (10,000, 64)		E_3 (153,600, 2145)		E_4 (10,000, 128)		
	PR	SVM	PR	SVM	PR	SVM	PR Plain Text	PR Ciphertext	SVM
Accuracy	57.34	87.46	57.27	79.33	83.92	84.27	68.52	68.52	79.08
Sensitivity	57.26	89.06	61.20	80.15	83.81	83.97	68.45	68.45	79.22
Specificity	57.43	85.86	53.35	78.51	84.03	84.57	68.59	68.59	78.94
PPV	57.35	86.30	56.74	78.86	84.00	84.48	68.55	68.55	79.00
NPV	58.85	88.70	57.90	79.82	83.85	84.07	68.49	68.49	79.16

Table 5. Performance comparison when using different input data configurations in the EEG-based predisposition to alcoholism detection use case after applying the voting scheme.

	E_1 (153,600, 64)		E_2 (10,000, 64)		E_3 (153,600, 2145)		E_4 (10,000, 128)		
	PR	SVM	PR	SVM	PR	SVM	PR Plain Text	PR Ciphertext	SVM
Accuracy	59.66	96.66	60.16	94.16	95.50	93.66	79.83	79.83	96.16
Sensitivity	63.00	98.00	65.33	95.66	97.33	95.33	77.66	77.66	97.66
Specificity	56.33	95.33	55.00	92.66	93.66	92.00	82.00	82.00	94.66
PPV	50.06	95.45	59.21	92.88	93.89	92.25	81.18	81.18	94.82
NPV	60.35	97.94	61.33	95.53	97.23	95.17	78.59	78.59	97.59

Table 6. Training and inference runtime comparison between plaintext and ciphertext versions in the EEG-based predisposition to alcoholism detection use case.

	Plaintext	Ciphertext	Ciphertext Parallelized
Training	0.0286 s	59.69 h	32.28 h
Inference	3.46×10^{-6} s	0.018 s	

Table 7. Influence of the precision parameter on the loss of accuracy and on the computational time in the EEG-based predisposition to alcoholism detection use case.

	Precision Parameter (Number of Terms)							
	35	32	29	26	23	20	17	14
Loss in accuracy	0	0	0	0	0	0	0	0.41
Runtime (hours)	32.28	26.53	22.17	17.91	13.02	10.90	8.16	5.75

4. Discussion and Conclusions

We proposed a method for performing privacy-preserving machine learning-based predictions on homomorphically encrypted medical data. Although there are many proposed HE methods and significant research has been performed on this topic, such methods are still relatively far from being usable in real-world scenarios. Besides existing limitations, such as computational complexity and noise accumulation, typical HE methods are only capable of performing addition and (at most) multiplication on integer values. This significantly limits their usability, especially in the medical field.

A typical usage scenario is when sensitive patient data need to be externalized to a cloud environment for developing machine learning-based models for various prediction tasks. To this extent, we applied the proposed methodology on two real-world use cases involving patient data with different privacy requirements: the prediction of predisposition to alcoholism and the detection of seizures, both using EEG signals. Although the computations on encrypted data are significantly impaired, the prediction performance is maintained at a level comparable to that of standard methods operating on unencrypted data. To achieve this, we proposed a set of methods targeting the inherent limitations in HE computations.

First, we proposed an encoding scheme that allows HE schemes to be applied to real-valued numbers. The encoding scheme relies on transforming real-valued numbers in a sequence of integer values, and for algebraic operations to be performed on the encoded form, using only addition and multiplication. Encoded values can then be encrypted, using existing HE schemes that support addition and multiplication, resulting in a method capable of performing arbitrary algebraic computations on homomorphically encrypted real values.

Another significant difficulty in HE computations is that the numerical values must remain in a predefined (and typically small) interval; exceeding this interval no longer allows for decryption and the information is lost. The initial values that are encrypted can be forced (e.g., through normalization) to be in a certain range; however, values derived from operations on input data can exceed the limit and become unusable. This is a major difficulty, especially in machine learning-based applications, as algorithms in most cases must perform an unknown number of iterations (e.g., training steps, and epochs). To this extent, we employed a polynomial model together with a direct (non-iterative) fitting method such that the required number of operations can be determined a priori. Knowing the input data range, along with the exact number of operations to be performed, allows for a choice of the right encryption scheme along with its parameters for a particular task, resulting in a guaranteed correct outcome.

As for the limitations of the proposed scheme, it is difficult to perform the division of two encoded values. Although several possibilities were explored, this still remains a difficult task to achieve within the current approach. One possibility would be to introduce an additional encoding layer by rewriting a real number as a fraction, and then encoding the numerator and denominator as in (1), but the computational overhead would increase drastically. Furthermore, these supplementary operations do not guarantee a correct result because of the increasing noise that some HE schemes are based on. Following the idea of computing division by multiplying the numerator with the inverse of the denominator, another question that emerges is how can the inverse of an encoded and encrypted number be calculated. One solution might be the Newton–Raphson iterative method for solving an equation. The equation that one would try to solve is $1/X - den = 0$, which provides the inverse of the denominator. The correctness of the result is dependent on the initial approximation. As the underlying information is secret, it is not possible to evaluate which initialization would guarantee convergence to the optimum value in a few iterations. On the other hand, because a comparison cannot be performed between encrypted numbers, a standard/fixed number of iterations needs to be performed. This approach is prone to errors that might occur because of an insufficient number of iterations, or because the plaintext domain is exceeded during the computations.

In this use case, each data point is considered an individual feature (no temporal meaning is considered). Furthermore, although the model achieves good prediction performance, a baseline EEG analysis might be invalidated by downsampling the signal to 40 Hz. As mentioned in [34], a band-pass filter (that allows for frequencies between 0.53 and 40 Hz) was used to filter the signal after data acquisition. According to the Nyquists' theorem and considering the highest frequency of the EEG (40 Hz), to obtain a valid signal, the sampling rate should be at least 80 Hz. The overall performance for the five folds in this experiment (160 features: 80 original features and their squares) on the plaintext data lies between the experiments with 356 and 80 features, respectively. The time required for training on ciphertext data was approximated to be 27 h per fold if 256-bit encryption, 23 encoding terms and 2 processes in the parallelized version were considered.

The computational overhead introduced through encoding and encryption represents another limitation of the method. As experiments have shown, a larger precision parameter and a larger encryption key lead to a higher computational time for the training phase. Regarding the precision parameter, we indicate that the number of terms required to ensure no loss in accuracy cannot be determined precisely based on the dataset dimensionality, as it also depends on the initial precision of the numbers contained within a certain dataset. The total number of operations performed on encoded and encrypted data is $N_{op} = 2sf(f+1)$, where s is the number of training samples and f is the number of features. The computational time can also be predicted according to the number of samples, the number of features and the number of encoding terms n by computing $T = T_{ad}\frac{N_{op}}{2}n + T_{mul}\frac{N_{op}}{2}n^2$, where T_{ad} and T_{mul} denote the average time in seconds required for an addition or a multiplication operation between two encrypted coefficients of the polynomial representation (1). The value of T is also expressed in seconds.

The proposed method can be successfully employed in real-world scenarios. It allows for homomorphic addition, subtraction and multiplication on real numbers, while only requiring an encryption scheme capable of performing the addition and multiplication of small integers. Distributing the information between multiple ciphertext values (polynomial coefficients) allows for the encoding of real numbers of arbitrary sizes, but at the same time, a trade-off between the computational accuracy and the computational time is made.

Author Contributions: Conceptualization, A.B.P. and C.I.N.; methodology, A.B.P., I.A.T., A.V., R.D., C.S. and L.M.I.; software, A.B.P. and I.A.T. ; validation, C.I.N., A.V. and L.M.I.; formal analysis, C.I.N.; investigation, A.B.P., I.A.T., C.I.N., A.V. and L.M.I.; resources, L.M.I.; data curation, A.B.P. and I.A.T.; writing—original draft preparation, A.B.P. and I.A.T.; writing—review and editing, C.I.N., A.V., R.D., C.S. and L.M.I.; visualization, A.B.P.; supervision, R.D., C.S. and L.M.I.; project administration, L.M.I.; funding acquisition, L.M.I. All authors have read and agreed to the published version of the manuscript.

Funding: This work was supported by a grant of the Romanian Ministry of Education and Research, CCCDI—UEFISCDI, project number PN-III-P2-2.1-PED-2019-2415, within PNCDI III. This work was also supported by a grant of the Ministry of Education and Research, CNCS/CCCDI—UEFISCDI, project number PN-III-P3-3.6-H2020-2020-0145/2021. This project received funding from the European Union's Horizon 2020 research and innovation program under grant agreement No. 875351.

Institutional Review Board Statement: Not applicable.

Informed Consent Statement: Not applicable.

Data Availability Statement: Publicly available datasets were analyzed in this study. Data can be accessed at: https://archive.ics.uci.edu/ml/datasets/Epileptic+Seizure+Recognition (accessed on 21 March 2021) and https://www.kaggle.com/chaditya95/epileptic-seizures-dataset (accessed on 21 March 2021).

Conflicts of Interest: Andreea Bianca Popescu and Ioana Antonia Taca received scholarships from Siemens SRL, Brasov, Romania. Cosmin Ioan Nita, Anamaria Vizitiu, Robert Demeter, Constantin Suciu, and Lucian Mihai Itu are employees of Siemens SRL, Brasov, Romania. The funders had no role in the design of the study; in the collection, analyses, or interpretation of data; in the writing of the manuscript, or in the decision to publish the results.

References

1. Miotto, R.; Wang, F.; Wang, S.; Jiang, X.; Dudley, J.T. Deep learning for healthcare: Review, opportunities and challenges. *Briefings Bioinform.* **2018**, *19*, 1236–1246. [CrossRef] [PubMed]
2. Shokri, R.; Shmatikov, V. Privacy-preserving deep learning. In Proceedings of the 22nd ACM SIGSAC Conference on Computer and Communications Security, Denver, CO, USA, 12–16 October 2015; pp. 1310–1321.
3. Mohassel, P.; Zhang, Y. Secureml: A system for scalable privacy-preserving machine learning. In Proceedings of the 2017 IEEE Symposium on Security and Privacy (SP), San Jose, CA, USA, 22–26 May 2017; pp. 19–38.
4. McMahan, H.B.; Ramage, D.; Talwar, K.; Zhang, L. Learning differentially private recurrent language models. *arXiv* **2017**, arXiv:1710.06963.
5. Phan, N.; Wang, Y.; Wu, X.; Dou, D. Differential privacy preservation for deep auto-encoders: an application of human behavior prediction. In Proceedings of the AAAI Conference on Artificial Intelligence, Phoenix, AZ, USA, 12–17 February 2016; Volume 30.
6. Abadi, M.; Chu, A.; Goodfellow, I.; McMahan, H.B.; Mironov, I.; Talwar, K.; Zhang, L. Deep learning with differential privacy. In Proceedings of the 2016 ACM SIGSAC Conference on Computer and Communications Security, Vienna, Austria, 24–28 October 2016; pp. 308–318.
7. Orlandi, C.; Piva, A.; Barni, M. Oblivious neural network computing via homomorphic encryption. *EURASIP J. Inf. Secur.* **2007**, *2007*, 1–11. [CrossRef]
8. Gilad-Bachrach, R.; Dowlin, N.; Laine, K.; Lauter, K.; Naehrig, M.; Wernsing, J. Cryptonets: Applying neural networks to encrypted data with high throughput and accuracy. In Proceedings of the International Conference on Machine Learning, New York, NY, USA, 20–22 June 2016; pp. 201–210.
9. Bos, J.W.; Lauter, K.; Loftus, J.; Naehrig, M. Improved security for a ring-based fully homomorphic encryption scheme. In Proceedings of the IMA International Conference on Cryptography and Coding, Oxford, UK, 17–19 December 2013; pp. 45–64.
10. Hesamifard, E.; Takabi, H.; Ghasemi, M. Cryptodl: Deep neural networks over encrypted data. *arXiv* **2017**, arXiv:1711.05189.
11. Fan, J.; Vercauteren, F. Somewhat practical fully homomorphic encryption. *IACR Cryptol. ePrint Arch.* **2012**, *2012*, 144.
12. *Microsoft SEAL (Release 3.6)*; Microsoft Research: Redmond, WA, USA, 2020. Available online: https://github.com/Microsoft/SEAL (accessed on 14 April 2021).
13. Cheon, J.H.; Kim, A.; Kim, M.; Song, Y. Homomorphic encryption for arithmetic of approximate numbers. In Proceedings of the International Conference on the Theory and Application of Cryptology and Information Security, Hong Kong, China, 3–7 December 2017; pp. 409–437.
14. Paillier, P. Public-key cryptosystems based on composite degree residuosity classes. In Proceedings of the International conference on the theory and applications of cryptographic techniques, Prague, Czech Republic, 2–6 May 1999; pp. 223–238.
15. ElGamal, T. A public key cryptosystem and a signature scheme based on discrete logarithms. *IEEE Trans. Inf. Theory* **1985**, *31*, 469–472. [CrossRef]
16. Chen, H.; Laine, K.; Player, R.; Xia, Y. High-precision arithmetic in homomorphic encryption. In Proceedings of the Cryptographers' Track at the RSA Conference, San Francisco, CA, USA, 16–20 April 2018; pp. 116–136.
17. Chen, H.; Laine, K.; Player, R. Simple encrypted arithmetic library-SEAL v2. 1. In Proceedings of the International Conference on Financial Cryptography and Data Security, Sliema, Malta, 3–7 April 2017; pp. 3–18.
18. Arita, S.; Nakasato, S. Fully homomorphic encryption for point numbers. In Proceedings of the International Conference on Information Security and Cryptology, Beijing, China, 4–6 November 2016; pp. 253–270.
19. Dowlin, N.; Gilad-Bachrach, R.; Laine, K.; Lauter, K.; Naehrig, M.; Wernsing, J. Manual for using homomorphic encryption for bioinformatics. *Proc. IEEE* **2017**, *105*, 552–567. [CrossRef]
20. Fouque, P.A.; Stern, J.; Wackers, G.J. Cryptocomputing with rationals. In Proceedings of the International Conference on Financial Cryptography, Southampton, Bermuda, 11–14 March 2002; pp. 136–146.
21. Moon, S.; Lee, Y. An efficient encrypted floating-point representation using HEAAN and TFHE. *Secur. Commun. Netw.* **2020**, *2020*, 1250295. [CrossRef]
22. Jäschke, A.; Armknecht, F. (Finite) field work: Choosing the best encoding of numbers for FHE computation. In Proceedings of the International Conference on Cryptology and Network Security, Hong Kong, China, 30 November–2 December 2017; pp. 482–492.

23. Jäschke, A.; Armknecht, F. Accelerating homomorphic computations on rational numbers. In Proceedings of the International Conference on Applied Cryptography and Network Security, London, UK, 19–22 June 2016; pp. 405–423.
24. Lin, C.F.; Shih, S.H.; Zhu, J.D. Chaos based encryption system for encrypting electroencephalogram signals. *J. Med. Syst.* **2014**, *38*, 1–10. [CrossRef] [PubMed]
25. Ahmad, M.; Farooq, O.; Datta, S.; Sohail, S.S.; Vyas, A.L.; Mulvaney, D. Chaos-based encryption of biomedical EEG signals using random quantization technique. In Proceedings of the 2011 4th International Conference on Biomedical Engineering and Informatics (BMEI), Shanghai, China, 15–17 October 2011; Volume 3, pp. 1471–1475.
26. Liu, Y.; Huang, H.; Xiao, F.; Malekian, R.; Wang, W. Classification and recognition of encrypted EEG data based on neural network. *J. Inf. Secur. Appl.* **2020**, *54*, 102567. [CrossRef]
27. Xiang, G.; Chen, X.; Zhu, P.; Ma, J. A method of homomorphic encryption. *Wuhan Univ. J. Nat. Sci.* **2006**, *11*, 181–184.
28. Sathya, S.S.; Vepakomma, P.; Raskar, R.; Ramachandra, R.; Bhattacharya, S. A review of homomorphic encryption libraries for secure computation. *arXiv* **2018**, arxiv:1812.02428.
29. Babyak, M.A. What you see may not be what you get: A brief, nontechnical introduction to overfitting in regression-type models. *Psychosom. Med.* **2004**, *66*, 411–421. [PubMed]
30. Alomari, M.H.; AbuBaker, A.; Turani, A.; Baniyounes, A.M.; Manasreh, A. EEG mouse: A machine learning-based brain computer interface. *Int. J. Adv. Comput. Sci. Appl.* **2014**, *5*, 193–198.
31. Shenoy, H.V.; Vinod, A.P.; Guan, C. Shrinkage estimator based regularization for EEG motor imagery classification. In Proceedings of the 2015 10th International Conference on Information, Communications and Signal Processing (ICICS), Singapore, 2–4 December 2015; pp. 1–5.
32. Pedregosa, F.; Varoquaux, G.; Gramfort, A.; Michel, V.; Thirion, B.; Grisel, O.; Blondel, M.; Prettenhofer, P.; Weiss, R.; Dubourg, V.; et al. Scikit-learn: Machine Learning in Python. *J. Mach. Learn. Res.* **2011**, *12*, 2825–2830.
33. Van Rossum, G.; Drake, F.L. *Python 3 Reference Manual*; CreateSpace: Scotts Valley, CA, USA, 2009.
34. Andrzejak, R.G.; Lehnertz, K.; Mormann, F.; Rieke, C.; David, P.; Elger, C.E. Indications of nonlinear deterministic and finite-dimensional structures in time series of brain electrical activity: Dependence on recording region and brain state. *Phys. Rev. E* **2001**, *64*, 061907. [CrossRef] [PubMed]
35. Dua, D.; Graff, C. UCI Machine Learning Repository. 2017. Available online: http://archive.ics.uci.edu/ml (accessed on 21 March 2021).
36. Bressert, E. *SciPy and NumPy: An Overview for Developers*; O'Reilly Media, Inc.: Sebastopol, CA, USA, 2012.
37. Sun, Y.; Ye, N.; Xu, X. EEG analysis of alcoholics and controls based on feature extraction. In Proceedings of the 2006 8th International Conference on Signal Processing, Guilin, China, 16–20 November 2006; Volume 1.
38. Snodgrass, J.G.; Vanderwart, M. A standardized set of 260 pictures: Norms for name agreement, image agreement, familiarity, and visual complexity. *J. Exp. Psychol. Hum. Learn. Mem.* **1980**, *6*, 174. [CrossRef]

Review

Current Challenges Supporting School-Aged Children with Vision Problems: A Rapid Review

Qasim Ali [1], Ilona Heldal [1,*], Carsten G. Helgesen [1], Gunta Krumina [2], Cristina Costescu [3], Attila Kovari [4], Jozsef Katona [5] and Serge Thill [6]

1. Department of Computing, Mathematics, and Physics, Western Norway University of Applied Sciences, Inndalsveien 28, 5063 Bergen, Norway; ali.qasim@hvl.no (Q.A.); carsten.gunnar.helgesen@hvl.no (C.G.H.)
2. Department of Optometry and Vision Science, University of Latvia, LV-1586 Riga, Latvia; gunta.krumina@lu.lv
3. Special Education Department, Faculty of Psychology and Educational Sciences, Babes-Bolyai University, Sindicatelor Street No 7, 400029 Cluj-Napoca, Romania; christina.costescu@gmail.com
4. Department of Natural Sciences and Environmental Protection, Institute of Engineering, University of Dunaujvaros, Tancsics M 1/A, 2400 Dunaujvaros, Hungary; kovari@uniduna.hu
5. Department of Software Development and Application, Informatics Institute, University of Dunaujvaros, Tancsics M 1/A, 2400 Dunaujvaros, Hungary; katonaj@uniduna.hu
6. Donders Institute for Brain, Cognition, and Behaviour, Radboud University, Thomas van Aquinostraat 4, 6525 GD Nijmegen, The Netherlands; serge.thill@donders.ru.nl
* Correspondence: ilona.heldal@hvl.no; Tel.: +47-5558-7524

Abstract: Many children have undetected vision problems or insufficient visual information processing that may be a factor in lower academic outcomes. The aim of this paper is to contribute to a better understanding of the importance of vision screening for school-aged children, and to investigate the possibilities of how eye-tracking (ET) technologies can support this. While there are indications that these technologies can support vision screening, a broad understanding of how to apply them and by whom, and if it is possible to utilize them at schools, is lacking. We review interdisciplinary research on performing vision investigations, and discuss current challenges for technology support. The focus is on exploring the possibilities of ET technologies to better support screening and handling of vision disorders, especially by non-vision experts. The data orginate from a literature survey of peer-reviewed journals and conference articles complemented by secondary sources, following a rapid review methodology. We highlight current trends in supportive technologies for vision screening, and identify the involved stakeholders and the research studies that discuss how to develop more supportive ET technologies for vision screening and training by non-experts.

Keywords: school children; functional vision; vision screening; vision training; eye-tracking; stakeholders

1. Introduction

Our vision system is essential for performing daily activities and interactions with the environment. The vision problems that can affect a child's vision are divided into two groups: eye disorders in which the eye does not focus the light that enters, resulting in blurred vision, such as myopia, amblyopia, strabismus, and eye diseases that are caused by changes in eye structure, such as a cataract or retinoblastoma. Uncorrected refractive error, one of the most common vision problems, and one of the major causes of impaired vision in both industrialized and developing countries is a significant cause of blindness [1,2]. Indeed, 99.2 million children around the world under 15 years of age have visual impairment (visual acuity < 6/10) [3]. A vision problem can occur even if eye health seems normal or does not show any signs of binocular disorders [4]. There are many children with cognitive impairments who also have vision problems [5,6]. Scheiman associates such problems with visual information processing due to physical or cognitive deficiencies [7]. ADHD and dyslexia are examples of cognitive disorders that are often associated with vision problems;

they have the same symptoms, but need to be treated separately [5,8,9]. The early detection of vision disorders is also important for minimizing upcoming problems associated with academic performance and social life [10–12].

Vision screening is compulsory in many European countries, and is generally performed on children prior to the commencement of school education at the age of 4–5 years [13]. After this age, the parent or legally responsible person has to follow up on their child's vision health. Since children's vision continues to develop, it may change and they may experience difficulties, especially when they begin at school [6,14,15]. Teachers or parents can detect and help the problems, but the most important stakeholders for screening and helping with treatments are the responsible vision specialists from healthcare institutions. Visual functional assessments (e.g., visual acuity, color vision testing, and contrast sensitivity) can fail to diagnose the possible functional problems of the vision system [16]. These not only relate to reading difficulties, but have other symptoms, such as deficient attention span, fatigue, headache, dizziness, poor academic performance, and productivity [17–19]. A child with some of these problems can often control eye movements for a few minutes before experiencing difficulties. Therefore, during typical tasks with vision screenings—which often have a short duration—these problems may not be identified [20]. Some of these problems can be due to various concrete functional problems of the vision system, and are alleviated through vision training [21,22].

In recent years, promising options supporting necessary vision tests have begun to appear that use technologies and computerized tasks [20]. Several eye-tracking (ET) technologies can measure the movements of the left and right eye separately, and therefore ET technologies are promising for producing a measurement that can help to indicate functional vision problems [23]. ET technologies are often integrated with computer programs to make screening and training more flexible by incorporating serious games (SG), or utilizing virtual reality (VR) technologies to use the surrounding space [24]. However, it is difficult to keep track of and understand the role of these new solutions.

The motivation behind this paper is to enhance the understanding of what technology development can contribute to screening the vision of school-aged children, especially for non-vision expert users. For this, one needs to know who can be involved in vision screening, their roles, competencies, and willingness. Their work needs to complement or be aligned with professional vision experts' work, and possibly be assisted with tools and technologies providing evidence for further actions if needed. For developing new supporting technologies, it is not enough to understand who can use it (i.e., the possible non-vision experts) and for whom (i.e., the school-aged children). The knowledge of vision experts is necessary to determine how screening by non-experts can be quality assured.

The overall aim of this paper is to identify current trends in the research that is needed to enhance technological support for vision screening and training and involve non-vision expert stakeholders. This will be completed via a rapid review [25] based on an iterative approach to mapping evidence with knowledge gaps from highly multidisciplinary domains, such as those practicing vision science, special education, cognitive science, and technology development. While it is difficult to perform a full and systematic review at this stage, some basic ideas originating from mapping studies [26], used for answering the questions defined in Table 1, also influenced this research.

Table 1. Questions defined for the rapid review.

ID	Research Question	Rationale (Is To)
RQ1	Who are the stakeholders influencing vision screening at schools, and what are their roles?	Identifying the involved vision experts and non-experts influencing vision screening in children.
RQ2	What is known about children's vision screening in schools?	Exploring current methodologies and procedures that exist for children in school.
RQ3	How can ET better support screening and training vision?	Investigating the evidence of eye-tracking technologies to support stakeholders in vision screening and training.

This paper reviews literature about screening and helping school-aged, 6–15 year old children. The main focus is not on vision experts from healthcare (e.g., ophthalmologists or optometrists) or special pedagogy, but on other stakeholders without vision competence who can help to identify vision problems in children (e.g., school nurses, curators, teachers). While non-expert stakeholders can seldom make a diagnosis, they can produce clear indicators about possible issues, and communicate these to a child's parents or legally responsible person. As the next step, these indicators will support communication with healthcare institutions.

While ET technologies have existed since the last century [27], affordable ET technologies are available only during the last two decades, and therefore we do not limit the search to a specific period. To provide a context for better potential vision care in school, we aimed to position functional vision disorders on the map of broader vision problems and at schools [28].

2. Methodology

The intention here is to define the necessary preconditions, such as who influences the handling of vision problems, and how technologies can better support their work. To map all involved stakeholders and technologies, we performed this rapid review, following the eight phases defined by Khangura et al. [25] and the process described in Figure 1. The main reason for this approach was the need to bring together knowledge from health, education and technology development, three broad domains and identify essential gaps that needed to be overcome during a limited timeframe. This approach considered the strengths and limitations defined by Gant [29]. One of the main strengths is the possibility of considering these three broad domains in the review. The primary limitations concern possible missed studies and difficulty contrasting them.

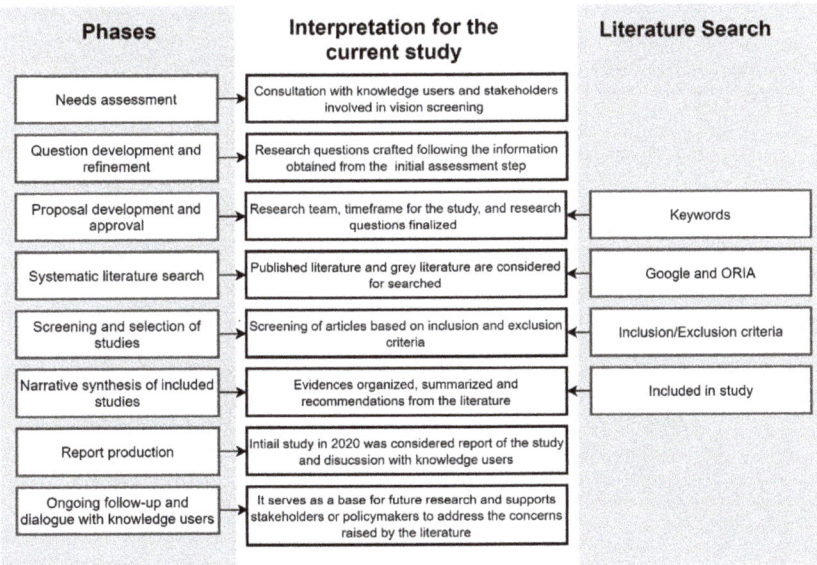

Figure 1. The eight steps considered for rapid review and their relationship within the context of current study and literature search.

While the timeframe for a rapid review is much shorter than for systematic reviews, the process is still based on clear questions, chosen criteria, and is rigorous. Instead of quantitative results, it ends with a descriptive summary of categories important for the research questions. This paper utilizes the rapid review methodology suggested by

Khangura et al. [25], extracts and summarizes evidence from different domains, and brings knowledge to practice in eight steps. The eight phases are defined as follows: (1) needs assessment, (2) question development and refinement, (3) proposal development and approval, (4) systematic literature search, (5) screening and selection of studies, (6) narrative synthesis of included studies (including evidence), (7) report production, and (8) ongoing follow-up and dialogue with knowledge representatives from the involved domains (see Figure 1).

The first two phases were completed in the last five months of 2020 [30], resulting in the three research questions that are synthesized in Table 2. The second phase for proposal development was also influenced by how mapping studies investigates literature and grey literature to map already existing research to research questions [31]. Phase (3) is realized through discussions with clinician vision experts, experts in special pedagogy, cognitive- and neuro-psychology for school beginners, and technology developers; this defines the search strategy and primary sources for research works (see Section 2.1). Phase (4) identifies and delimits the research studies, and presents the narrative in the flowchart in Figure 1 (for presenting phases (5) and (6), see Section 2.2). Phase (7) resulted in an earlier initial report from this study [30], which was followed by a discussion with knowledge users (phase (8)). For this rapid review, we use the term "phase" (not steps, as the original research by Khangura et al. [25] calls these) to distinguish from the major steps guiding the literature search (last column in Figure 1).

Table 2. Keywords guided the search of the literature for each research question.

Research Questions	Main Keywords
RQ1	"Vision screening" AND "school" AND "stakeholders"
RQ2	("vision screening" OR "vision assessment") AND "school children" AND "oculomotor dysfunction"
RQ3	("vision screening" OR "vision assessment") AND "oculomotor dysfunction" AND "eye-tracking"

2.1. Search Strategy

Google Scholar and HVL Library (ORIA) [32] search engines were used to search the articles from domains answering the research questions. The HVL Library has access to more than 160 resources and databases, including MEDLINE, BMC, ACM, IEEE, Web of Science, and SCOPUS. We identified literature on vision screening from 15 multidisciplinary sources of clinical, engineering, and education journals and conferences, including "*Neural Syst Rehabil Eng*", "*Teaching Exceptional Children*", "*E-Health and Bioengi-neering Conference (EHB)*", "*Virtual Rehabilitation (ICVR)*", "*Child: care, health, and development*", "*American Orthoptic Journal*", "*Pediatrics*", "*BMC Medical Education*", "*Software Engineering (JCSSE)*", "*Australian Journal of Education*", "*Survey of Oph-thalmology*", "*Conference on Cognitive Info-communications*", "*The Future of Educa-tion*", "*Journal of Behavioral Optometry*", and "*Serious Games and Applications for Health (SeGAH)*".

The keywords used in the search process are defined in Table 2. After the search, we checked the title, keywords, and abstract to examine the cited papers of relevant articles. The search strategy also includes the selected papers' reference lists. If we found an author with important papers, such as [33], we checked related publications from their environment. Examples of such manual selection included recently published articles involving practitioners [34,35], while the EU database for current vision screening projects [11,13] was further explored for grey literature. Since the authors are experts in different domains, they suggested literature or checked others for representativeness of their own domains.

2.2. Inclusion and Exclusion Criteria, the Narrative and Study Selection

The following were the inclusion criteria for selected articles:

(1) Studies that included vision screening and vision rehabilitation.
(2) Studies about stakeholders involved in vision screening.

(3) Studies considering functional vision problems (oculomotor dysfunction assessment).
(4) Written in the English language.

Exclusion criteria were as follows:

(1) Studies intended for hearing, cognitive, or related to any other disabilities.
(2) Severe vision problems such as blindness and retinoblastoma.
(3) Use of eye-tracking technologies for purposes other than vision screening or rehabilitation.

The results included a large number of articles, including review papers, mainly from the fields of medicine, education, applied sciences, optometry, and ophthalmology. The initial number of papers was 1049 from Google-Scholar, and 185 from O-HVL. The number of examined papers that we decided to further examine was 192, including eight review papers and four survey papers based on analyzing their title and reviewing the abstract by two authors. After the initial selection, 142 articles were considered relevant for this review. The descriptive results begin with Sections 3 and 4, positioning the problem area while bridging the domains that handle practical vision screening and training to technology development, and connecting to the literature findings and answering the three research questions in Section 5 (RQ1), Section 6 (RQ2), and Section 7 (RQ3).

3. Overall Context of Vision Screening

As we described earlier, there are vision problems influencing the quality of life and development of individuals. Presently, vision problems are detected by traditional and instrumental screening, often by professionals. Most detected vision problems can be resolved by corrective lenses (spectacles or contact lenses), vision training, or eye surgery [36]. Children's vision is essential when starting school; 70% to 80% of all tasks in an educational setting require good vision, yet children's vision is seldom screened at this stage [37,38]. Many children may be disadvantaged by not enjoying the same learning conditions as children with no vision problems, or those having earlier recognized vision problems. The level of income, parental education, the policies of countries, and ethnicities are identified as factors influencing whether school-aged children benefit from vision screening or not [28,35,39].

Vision screening is usually carried out using a visual attention task that requires children to judge how much of the visual world they see in clinical and non-clinical settings. During the task, the children's visual attention may drop, and they may not be able to maintain their focus [40]. If the children cannot keep their attention, they may stop looking at the given task and become bored. Consequently, their responses become less accurate and sometimes they can be misdiagnosed [41]. Occasionally, attention may drop due to other problems, such as ADHD or dyslexia [42], problems associated with vision difficulties [43–45]. Vision problems are often associated with hearing or other motoric problems, such as balance [24,46].

Generally, vision screening tests are carried out in primary care clinics by practitioners for checking eye health in children. Some of the tests can be supplemented by technologies. Metsing et al. [47] identified three categories for vision screening, namely, conventional (traditional or manual), instrumental, and computer software (see Figure 2). Her categories can be extended for vision training as well, and the second and the third categories can also overlap each other. In this paper, an additional category is differentiated (Section 3.4.) for web and smartphone-based software tools; the discrimination of this category often overlaps with other computer-based programs. However, this category points to newer, more affordable, and portable possibilities that appear today.

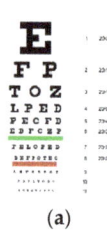 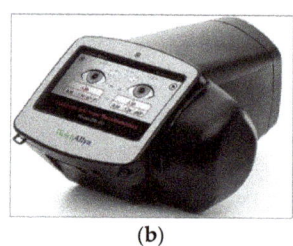 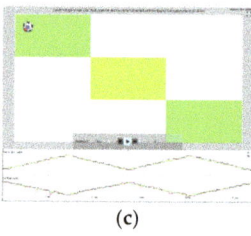

(a) (b) (c)

Figure 2. Different tools and computer software programs supporting different phases in a vision screening: (**a**) Snellen chart used for traditional vision screening (image downloaded from https://en.wikipedia.org/wiki/Snellen_chart#/media/File:Snellen_chart.svg, accessed date 12 October 2021); (**b**) Welsch Allyn Spot Vision Screener is an instrumental device used for measuring refractive error (image downloaded from https://www.gocheckkids.com/welch-allyn-spot-vision-screener, accessed date 12 October 2021); (**c**) C&Look is an eye-tracking computer software for screening functional vision problems.

3.1. Conventional (Traditional, Manual) Screening

The traditional way of diagnosing vision problems among children and adults is by "manual" vision screening during which vision specialists examine vision problems through several tests. Specialists define the procedure (the included tests and how the chosen tests are combined in a test battery). Already, these procedures use several aids, e.g., the Snellen chart (see Figure 2a), the plus lens test, and stereo test [48]. These are aids helping to investigate problems such as reduced visual acuity, hypermetropia, and binocular vision. Traditional methods of vision screening are considered gold standard procedures. As such, they have the advantage of having been tested and approved extensively by now. However, administering them requires expert knowledge from vision specialists, or specially trained school nurses, or educated volunteers.

In most countries, systematic screening is implemented around the age of four or five [13] as a last recommended and state-financed vision screening. This screening seldom includes functional vision screening, e.g., screening for identifying ocular motility disorders or other problems with the vision system in one's life [6,49]. Common vision problems occur in preschool- and primary school-aged children, and the main focus of most screening programmes is to detect amblyopia, refractive errors, binocular vision, and ocular motility disorders (e.g., strabismus) [47].

Meanwhile, the most common vision problems in primary school- and secondary school-aged children are uncorrected refractive error [50], oculomotor dysfunction (an anomaly in fixation, saccades, or pursuit eye movements) [51], and accommodative and non-strabismic binocular disorders (e.g., accommodative insufficiency, convergence insufficiency) [52,53], especially in learning disabilities and children with ADHD [54]. However, visual screenings to assess this visual functionality do not exist.

3.2. Instrumental Screening

Instrumental screening methods are already appropriate for children 1 or 2 years old, and are recommended for children older than 2 years old under certain conditions [55]. Instrumental vision screening includes supported instruments from the available automatic vision screening devices that use state-of-the-art technologies, e.g., autorefraction, retinal birefringence, or a photo screener [56]. Automated vision screening devices are used primarily for diagnosing refractive errors, media opacities, and misalignments of eyes in children and adults [57] (see Figure 2b).

Instrumental technologies show high accuracy in identifying vision problems in children. Silverstein et al. [58] showed that photoscreener-based instrumental technology could detect refractive errors in children (particularly useful in younger children (ages 3–5), preverbal children (under age 3), and nonverbal children) that may be missed in traditional screening methods. The speed of vision screening is almost three times faster than the traditional method. Vaughan et al. [59] compared the results obtained via photoscreening technology with traditional chart methodology to identify amblyopia and reflective errors in preschool children. The results showed 82.9% of the children passed instrumental eye exams, while this rate was 64% through LEA chart-based methodology [59]. The LEA chart has the same function for preschool and kindergarten children as the Snellen chart (Figure 2a) but instead of letters, it uses symbols.

The acquisition of useful visual field measurements in children, particularly younger children, is difficult and sometimes impossible. Children with moderate and severe visual impairment, glaucoma, damaged brain, or a central nervous system abnormality, such as perinatal stroke or optic pathway glioma, should undergo visual field testing. The biggest challenge in visual field testing is recognizing when a child may have a visual field defect, especially if it combines with ADHD. Vision specialists and researchers commonly use standard automated perimetry (SAP) in clinical settings for visual field testing [60]. The accuracy of visual field testing can be influenced by several factors, including time to respond, attention, fatigue, and stimulus size and duration [61]. Akar et al. [62] investigated the visual field testing strategy for healthy school-aged children with the Goldmann perimeter and Humphrey Field Analyzer. In order to avoid fixation loss, different rewards, verbal instruction, and gamified response buttons were given to children. The results showed that children faced difficulties with Goldmann kinetic perimetry tests. They concluded that the visual field of children could be examined using static perimetry at the age of six. However, visual field assessment is harder to perform in children with AHDH due to the lack of comprehension of the testing method, loss of gaze on the fixation target, rapid boredom, fatigue, and distraction [63]. In this case, eye-tracking technology is most suitable to compensate for inadequate eye fixation control during perimetry [64].

The use of photo screeners allows for the quick, non-invasive measurement of refraction and ocular alignment in both eyes, and would be of great value in refractive error screening, early detection of amblyopia, and in eye care practice and research [65]. Unfortunately, there is no possibility to assess the visual function (visual acuity, color vision, stereovision etc.)

3.3. Computerized Vision Screening Programs

Computerized vision screening programs offer a broad range of test batteries for vision screening, including visual acuity, color vision, contrast vision, stereovision, visual efficiency skills, visual field, test, and oculomotor behavior of the eyes [66,67]. For example, Visual Efficiency Rating (VERA) is a computerized software developed for school nurses to screen visual acuity, saccades, and the accommodative and vergence facility in children [66]. VERA simulates reading by placing numbers or text sequentially in a pattern while performing the vision test on children [68]. However, the disadvantage of VERA is the low range of sensitivity, which could miss children with vision problems. Gallaway et al. [69] observed that the sensitivity of VERA ranges from 45% to 64% on school children with visual acuity of 20/25 or better.

In recent years, there has been an increasing interest in using eye-tracking computer programs to screen oculomotor behaviors of the eyes by reading or looking at the animated objects on a computer screen. RightEye [35] and C&Look [23] are the latest developments in the quest to create a new generation of software that can detect the signs and patterns of oculomotor problems (see Figure 2c).

As mentioned earlier, the accuracy of visual field testing can be lowered due to several factors, especially alertness and the inappropriate responses of children in the perimeter. Miranda et al. [70] developed a computerized solution for pediatric visual field tests using computer games and an organic light-emitting diode (OLED) to address such challenges. This approach was adopted to provide an enjoyable test experience, and results indicated that this solution is feasible for children.

3.4. Web and Smartphone-Based Tools

With the advent of modern technologies, several improvements have been made in visual function testing, leading to robust, time-efficient, reliable, and evidence-based screening tools [71]. Studies show the reliability of automated vision testing tools for objective measurements of eyes, including autorefractors, photo screeners, and smartphones-based devices [56]. Such tools can offer a broad range of vision tests, such as refractive errors, visual acuity, visual efficiency skills, color vision, stereovision, contrast vision, visual field test, and testing the oculomotor behavior of the eyes.

Today, health communities often use smartphone technologies in telemedicine for vision screening. Due to COVID-19, teleophthalmolog is being used by researchers for measuring visual acuity remotely [72]. Most of the available applications use optotypes for vision screening. "Peek Acuity" [73] uses the optotype letter "E" (different sizes, and rotated at different angles) for the visual acuity test. The Freiburg Vision Test (FrACT) [74] is a corresponding web-based application that also uses similar optotypes (often "E" or "C" of different sizes, and rotated at different angles) to evaluate visual acuity, and has significant sensitivity among preschool-aged children [75].

The use of "E" or "C" optotypes in web-based applications increases the objectivity of the test when the child cannot recognize the letter and there is no learning effect from one eye to the other. The web-based tests are easily accessible, validated, store data to electronic patient record, and can be shared remotely. Several computerized vision screening systems are portable and often provide a self-assessment environment without a need for the supervision of vision experts [76].

The main disadvantage of the tests is the presentation of one optotype without visual crowding effect (added distractors around a central target), which is very important in the assessment of visual acuity, the calculation of crowding ratio, and the diagnostics of amblyopia, dyslexia, learning disabilities, and ADHD [77,78]. Figure 3 shows the different types of technologies supporting vision screening.

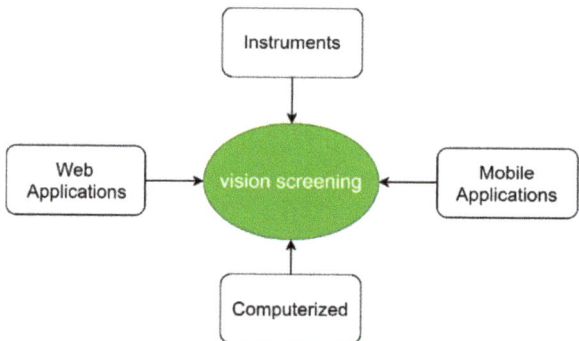

Figure 3. Different paradigms of technologies assisting experts in vision screening.

4. Visual Information Processing and Visual Cognition

Visual information processing is a cognitive skill to extract, organize, and analyze the visual information acquired from the environment [79]. Researchers believe that visual cognition is a complex process that involves visual information processing and cognitive factors. These work together to recognize the object from the scene using prior knowledge and retina input [80,81]. Scheiman [7] divides the information processing mechanism into three components: visual analysis, visual-spatial, and visual-motor processes. Figure 4 provides a basic description of these three components. These components represent specific skills of the individuals involved in performing tasks or activities. The literature argues that poor information processing skills such as visual-spatial and visual-motor integration, and visual analysis skills, are critical for school children [82].

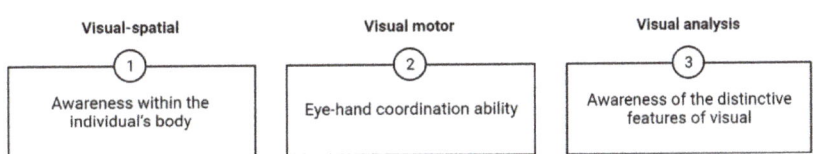

Figure 4. Visual information processing skills required for awareness and eye–hand coordination.

Vision experts test visual information processing skills in children using various objective or subjective standardized tests such as the Beery–Buktenica Developmental Test of Visual–Motor Integration (VMI), Motor-Free Visual Perceptual Tests (MVPT), and Rapid Automatized Naming (RAN) with the subset of the Developmental Eye Movement (DEM) test [82,83]. Studies have shown that children have visual processing disorders if they fail in such tests. Insufficient visual information processing is further associated with lower academic outcomes [79,84].

While eye-tracking technologies are helpful in recording eye movements such as fixations and saccades [85], which contribute to examining gaze movement in relation to visual representations, these relations are not necessarily enough to understand a whole picture of problems. For a better interpretation of the results, the relationships between cognitive skills are also important. Furthermore, handling the direct interpretations depends on prior knowledge [86] or usage of the context [87,88]. Other studies that may contribute to improved results would need a better understanding and interpretation of individual capabilities [89,90], hand–eye coordination [91], or children's adaptability [92] to different situations.

Information about the systems that observe brain activities [93,94], or advanced methodologies that create new technologies to support health [95], requires a high level of knowledge about information technologies and human cognition. According to what we understand, the use of eye-tracking technology is not widespread for diagnosing vision problems. However, it has great potential for complementing the work of vision experts by recording and analyzing the eye movements, and comparing the stored data of the subjects. The newest eye-tracking technology allows for controlling the distance from the screen, adjusting the position of the child's head to the test, and informing the assistant about a wrong position, which is very helpful for collecting valid data.

5. Stakeholders Influencing Children's Vision Screening (RQ1)

This section identifies the main stakeholders in order to answer the RQ1: "Who are the stakeholders influencing vision screening at schools, and what are their roles?" Professional vision experts are not the only ones who perform the vision screening or help with managing vision problems. Other stakeholders that are involved or influence vision treatments include trained laypersons, school teachers, other educationalists, or parents [95]. In general, vision screening includes stakeholders from different domains who may have different routines and roles.

5.1. Vision Experts from Clinical Settings

Matta et al. [96] surveyed eighteen countries to examine the current state of international vision screening efforts worldwide. Their results show that 22% of vision screening is performed by orthoptists, 50% by ophthalmologists, optometrists, opticians, and parents, 17% by nurses, and 6% by doctors only. The primary stakeholders responsible for vision screening in children are qualified vision experts in clinical settings. The experts involved in children's vision screening, such as optometrists, ophthalmologists, and nurses, have the required expertise in clinical settings [97]. Their primary role is to diagnose vision problems in children that make them vulnerable during their development in general, and to vision impairments in particular [66].

5.2. Nurses, Parents, and Laypersons

A recent study by Sabri et al. [98] showed the result of non-healthcare professionals for vision screening. The accuracy of volunteers was 75% compared to that of optometrists, and they could be trained to perform vision screening on children. The literature review of Sharma et al. [14] identified many professionals and non-professionals such as nurses, teachers, health technicians, parents, and volunteers for vision screening for school-aged children.

In the United States, the results of school vision screening are often reported to parents or guardians, and they were instructed to follow up with an appropriate local eye-care professional [38]. Involvement from parents is crucial for identification and diagnosis, as well as for having a role in assisting the rehabilitation of children's functional vision.

5.3. Vision Teachers

Vision teachers (VT) are usually educated to Master of Science level in special pedagogy that focuses on children with visual impairment (low vision and blindness). A survey [99] found that the vision skills identified by VTs are mainly near and distance visual acuity, tracking, peripheral visual field, color perception, fixation, shift of gaze, contrast vision, and depth perception. The most common vision assessment method is a visual acuity test, which cannot detect a number of visual problems [39].

5.4. Other Specialists

Eye-care providers and primary care physicians have created several test batteries for vision screening in children, including cover tests, fixation, red-flex, acuity, color vision, and another test to identify oculomotor dysfunction [37,100,101]. As we have argued, several of these can be challenging to detect, in particular in relation to functional vision problems [100]. Recent studies indicate that 22% to 24% of school children in total, and 96% of children with a learning disability, have oculomotor dysfunction [102,103]. If the numbers are accurate, and the symptoms of oculomotor dysfunction and dyslexia can be the same, many children may receive the wrong diagnoses. Table 3 summarizes the stakeholders and their roles in children's vision screening.

Table 3. Papers grouped based on stakeholders and their roles.

Studies	Stakeholders	Roles
[96,97]	Ophthalmologists, Optometrists, Orthoptists, Opticians, Nurses	Vision screening in clinical setting.
[14,44,104–107]	Educationalist, Vision Teachers	Vision screening in schools.
[14,96,98]	Volunteers, Parents	Non-health professionals supporting vision screening directly or indirectly.

6. School-Aged Children's Vision Screening (RQ2)

To answer Research Question 2, "What is known about children's vision screening in schools?", we need to understand the cases where school children are screened and how studies report these results.

6.1. Vision Screening Programs for Schools

There are several countries with school vision screening programs, such as parts of the US, South Africa [39], and a county in Norway [18]. Across different countries, there is a substantial variation in screening programs focusing on vision and hearing. Studies show that only 21–36% of children younger than six years of age have received vision screening in the US [108].

A literature review [39] examining vision screening in 18 countries shows that many countries only perform partial vision screening; 88% are missing necessary tests, such as those indicating a need for eyeglasses (e.g., tests with Snellen or Tumbling "E" charts). In another study, Ciner et al. [109] highlighted the potential pitfalls in the screening process for preschool children, such as the agreement of effective methods of screening and the variability of test batteries in US states. Some studies reported the lack of professionals for vision screening in countries such as Turkey and Tanzania, and documented in reports from European countries [110–112]. Hopkins et al. [101] also argued that refractive error, strabismus, and amblyopia are most likely considered for screening in preschool and school entry children, whereas there is no agreed policy for the screening of binocular vision problems, functional vision, and other eye conditions in Australia and worldwide. While the work of non-professionals can be supportive and complimentary, their duties and roles are unclear at the moment. They supplement professionals in different ways and with different, often limited tests [39,42,112]. Sabri et al. [98] found that the accuracy of non-healthcare workers is 75% compared to professionals. However, non-professional volunteers can be trained to detect vision problems in children, especially if they have the right tools to support them.

6.2. Opinions on Vision Screening Programs at Schools

Screening for all schools may help identify children with vision problems as early as possible and increase awareness among school staff and parents about vision problems [113]. Among the benefits of school vision screening programs is improving the teacher–child relationship, and enabling teachers to know which support they need to provide [114]. This support also depends on how children can accept instructions, and their cognitive abilities. A basic vision screening can be performed in children's classrooms, which means that their school routine is not necessarily disturbed [111,115]. On the contrary, the training for school teachers may be perceived as an additional burden or a need to require additional resources [116]. For some schools, obtaining permission from the school authorities for holding an eye screening activity is a major challenge; school authorities often feel that the screening program interferes with their regular academic calendar and does not add value to the school or the children [117,118].

Teachers should collaborate with vision specialists, vision teachers, or occupational therapists to avoid misidentifying children with vision problems [110]. Regarding parental involvement, there should be announcements and discussions during the parent–teacher meeting in schools about the benefits and screening program structure prior to the screening [95]. After the screening, parents should be advised regarding the problems identified and the possibilities for intervention.

6.3. Opinions on Vision Screening for Many, and by Non-Professionals at Schools

Sudhan et al. [113] examined 530 schools with 77,778 children in India. While the teachers involved in the study may have complemented the screening of health professionals, the accuracy of their identification was not recognized [113]. Since the data collection mainly consisted of manual screening of various quality, it is not straightforward to relay or compare certain screening results from this study.

Collecting vision measurements would improve vision assessment protocols and support progress evaluation during rehabilitation, training, or visual therapy. Objective measurements are possible using ETs [33,119,120], but ETs alone are not enough [23]. We must consider new user-friendly, accessible, and cost-effective specialized software tools to support the right interventions from stakeholders contributing to a broader investigation. However, as pointed out earlier, there is also a lack of stakeholders who can perform vision screening and assist rehabilitation.

Figure 5 depicts current recommendations suggested by the literature to strengthen the school vision screening programs. The purple box highlights the need for uniform screening methods [109,121], the red box shows that including volunteers would be more helpful in screening programs [98], the yellow box represents the recommendation to consider that functional vision screening should be included in screening programs [101]. The blue box recommends that school teachers could be trained to support vision screening [116]. Last, the green box shows all stakeholders involved in vision screening should communicate clearly and provide the information necessary to help vision screening [38,101].

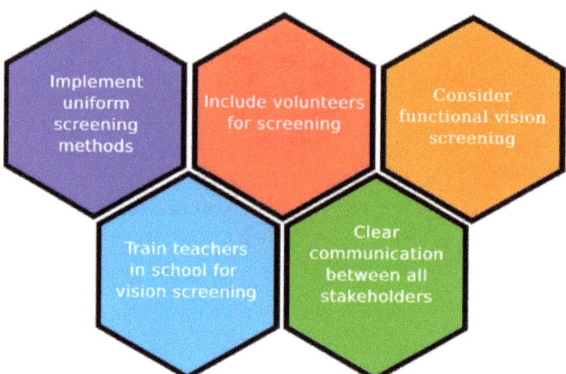

Figure 5. Current recommendations from the literature to fill the missing gaps in school vision screening programs. The colors in the figure depict recommendations for various methods and from various stakeholder groups.

7. ET Support for Vision Screening and Training (RQ3)

This section discusses Research Question 3: "How can ET better support screening and training vision?"

7.1. Usage and Types of ET Technologies

The use of eye-tracking technologies has significantly increased in cognitive neuroscience, the assessment of functional vision, applied engineering, education, and customer visual attention research [122–124]. In recent years, research studies have shown a potential use of eye-trackers to quantify visual information processing, visual function assessment (VFA), and diagnose nystagmus and strabismus vision problems [33,125].

The eye-tracking method aims to record the gaze direction at a specific time and understand visual attention. To achieve this aim, four types of eye-tracking techniques have developed over time. The classifications of four general ET techniques are: scleral search coil, electro-oculography (EOG), infrared oculography (IOG), and video oculography (VOG) [126]. Most modern eye-trackers use the video oculography (VOG) technique to record eye movements by applying the pupil center corneal reflection (PCCR) method [127]. The advanced image processing algorithms are also applied to the subject's eye position, relevant to stimuli [128]. The commercial eye-trackers are mobile, remote, and embedded in a head-mounted display (HMD), or add-ons for HMD. Remote eye-trackers require the participants to maintain head position throughout the entire process, while mobile or HMD eye-trackers do not have such conditions.

Generally, the refresh rate of eye-trackers varies from 30 Hz to 1200 Hz for recording eye movements, such as fixations and saccades. Eye trackers with low-frequency sampling have high latency in evaluating the point of gaze. The eye-trackers with a high-frequency sampling rate and low latency come with high-resolution cameras and they cost.

Several research studies showed the use of ET for vision screening, especially the visual functional assessment (VFA). Kooiker et al. [33] observed and quantified ET data to examine the oculomotor control and visual function assessment of visually impaired children. Murray et al. [35] demonstrated the reliability of ET by cluster analysis to determine the normative values of oculomotor metrics, such as saccades and fixations. Their results showed 85% of eye movement metrics were highly reliable and acceptable. Eide et al. [23] developed C&Look software that uses ET for complementing VTs and

VT students to diagnose vision problems associated with oculomotor dysfunction in school children [111].

Furthermore, some research studies have shown the advantages of eye-tracking technologies for diagnosing strabismus and pathological nystagmus [129,130]. Table 4 shows the directions of ET used by researchers in the literature for vision screening.

Table 4. Overview of the studies that used ET for vision screening. N/R = not reported.

Studies	Origin	Method	Vision Problem	Stakeholders	Subjects	Age
[129]	Italy	Eye-tracker	Nystagmus	Ophthalmologists	15	N/A
[131]	India	Eye-tracker	OMD	N/R	16	45–72
[33]	Netherlands	Eye-tracker	OMD, visual functions	Orthoptists, Optometrists, Ophthalmologists	126	1–14
[130]	Thailand	Eye-tracker	Strabismus	N/R	50	7–50
[35]	U.S	Eye-tracker	OMD	Optometrists	2993	5–62
[132]	Eastern Europe	Eye-tracker	OMD	Ophthalmologists	58	4–19
[133]	South Africa	Eye-tracker	OMD	Clinicians	33	21–24
[79]	Australia	Eye-tracking, traditional	Refractive errors, binocular vision anomalies, OMD	Optometrist	108	8–9

7.2. Recommendations for Challenges in ET Screening

While performing calibration or conducting the experiment with an eye-tracker, several factors can influence the accuracy and quality of the data being recorded, such as sitting position, body or head movements, and the person responsible for concluding the experiment [134]. Such factors are even more challenging for children and people with neurological disorders such as Alzheimer's disease, schizophrenia, stroke or autism [135]. Eye-tracking has been one of the main research methods in the last decades, not only for the diagnosis of eye movement disorders, but additionally for gaining insights into early (neuro)cognitive development. Through eye-tracking, gaze location can be objectively measured from children as young as a few days old, and up to adulthood.

Several research studies have recommended protocols and introduced new algorithms to mitigate the potential difficulties in the calibration or eye movements recordings process. Table 5 illustrates the recommendations extracted from the literature.

Table 5. Recommendations for calibration and screening for children and infants.

Studies	Recommendations
[136]	Play cartoons to attract the attention of the child, use Rifton chair, caregiver support.
[137]	Apply linear transformation on fixation coordinates.
[134]	Use of animated calibration targets (looming or twisting), gaze coordinates should be accepted within first 4 s, keep changing background screen color of calibration scene.
[138]	Consider monocular calibration of each eye.

7.3. Microsaccades

Microsaccades are small involuntary movements, including drift, tremor, and fixational saccades [139], and ranging from 0.5 to 2.5 Hz [140]. Studies reported that microsaccades help attain visual attention, processing cognitive tasks. In the non-visual cognition process, the frequency rate of microsaccade decreases and the magnitude of microsaccade increases with the task difficulty level [141]. Microsaccades can be detected with an eye-tracker, but with the cost of a high-frequency rate. Marcus et al. [140] analyzed the data quality of microsaccade measurements collected by Tobii Pro Spectrum 1200 Hz and EyeLink 600 Hz eye-trackers. The results of their study showed that the high-frequency eye-tracker attained the microsaccades rate without missing any microsaccades.

7.4. Eye-Tracking Data Analysis

Eye-trackers offer high-frequency sampling rates. Therefore, investigating different parameters and conducting a statistical analysis of extensive data sets requires a set of tools. Many open-source and commercial software solutions are available (see Table 6) for generating heat maps, gaze points, blinking, fixation, microsaccade, and calculating the pupil size over time in an experiment.

Table 6. Names of open-source and commercial tools for recording and analyzing the experimental study using eye-trackers.

Tool Name	Free	Paid
EventIDE [142]		✓
Tobii Pro Lab [143]		✓
SMI BeGaze [144]		✓
iMotions [145]		✓
Psychopy [143]	✓	
PyGaze [143]	✓	
OGAMA [146]	✓	
EMA Toolbox [147]	✓	
Gazepath [148]	✓	
PyTrack [149]	✓	

8. Synthetizing the Results

8.1. Main Findings and Future Needs

The goal of this study is to gain an understanding of the current state of research in stakeholders involved in vision screening for children and the use of state-of-art eye-tracking technology to support vision experts, and to fill the gaps in the present literature on the use of eye-tracking technology for new possibilities and future lines of research. After answering the research questions, the following points synthesize the sometimes contradictory findings and serve as a basic departure for further research:

- For the first research question (i.e., who are the stakeholders influencing vision screening at schools?), seven primary vision experts and non-experts emerged from the literature: (i) orthoptists [33] (ii) optometrists [35] (iii) ophthalmologists [129] (iv) vision teachers [44] (v) nurses [150] (vi) parents (vii) volunteers [98].
- In most countries, routine vision screening starts at the age of 4 to 5 [49,151] or from 3 to 5 [95]. However, there are no uniform methodologies for vision screening [101,109,121].
- Many factors influence the low number of follow-up vision screenings for children, including parents' education, awareness, and social and economic factors [38,152,153]. Amblyopia (lazy eye) is often neglected in children because of the mentioned factors and the local medical system [154].
- Basic terms such as vision, screening, and technology mean different things in different fields. To find correct and accurate search terms is difficult and time-consuming.
- Vision problems affect children's social and academic performances, and can be a reason for students to drop out from educational institutes [19,28]. Therefore, it is crucial to assess children's vision throughout the school period at different stages [155].
- The examined documents reflect the current obstacles faced by children in their schools. At this age, the guardians of students, the teachers, and the child may not be aware of vision deficits. Consequently, concerns related to children's vision cannot be communicated well between the children and the guardians, and after that, if the children have possible vision problems, between the guardians and other involved stakeholders.
- An important observation from the literature and whitepapers is that vision care is often limited to the screening or handling of a specific vision problem. The support for treatment approaches and the responsibilities of stakeholders to aid visual deficits are lacking. Often, many stakeholders need to be involved in the treatment of vision problems, but their collaboration is fragmented [28,39].

- Stakeholders do not widely use eye-tracking technology, even though the literature has shown the promising applications and benefits of ET in diagnosing OMD, nystagmus, and strabismus [33,111,129,130].
- ET can collect eye movement metrics such as fixations, the number of fixations, saccades, microsaccades, and smooth pursuits [85,140]. The objective data collected from ET are reliable, and studies demonstrate that such metrics are helpful in functional vision assessment (FVA) [33,35].

8.2. Limitations

Several limitations have been identified in this study:

- We used Google Scholar and HVL Library as sources to investigate the literature. We may have missed some actual results that were not indexed in these two sources.
- The participants in the included studies were from different geographic locations, languages, and cultural backgrounds. The laws and regulations related to vision screening in schools and clinical may vary for each study, so comparing them is difficult.
- Most of the examined papers refer to vision screening activities, while treatment of certain vision problems also utilizes technologies. Existing studies for vision treat-ments often focus on improving children's attention for visual tasks. Today, it is challenging to differentiate between attention problems and problems with the vision system.
- It was not systematic enough to identify more profound related domain (health education, involved policy developments, and technology management knowledge) gaps for helping all school-aged children.
- This study used rapid review; therefore, results can be limited compared to systematic review where quantitative results can be provided.

8.3. Future Work

We used Google Scholar and HVL Library (ORIA) for the literature search. However, a systematic literature review is still needed to ensure the precise roles of stakeholders in screening and rehabilitation of children's vision. While this work has begun with the ambition to conduct a systematic literature review involving the domains promising better help for school-aged children with possible vision problems (e.g., education, health, and technology development and management), to perform such a review is difficult. The three domains, on their part focusing on vision disorders, are fragmented silos. Critical studies are complex to compare and contrast. This study identified several issues influencing the use of technologies supporting school-aged children's vision. To develop and use these technologies further systematic studies producing quantitative evidences on the benefits and limitations of the available technologies are needed (e.g., [17,33,37,156]).

Figure 6 shows the stakeholders scattered across different areas of research that are directly or indirectly involved in the general or functional vision screening and rehabilitation of children. A systematic literature review can provide evidence to support vision screening from such domains and stakeholders, and overcome the disconnect between such entities. However, such reviews must handle the different content, and compare and contrast. This will be difficult due to the many different sources and different types of evidence presented. Even the basic terms to use, such as vision, visual, screening, and technologies, have different meanings in the various associated fields. Vision (after relating the word to seeing or the eyes) can be much more than seeing or vision systems, especially in technology development and use, or in education. Screening automatically means examination by clinical experts in medical journals, but can additionally refer to data collection in engineering papers. Screening technologies can collect different types of information and data. Technologies can vary from tools such as using pens to examine vision to complex instruments or software programs helpful for clinical professionals.

Figure 6. Research domains for further investigation.

9. Discussion

This paper investigated vision disorders in children, the stakeholders involved in vision screening for school-aged children, and focuses on identifying state-of-the-art eye-tracking technology for supporting vision screening. We identified important stakeholders contributions to children's vision health and discussed their role for collaboration. One of the main results is that many stakeholders with potential influences on managing children's vision are not systematically communicating with each other. Many non-specialist observers of possible vision problems, and the specialists diagnosing these problems, have difficulties initiating help for the children since they are not communicating. The potential for better communication needs to involve schools, where possible observers are, define the roles of responsibility, and involve other stakeholders, parents and professional vision experts to help children with vision impairment. The first step for better eye health, starting at schools, can be the recommendations summarized in Figure 5, according to several earlier literature (e.g., [6,15,111,115,117]).

Vision care is more than only diagnosing and screening children's vision. While research is pointing to training as an important part of treating vision disorders, training is often neglected and not connected to screening. This issue is reflected in the research on applying technologies. We exemplified the most common vision screening methods today, grouped into traditional, instrumental, and computerized tools, to better identify the potential of available tools and techniques in vision screening. However, we do not discuss a connection between screening and treatments, and if tools and instruments can be involved.

While the interest to use ET technology for diagnosing different vision impairments is growing, many of these technologies need to be improved to produce more understandable evidence regarding the different tests for non-specialists and for specialists. There are ET-based methodologies for analyzing eye movements; however, what exactly triggers the recorded eye movements, and what the relation is between attention and focus and the different parts of the vision system, should be further examined. The concrete measurements that ET can produce, including the fixation points, areas, number of fixations, saccades, smooth pursuits, and microsaccades from eye data, can be recorded and examined. Still, its role for visual information processing and the vision system needs better clarifications. This can only be achieved through the collaboration of vision experts and technology developers.

This study used Google Scholar and HVL Library (ORIA) [32] was used to find a broad range of published literature (peer-reviewed journals) and grey literature, including white papers and conference proceedings focusing on the areas of vision screening, education, technology, and cognition. In this study, we used a rapid review approach following the eight phases defined by Khangura et al. [25] to collect and synthesize the pieces of evidence that provide a high level of understanding of current challenges in school-aged children's vision screening programs. Defining research questions and performing the literature search were performed using those eight phases iteratively.

Across the world, many screening programs exist for preschool and school children. The literature demonstrates that there are still several gaps and limitations in the current screening procedure, including no adequate information on the ideal age of children for screening, inconsistent methods of screening in different states and regions, little agreement on effective methods of screening, insufficient resources and vision experts and an absence of functional vision assessments [101,109–112]. Intensive research is needed to see how to solve these problems, and the possibilities need to be investigated. A possibility that was investigated in this paper is empowering non-health-specialist volunteers and school teachers to perform vision screening where the intervention can be performed without a vision specialist, achieved by using technologies [14,98]. Professionals utilize traditional, instrumental, and computerized tools and technologies to conduct vision screening. Although traditional methods of screening are the gold standard, instrumental screening is gaining the attention of vision experts and is recommended by the American Academy of Ophthalmology (AAO), the American Academy of Pediatrics (AAP), the American Association of Certified Orthoptists (AACO), and the American Association for Pediatric Ophthalmology and Strabismus (AAPOS) for children under certain conditions [48,55]. Computerized technology, especially eye-tracker devices integrated with computer software, shows promising applications for collecting objective evidence of the eye movements that can be utilized to assess children's functional vision [33,132]. While using ET, some factors such as calibration setup, the sitting position of children, and the adjustment for children with special needs can influence data quality. Still, research studies have demonstrated that adopting various techniques and algorithms could optimize data quality and minimize the mentioned challenges [136].

10. Conclusions

This paper revealed how stakeholders, improved communication, and technologies could better support vision screening for school-aged children. One of the most important current challenges is to involve schools in these activities. This involvement requires uniform screening methods that include functional vision screening and possibilities to follow up vision problems. Since the number of vision experts performing the screening is low, it would be necessary to involve non-experts. Current technologies, including eye-tracking technologies, have great potential to support non-experts and these screening and training methods, providing measurable results, and bridging the fragmented communication between the involved stakeholders.

Author Contributions: Q.A. was the primary author responsible for investigation, data curation, visualization, and writing—original draft preparation. Supervision, conceptualization, methodology, writing—review and editing, I.H.; supervision, writing—review and editing, and methodology, C.G.H.; validation, writing—review and editing, G.K.; validation from vision science and optometry, C.C.; validation from cognitive science, writing—review and editing, A.K.; writing—review and editing, J.K.; writing—review and editing, S.T. All authors have read and agreed to the published version of the manuscript.

Funding: The project is sponsored by SECED, funded by the Research Council of Norway, project no: 267524/H30. The project is also sponsored by EFOP-3.6.1-16-2016-00003 funds, consolidatation of long-term R and D and I processes at the University of Dunaujvaros.

Institutional Review Board Statement: Not applicable.

Informed Consent Statement: Not applicable.

Conflicts of Interest: The authors declare no conflict of interest.

References

1. Dandona, L.; Dandona, R.; Naduvilath, T.J.; McCarty, C.A.; Srinivas, M.; Mandal, P.; Nanda, A.; Rao, G.N. Burden of moderate visual impairment in an urban population in southern India. *Ophthalmology* **1999**, *106*, 497–504. [CrossRef]
2. Taylor, H.R.; Livingston, P.M.; Stanislavsky, Y.L.; McCarty, C.A. Visual impairment in Australia: Distance visual acuity, near vision, and visual field findings of the Melbourne Visual Impairment Project. *Am. J. Ophthalmol.* **1997**, *123*, 328–337. [CrossRef]
3. Pascolini, D.; Mariotti, S.P. Global estimates of visual impairment: 2010. *Br. J. Ophthalmol.* **2012**, *96*, 614–618. [CrossRef]
4. Goldstand, S.; Koslowe, K.C.; Parush, S. Vision, visual-information processing, and academic performance among seventh-grade schoolchildren: A more significant relationship than we thought? *Am. J. Occupat. Ther.* **2005**, *59*, 377–389. [CrossRef] [PubMed]
5. Stein, J. Reply to:"The Relationship between eye movements and reading difficulties", Blythe, Kirkby & Liversedge. *Brain Sci.* **2018**, *8*, 99.
6. Wilhelmsen, G.B. School starters' vision—An educational approach. *Improv. Sch.* **2016**, *19*, 141–153. [CrossRef]
7. Scheiman, M. *Understanding and Managing Vision Deficits: A Guide for Occupational Therapists*; SLACK: Thorofare, NJ, USA, 2011.
8. Kaul, F.Y.; Johansson, M.; Månsson, J.; Stjernqvist, K.; Farooqi, A.; Serenius, F.; Thorell, B.L. Cognitive profiles of extremely preterm children: Full-Scale IQ hides strengths and weaknesses. *Acta Paediatr.* **2021**, *110*, 1817–1826. [CrossRef]
9. Fabian, I.D.; Kinori, M.; Ancri, O.; Spierer, A.; Tsinman, A.; Ben Simon, G.J. The possible association of attention deficit hyperactivity disorder with undiagnosed refractive errors. *J. AAPOS* **2013**, *17*, 507–511. [CrossRef]
10. Zaba, J.N. Children's vision care in the 21 st century & its impact on education, literacy, social issues, & the workplace: A call to action. *J. Behav. Optom.* **2011**, *22*, 39–41.
11. Eurostat. *Correction of Vision Problems by Sex, Age and Degree of Urbanisation*; Eurostat: Luxembourg, 2014.
12. Vaughn, W.; Maples, W.C.; Hoenes, R. The association between vision quality of life and academics as measured by the College of Optometrists in Vision Development Quality of Life questionnaire. *Optom. J. Am. Optom. Assoc.* **2006**, *77*, 116–123. [CrossRef]
13. EUScreen. *Summary Vision Screening Data*; EUScreen: Harderwijk, The Netherlands, 2019.
14. Sharma, A.; Congdon, N.; Patel, M.; Gilbert, C. School-based approaches to the correction of refractive error in children. *Surv. Ophthalmol.* **2012**, *57*, 272–283. [CrossRef]
15. Johnson, C.; Majzoub, K.; Lyons, S.; Martirosyan, K.; Tattersall, P. Eyes that thrive in school: A program to support vision treatment plans at school. *J. Sch. Health* **2016**, *86*, 391–396. [CrossRef]
16. Bennett, C.R.; Bex, P.J.; Bauer, C.M.; Merabet, L.B. The assessment of visual function and functional vision. *Semin. Pediatr. Neurol.* **2019**, *31*, 30–40. [CrossRef]
17. Thorud, H.-M.S.; Aurjord, R.; Falkenberg, H.K. Headache and musculoskeletal pain in school children are associated with uncorrected vision problems and need for glasses: A case–control study. *Sci. Rep.* **2021**, *11*, 2093. [CrossRef]
18. Falkenberg, H.K.; Langaas, T.; Svarverud, E. Vision status of children aged 7–15 years referred from school vision screening in Norway during 2003–2013: A retrospective study. *BMC Ophthalmol.* **2019**, *19*, 180. [CrossRef]
19. Bonilla-Warford, N.; Allison, C. A review of the efficacy of oculomotor vision therapy in improving reading skills. *J. Optom. Vis. Dev.* **2004**, *35*, 108–115.
20. Eide, G.M.; Watanabe, R.; Heldal, I.; Helgesen, C.; Geitung, A.; Soleim, H. Detecting oculomotor problems using eye tracking: Comparing EyeX and TX300. In Proceedings of the 10th IEEE Conference on Cognitive Infocommunication, Neaples, Italy, 23–25 October 2019.
21. Boon, M.Y.; Asper, L.J.; Chik, P.; Alagiah, P.; Ryan, M. Treatment and compliance with virtual reality and anaglyph-based training programs for convergence insufficiency. *Clin. Exp. Optom.* **2020**, *103*, 870–876. [CrossRef]
22. Colenbrander, A. Assessment of functional vision and its rehabilitation. *Acta Ophthalmol.* **2010**, *88*, 163–173. [CrossRef]
23. Eide, M.G.; Heldal, I.; Helgesen, C.G.; Wilhelmsen, G.B.; Watanabe, R.; Geitung, A.; Soleim, H.; Costescu, C. Eye tracking complementing manual vision screening for detecting oculomotor dysfunction. In Proceedings of the 2019 E-Health and Bioengineering Conference (EHB), Iasi, Romania, 21–23 November 2019; pp. 1–5.
24. Linehan, C.; Waddington, J.; Hodgson, T.L.; Hicks, K.; Banks, R. Designing games for the rehabilitation of functional vision for children with cerebral visual impairment. In Proceedings of the CHI'14 Extended Abstracts on Human Factors in Computing Systems, Toronto, ON, Canada, 26 April–1 May 2014; pp. 1207–1212.
25. Khangura, S.; Konnyu, K.; Cushman, R.; Grimshaw, J.; Moher, D. Evidence summaries: The evolution of a rapid review approach. *Syst. Rev.* **2012**, *1*, 10. [CrossRef] [PubMed]
26. Petersen, K.; Feldt, R.; Mujtaba, S.; Mattsson, M. Systematic mapping studies in software engineering. In Proceedings of the 12th International Conference on Evaluation and Assessment in Software Engineering, Bari, Italy, 26–27 June 2008; pp. 68–77.
27. Yarbus, A.L. *Role of Eye Movements in the Visual Process*; Nauka: Oxford, UK, 1965; p. 167.
28. Vågenes, V.; Ludvigsen, K.; Heldal, I.; Helgesen, C. Supporting vision competence creation for sustainable education in Tanzania. In Proceedings of the ECKM 21st European Conference on Knowledge Management, London, UK, 2–4 December 2020.
29. Grant, M.J.; Booth, A. A typology of reviews: An analysis of 14 review types and associated methodologies. *Health Info Libr. J.* **2009**, *26*, 91–108. [CrossRef] [PubMed]
30. Ali, Q.; Heldal, I.; Helgesen, C.G.; Costescu, C.; Kovari, A.; Katona, J.; Thill, S. Eye-tracking Technologies Supporting Vision Screening in Children. In Proceedings of the 2020 11th IEEE International Conference on Cognitive Infocommunications (CogInfoCom), Online, 23–25 September 2020; pp. 000471–000478.

31. Kitchenham, B.; Charters, S. *Guidelines for Performing Systematic Literature Reviews in Software Engineering*; CiteSeer: State College, PA, USA, 2007.
32. Library—Western Norway University of Applied Sciences. Available online: https://www.hvl.no/en/library/ (accessed on 31 May 2021).
33. Kooiker, M.J.; Pel, J.J.; Verbunt, H.J.; de Wit, G.C.; van Genderen, M.M.; van der Steen, J. Quantification of visual function assessment using remote eye tracking in children: Validity and applicability. *Acta Ophthalmol.* **2016**, *94*, 599–608. [CrossRef]
34. Hunfalvay, M.; Tyagi, A.; Whittaker, J.; Noel, C. Oculomotor Training Improves Binocular Vision with Improved Symptoms. Oculomotor Training & Binocular Vision. 2019. Available online: https://www.researchgate.net/publication/336530649_Oculomotor_Training_for_Poor_Pursuits_Improves_Functional_Vision_Scores_and_Neurobehavioral_Symptoms (accessed on 12 October 2021).
35. Murray, N.; Kubitz, K.; Roberts, C.-M.; Hunfalvay, M.; Bolte, T.; Tyagi, A. An examination of the oculomotor behavior metrics within a suite of digitized eye tracking tests. *IEEE J. Transl. Eng. Health Med.* **2018**, *5*, 1–5.
36. Sakimoto, T.; Rosenblatt, M.I.; Azar, D.T. Laser eye surgery for refractive errors. *Lancet* **2006**, *367*, 1432–1447. [CrossRef]
37. Kerr, N.C.; Arnold, R.W. Vision screening for children: Current trends, technology, and legislative issues. *Curr. Opin. Ophthalmol.* **2004**, *15*, 454–459. [CrossRef] [PubMed]
38. Shakarchi, A.F.; Collins, M.E. Referral to community care from school-based eye care programs in the United States. *Surv. Ophthalmol.* **2019**, *64*, 858–867. [CrossRef]
39. Chen, A.-H.; Bakar, N.F.A.; Arthur, P. Comparison of the pediatric vision screening program in 18 countries across five continents. *J. Curr. Ophthalmol.* **2019**, *31*, 357–365. [CrossRef]
40. Bortoli, A.D.; Gaggi, O. PlayWithEyes: A new way to test children eyes. In Proceedings of the 2011 IEEE 1st International Conference on Serious Games and Applications for Health (SeGAH), Braga, Portugal, 16–18 November 2011; pp. 1–4.
41. Gaggi, O.; Ciman, M. The use of games to help children eyes testing. *Multimed. Tools Appl.* **2016**, *75*, 3453–3478. [CrossRef]
42. Li, A. Classroom strategies for improving and enhancing visual skills in students with disabilities. *Teach. Except. Child.* **2004**, *36*, 38–46. [CrossRef]
43. Stein, J. Dyslexia: The role of vision and visual attention. *Curr. Dev. Disord. Rep.* **2014**, *1*, 267–280. [CrossRef] [PubMed]
44. Wilhelmsen, G.B.; Aanstad, M.L.; Leirvik, E.I.B. Implementing vision research in special needs education. *Support Learn.* **2015**, *30*, 134–149. [CrossRef]
45. Hjartarson, K.H. High-Level Vision in Dyslexia: The Possibility of Inefficient Visual Expertise as a Contributing Factor to Reading Problems. 2017. Available online: https://skemman.is/handle/1946/27659 (accessed on 12 October 2021).
46. Wiley, S.; Parnell, L.; Belhorn, T. Promoting early identification and intervention for children who are deaf/hard of hearing, children with vision impairment, and children with deaf-blind conditions. *J. Early Hear. Detect. Interv.* **2016**, *1*, 26–33.
47. Metsing, I.T.; Jacobs, W.; Hansraj, R. A review of vision screening methods for children. *Afr. Vis. Eye Health* **2018**, *77*, a446. [CrossRef]
48. Metsing, I.T.; Hansraj, R.; Jacobs, W.; Nel, E.W. Review of school vision screening guidelines. *Afr. Vis. Eye Health* **2018**, *77*, a444. [CrossRef]
49. Evans, J.R.; Morjaria, P.; Powell, C. Vision screening for correctable visual acuity deficits in school-age children and adolescents. *Cochrane Database Syst. Rev.* **2018**, *2*, CD005023. [CrossRef]
50. Atowa, U.C.; Hansraj, R.; Wajuihian, S.O. Vision problems: A review of prevalence studies on refractive errors in school-age children. *Afr. Vis. Eye Health* **2019**, *78*, a461. [CrossRef]
51. Simmons, K. Oculomotor Dysfunction: Where's the Evidence? *Optom. Vis. Perform.* **2017**, *5*, 198–202.
52. Wajuihian, S.O.; Hansraj, R. A review of non-strabismic accommodative-vergence anomalies in school-age children. Part 1: Vergence anomalies. *Afr. Vis. Eye Health* **2015**, *74*, a32. [CrossRef]
53. Wajuihian, S.O.; Hansraj, R. A review of non-strabismic accommodative and vergence anomalies in school-age children. Part 2: Accommodative anomalies. *Afr. Vis. Eye Health* **2015**, *74*, a33. [CrossRef]
54. Handler, S.M.; Fierson, W.M.; Section on, O. Learning disabilities, dyslexia, and vision. *Pediatrics* **2011**, *127*, e818–e856. [CrossRef]
55. Nottingham Chaplin, P.K.; Baldonado, K.; Cotter, S.; Moore, B.; Bradford, G.E. An eye on vision: Five questions about vision screening and eye health—Part 3. *NASN Sch. Nurse* **2018**, *33*, 279–283. [CrossRef]
56. Kassem, A.M. Automated vision screening. *Adv. Ophthalmol. Optom.* **2018**, *3*, 87–100. [CrossRef]
57. Nottingham Chaplin, P.K.; Baldonado, K.; Hutchinson, A.; Moore, B. Vision and eye health: Moving into the digital age with instrument-based vision screening. *NASN Sch. Nurse* **2015**, *30*, 154–160. [CrossRef]
58. Silverstein, E.; McElhinny, E.R. Traditional and instrument-based vision screening in third-grade students. *J. Am. Assoc. Pediatr. Ophthalmol. Strabismus* **2018**, *22*, e69. [CrossRef]
59. Vaughan, J.; Dale, T.; Herrera, D. Comparison of photoscreening to chart methodology for vision screening. *J. Sch. Nurs.* **2020**, 1–5. [CrossRef]
60. Phu, J.; Khuu, S.K.; Yapp, M.; Assaad, N.; Hennessy, M.P.; Kalloniatis, M. The value of visual field testing in the era of advanced imaging: Clinical and psychophysical perspectives. *Clin. Exp. Optom.* **2017**, *100*, 313–332. [CrossRef]
61. Parrish, R.K., 2nd; Schiffman, J.; Anderson, D.R. Static and kinetic visual field testing. Reproducibility in normal volunteers. *Arch. Ophthalmol.* **1984**, *102*, 1497–1502. [CrossRef] [PubMed]

62. Akar, Y.; Yilmaz, A.; Yucel, I. Assessment of an effective visual field testing strategy for a normal pediatric population. *Ophthalmologica* **2008**, *222*, 329–333. [CrossRef]
63. Tschopp, C.; Safran, A.B.; Viviani, P.; Reicherts, M.; Bullinger, A.; Mermoud, C. Automated visual field examination in children aged 5–8 years: Part II: Normative values. *Vis. Res.* **1998**, *38*, 2211–2218. [CrossRef]
64. Leitner, M.C.; Hutzler, F.; Schuster, S.; Vignali, L.; Marvan, P.; Reitsamer, H.A.; Hawelka, S. Eye-tracking-based visual field analysis (EFA): A reliable and precise perimetric methodology for the assessment of visual field defects. *BMJ Open Ophthalmol.* **2021**, *6*, e000429. [CrossRef]
65. Sanchez, I.; Ortiz-Toquero, S.; Martin, R.; de Juan, V. Advantages, limitations, and diagnostic accuracy of photoscreeners in early detection of amblyopia: A review. *Clin. Ophthalmol.* **2016**, *10*, 1365–1373. [CrossRef]
66. Atowa, U.C.; Wajuihian, S.O.; Hansraj, R. A review of paediatric vision screening protocols and guidelines. *Int. J. Ophthalmol.* **2019**, *12*, 1194–1201. [CrossRef] [PubMed]
67. Tekavi Pompe, M. Color vision testing in children. *Color Res. Appl.* **2020**, *45*, 775–781. [CrossRef]
68. Gallaway, M. The need for better school vision screening: The use of VERA vision screening in a community setting. *Optom. Vis. Dev.* **2010**, *41*, 232–239.
69. Gallaway, M.; Mitchell, G.L. Validity of the VERA visual skills screening. *Optom. J. Am. Optom. Assoc.* **2010**, *81*, 571–579. [CrossRef]
70. Miranda, M.A.; Henson, D.B.; Fenerty, C.; Biswas, S.; Aslam, T. Development of a pediatric visual field test. *Transl. Vis. Sci. Technol.* **2016**, *5*, 13. [CrossRef] [PubMed]
71. Schmidt, P.; Baumritter, A.; Ciner, E.; Cyert, L.; Dobson, V.; Haas, B.; Kulp, M.T.; Maguire, M.; Moore, B.; Orel-Bixler, D.; et al. Predictive value of photoscreening and traditional screening of preschool children. *J. Am. Assoc. Pediatr. Ophthalmol. Strabismus* **2006**, *10*, 377–378. [CrossRef]
72. Silverstein, E.; Williams, J.S.; Brown, J.R.; Bylykbashi, E.; Stinnett, S.S. Teleophthalmology: Evaluation of phone-based visual acuity in a pediatric population. *Am. J. Ophthalmol.* **2021**, *221*, 199–206. [CrossRef]
73. Bastawrous, A.; Rono, H.K.; Livingstone, I.A.T.; Weiss, H.A.; Jordan, S.; Kuper, H.; Burton, M.J. Development and validation of a smartphone-based visual acuity test (peek acuity) for clinical practice and community-based fieldwork. *JAMA Ophthalmol.* **2015**, *133*, 930–937. [CrossRef]
74. Bach, M. The freiburg visual acuity test-variability unchanged by post-hoc re-analysis. *Graefe's Arch. Clin. Exp. Ophthalmol.* **2006**, *245*, 965–971. [CrossRef] [PubMed]
75. Zhao, L.; Stinnett, S.S.; Prakalapakorn, S.G. Visual acuity assessment and vision screening using a novel smartphone application. *J. Pediatr.* **2019**, *213*, 203–210.e1. [CrossRef]
76. Huang, Y.; Ropelato, S.; Menozzi, M. Fully automatic and computerized self-vision-screening system: Vision at own—An E-health service of self vision examination and screening. In Proceedings of the 2015 Fifth International Conference on Digital Information Processing and Communications (ICDIPC), Sierre, Switzerland, 7–9 October 2015; pp. 262–266.
77. Bellocchi, S. Developmental Dyslexia, Visual Crowding and Eye Movements. In *Eye Movement: Developmental Perspectives, Dysfunctions and Disorders in Humans*; Stewart, L.C., Ed.; Nova Science Publishers: New York, NY, USA, 2013; pp. 93–110.
78. Huurneman, B.; Boonstra, F.N.; Cox, R.F.; Cillessen, A.H.; van Rens, G. A systematic review on 'Foveal Crowding' in visually impaired children and perceptual learning as a method to reduce Crowding. *BMC Ophthalmol.* **2012**, *12*, 27. [CrossRef]
79. Wood, J.M.; Black, A.A.; Hopkins, S.; White, S.L.J. Vision and academic performance in primary school children. *Ophthal. Physiol. Opt.* **2018**, *38*, 516–524. [CrossRef]
80. Pinker, S. Visual cognition: An introduction. *Cognition* **1984**, *18*, 1–63. [CrossRef]
81. Cavanagh, P. Visual cognition. *Vis. Res.* **2011**, *51*, 1538–1551. [CrossRef]
82. Hopkins, S.; Black, A.A.; White, S.L.J.; Wood, J.M. Visual information processing skills are associated with academic performance in Grade 2 school children. *Acta Ophthalmol.* **2019**, *97*, e1141–e1148. [CrossRef]
83. Valarmathi, A.; Suresh, K.; Venkatesh, L. Visual-perceptual function of children using the developmental test of visual perception-3. *Clin. Exp. Optom.* **2021**, 1–5. [CrossRef]
84. Webber, A.; Wood, J.; Gole, G.; Brown, B. DEM test, visagraph eye movement recordings, and reading ability in children. *Optom. Vis. Sci.* **2011**, *88*, 295–302. [CrossRef]
85. Török, Á.; Török, Z.G.; Tölgyesi, B. Cluttered centres: Interaction between eccentricity and clutter in attracting visual attention of readers of a 16th century map. In Proceedings of the 2017 8th IEEE International Conference on Cognitive Infocommunications (CogInfoCom), Debrecen, Hungary, 11–14 September 2017; pp. 000433–000438.
86. Ujbanyi, T.; Katona, J.; Sziladi, G.; Kovari, A. Eye-tracking analysis of computer networks exam question besides different skilled groups. In Proceedings of the 2016 7th IEEE International Conference on Cognitive Infocommunications (CogInfoCom), Wroclaw, Poland, 16–18 October 2016; pp. 000277–000282.
87. Walenstein, A. Observing and measuring cognitive support: Steps toward systematic tool evaluation and engineering. In Proceedings of the 11th IEEE International Workshop on Program Comprehension, Portland, OR, USA, 10–11 May 2003; pp. 185–194.
88. Berki, B. Desktop VR as a virtual workspace: A cognitive aspect. *Acta Polytech. Hung.* **2019**, *16*, 219–231.

89. Sik-Lanyi, C.; Shirmohammmadi, S.; Guzsvinecz, T.; Abersek, B.; Szucs, V.; Isacker, K.V.; Lazarov, A.; Grudeva, P.; Boru, B. How to develop serious games for social and cognitive competence of children with learning difficulties. In Proceedings of the 2017 8th IEEE International Conference on Cognitive Infocommunications (CogInfoCom), Debrecen, Hungary, 11–14 September 2017; pp. 000321–000326.
90. Csapó, G. Placing event-action based visual programming in the process of computer science education. *Acta Polytech. Hung.* **2019**, *16*, 35–57.
91. Kővári, A.; Katona, J.; Pop, C. Quantitative analysis of relationship between visual attention and eye-hand coordination. *Acta Polytech. Hung.* **2020**, *17*, 77–95. [CrossRef]
92. Péntek, Á.G.I. Adaptive services with cloud architecture for telemedicine. In Proceedings of the 2015 6th IEEE International Conference on Cognitive Infocommunications (CogInfoCom), Gyor, Hungary, 19–21 October 2015; pp. 369–374.
93. Singh, H.; Giardina, T.D.; Meyer, A.N.D.; Forjuoh, S.N.; Reis, M.D.; Thomas, E.J. Types and origins of diagnostic errors in primary care settings. *JAMA Intern. Med.* **2013**, *173*, 418–425. [CrossRef]
94. Katona, J.; Kovari, A. Examining the learning efficiency by a brain-computer interface system. *Acta Polytech. Hung.* **2018**, *15*, 251–280.
95. Akuffo, K.O.; Abdul-Kabir, M.; Agyei-Manu, E.; Tsiquaye, J.H.; Darko, C.K.; Addo, E.K. Assessment of availability, awareness and perception of stakeholders regarding preschool vision screening in Kumasi, Ghana: An exploratory study. *PLoS ONE* **2020**, *15*, e0230117. [CrossRef]
96. Matta, N.S.; Silbert, D.I. Vision screening across the world. *Am. Orthopt. J.* **2012**, *62*, 87–89. [CrossRef]
97. Flodgren, G.M.; Ding, Y. *Vision Screening in Children under the Age of 18: A Systematic Review*; Norwegian Institute of Public Health: Oslo, Norway, 2018.
98. Sabri, K.; Easterbrook, B.; Khosla, N.; Davis, C.; Farrokhyar, F. Paediatric vision screening by non-healthcare volunteers: Evidence based practices. *BMC Med. Educ.* **2019**, *19*, 65. [CrossRef]
99. Kaiser, J.T.; Herzberg, T.S. Procedures and tools used by teachers when completing functional vision assessments with children with visual impairments. *J. Vis. Impair. Blind.* **2017**, *111*, 441–452. [CrossRef]
100. Wilhelmsen, G.B. Barns Funksjonelle Syn. Gir Synsvansker Som Ikke Klassifiseres Etter ICD-10, Behov for Tiltak? 2012. Available online: https://hvlopen.brage.unit.no/hvlopen-xmlui/handle/11250/2481338 (accessed on 12 October 2021).
101. Hopkins, S.; Sampson, G.P; Hendicott, P.; Wood, J.M. Review of guidelines for children's vision screenings. *Clin. Exp. Optom.* **2013**, *96*, 443–449. [CrossRef]
102. Kelly, K.R.; Jost, R.M.; Cruz, A.D.L.; Dao, L.M.; Beauchamp, C.L.; Stager, D.; Birch, E.E. Slow reading in children with anisometropic amblyopia is associated with fixation instability and increased saccades. *J. AAPOS Off. Publ. Am. Assoc. Pediatr. Ophthalmol. Strabismus* **2017**, *216*, 447–451.e441. [CrossRef] [PubMed]
103. Sherman, A. Relating vision disorders to learning disability. *J. Am. Optom. Assoc.* **1973**, *44*, 140–141.
104. D'Andrea, F.M.; Farrenkopf, C. *Looking to Learn: Promoting Literacy for Students with Low Vision*; American Foundation for the Blind: New York, NY, USA, 2000.
105. Holbrook, M.C.; Koenig, A.J. *Foundations of Education: Instructional Strategies for Teaching Children and Youths with Visual Impairments*; American Foundation for the Blind: New York, NY, USA, 2000; Volume 2.
106. Kauffman, J.M.; Hallahan, D.P. *Handbook of Special Education*; Routledge: Oxford, UK, 2011.
107. Lueck, A.H. *Functional Vision: A Practitioner's Guide to Evaluation and Intervention*; American Foundation for the Blind: New York, NY, USA, 2004.
108. Hered, R.W.; Wood, D.L. Preschool vision screening in primary care pediatric practice. *Public Health Rep.* **2013**, *128*, 189–197. [CrossRef]
109. Ciner, E.B.; Schmidt, P.P.; Orel-Bixler, D.; Dobson, V.; Maguire, M.; Cyert, L.; Moore, B.; Schultz, J. Vision screening of preschool children: Evaluating the past, looking toward the future. *Optom. Vis. Sci.* **1998**, *75*, 571–584. [CrossRef] [PubMed]
110. Mathers, M.; Keyes, M.; Wright, M. A review of the evidence on the effectiveness of children's vision screening. *Child Care Health Dev.* **2010**, *36*, 756–780. [CrossRef] [PubMed]
111. Wilhelmsen, G.B. Better vision—Better reading teachers with vision competence change pupils reading skills. In Proceedings of the Conference—Future of Education, Florence, Italy, 11–12 June 2015; p. 442.
112. Gursoy, H.; Basmak, H.; Yaz, Y.; Colak, E. Vision screening in children entering school: Eskisehir, Turkey. *Ophthal. Epidemiol.* **2013**, *20*, 232–238. [CrossRef]
113. Sudhan, A.; Pandey, A.; Pandey, S.; Srivastava, P.; Pandey, K.P.; Jain, B.K. Effectiveness of using teachers to screen eyes of school-going children in Satna district of Madhya Pradesh, India. *Indian J. Ophthalmol.* **2009**, *57*, 455. [CrossRef] [PubMed]
114. Wilhelmsen, G.B.; Felder, M. Structured Visual Learning and Stimulation in School: An Intervention Study. 2021. Available online: https://hvlopen.brage.unit.no/hvlopen-xmlui/handle/11250/2738170 (accessed on 12 October 2021).
115. Narayanan, A.; Kumar, K. Strategy for mass vision screening of school children and its effectiveness. *Clin. Exp. Optom.* **2012**, *95*. [CrossRef]
116. Castanes, M. Major review: The underutilization of vision screening (for amblyopia, optical anomalies and strabismus) among preschool age children. *Binocul. Vis. Strabismus Quart.* **2003**, *18*, 217–232.
117. Hartmann, E.E.; Dobson, V.; Hainline, L.; Marsh-Tootle, W.; Quinn, G.E.; Ruttum, M.S.; Schmidt, P.P.; Simons, K. Preschool vision screening: Summary of a task force report. *Pediatrics* **2000**, *106*, 1105–1116. [CrossRef]

118. Zimmerman, D.R.; Ben-Eli, H.; Moore, B.; Toledano, M.; Stein-Zamir, C.; Gordon-Shaag, A. Evidence-based preschool-age vision screening: Health policy considerations. *Israel J. Health Policy Res.* **2019**, *8*, 70. [CrossRef]
119. Hunfalvay, M. Eye Tracking Technology for Clinical Practice: Benefits, Limitations, and Considerations. *Optom. Vis. Perform.* **2018**.
120. Ali, Q.; Heldal, I.; Mads Gjerstad, E.; Helgesen, C. Using eye-tracking technologies in vision teachers' work—A Norwegian perspective. In Proceedings of the 2020 E-Health and Bioengineering Conference (EHB), Iasi, Romania, 29–30 October 2020.
121. Shaw, R.; Russotti, J.; Strauss-Schwartz, J.; Vail, H.; Kahn, R. The need for a uniform method of recording and reporting functional vision assessments. *J. Vis. Impairm. Blindn.* **2009**, *103*, 367–371. [CrossRef]
122. Bobić, V.; Graovac, S. Development, implementation and evaluation of new eye tracking methodology. In Proceedings of the 2016 24th Telecommunications Forum (TELFOR), Belgrade, Serbia, 22–26 November 2016; pp. 1–4.
123. Kuo, Y.-L.; Lee, J.-S.; Kao, S.-T. Eye tracking in visible environment. In Proceedings of the 2009 Fifth International Conference on Intelligent Information Hiding and Multimedia Signal Processing, Kyoto, Japan, 12–14 September 2009; pp. 114–117.
124. Ehinger, B.V.; Groß, K.; Ibs, I.; König, P. A new comprehensive eye-tracking test battery concurrently evaluating the Pupil Labs glasses and the EyeLink 1000. *PeerJ* **2019**, *7*, e7086. [CrossRef]
125. Chen, Z.; Fu, H.; Lo, W.-L.; Chi, Z. Strabismus recognition using eye-tracking data and convolutional neural networks. *J. Healthc. Eng.* **2018**, *2018*, 7692198. [CrossRef]
126. Klaib, A.F.; Alsrehin, N.O.; Melhem, W.Y.; Bashtawi, H.O.; Magableh, A.A. Eye tracking algorithms, techniques, tools, and applications with an emphasis on machine learning and Internet of Things technologies. *Expert Syst. Appl.* **2021**, *166*, 114037. [CrossRef]
127. Blignaut, P.; Beelders, T.; Plessis, J.; Wium, D.; Brown, R. Demystifying the Black Box: From Raw Data to Applications. In Proceedings of the Conference on Eye Tracking South Africa, Cape Town, South Africa, 29–31 August 2013; pp. 1–18.
128. Lupu, R.G.; Ungureanu, F.; Siriteanu, V. Eye tracking mouse for human computer interaction. In Proceedings of the 2013 E-Health and Bioengineering Conference (EHB), Iasi, Romania, 21–23 November 2013; pp. 1–4.
129. Giordano, D.; Pino, C.; Spampinato, C.; Pietro, M.D.; Reibaldi, A. Eye tracker based method for quantitative analysis of pathological nystagmus. In Proceedings of the 2011 24th International Symposium on Computer-Based Medical Systems (CBMS), Bristol, UK, 27–30 June 2011; pp. 1–6.
130. Saisara, U.; Boonbrahm, P.; Chaiwiriya, A. Strabismus screening by Eye Tracker and games. In Proceedings of the 2017 14th International Joint Conference on Computer Science and Software Engineering (JCSSE), Nakhon Si Thammarat, Thailand, 12–14 July 2017; pp. 1–5.
131. Kumar, D.; Dutta, A.; Das, A.; Lahiri, U. SmartEye: Developing a novel eye tracking system for quantitative assessment of oculomotor abnormalities. *IEEE Trans. Neural Syst. Rehabil. Eng.* **2016**, *24*, 1051–1059. [CrossRef]
132. Pueyo, V.; Castillo, O.; Gonzalez, I.; Ortin, M.; Perez, T.; Gutierrez, D.; Prieto, E.; Alejandre, A.; Masia, B. Oculomotor deficits in children adopted from Eastern Europe. *Acta Paediatr.* **2020**, *109*, 1439–1444. [CrossRef]
133. Blignaut, P.; van Rensburg, E.J.; Oberholzer, M. Visualization and quantification of eye tracking data for the evaluation of oculomotor function. *Heliyon* **2019**, *5*, e01127. [CrossRef]
134. Schlegelmilch, K.; Wertz, A.E. The effects of calibration target, screen location, and movement type on infant eye-tracking data quality. *Infancy* **2019**, *24*, 636–662. [CrossRef]
135. Gavas, R.D.; Roy, S.; Chatterjee, D.; Tripathy, S.R.; Chakravarty, K.; Sinha, A. Enhancing the usability of low-cost eye trackers for rehabilitation applications. *PLoS ONE* **2018**, *13*, e0196348. [CrossRef] [PubMed]
136. Sasson, N.J.; Elison, J.T. Eye tracking young children with autism. *J. Vis. Exp.* **2012**, 3675. [CrossRef]
137. Vadillo, M.A.; Street, C.N.H.; Beesley, T.; Shanks, D.R. A simple algorithm for the offline recalibration of eye-tracking data through best-fitting linear transformation. *Behav. Res. Methods* **2015**, *47*, 1365–1376. [CrossRef] [PubMed]
138. Nyström, M.; Andersson, R.; Holmqvist, K.; van de Weijer, J. The influence of calibration method and eye physiology on eyetracking data quality. *Behav. Res. Methods* **2013**, *45*, 272–288. [CrossRef] [PubMed]
139. Sheynikhovich, D.; Bécu, M.; Wu, C.; Arleo, A. Unsupervised detection of microsaccades in a high-noise regime. *J. Vis.* **2018**, *18*, 19. [CrossRef] [PubMed]
140. Nyström, M.; Niehorster, D.C.; Andersson, R.; Hooge, I. The Tobii Pro Spectrum: A useful tool for studying microsaccades? *Behav. Res. Methods* **2020**. [CrossRef]
141. Krejtz, K.; Duchowski, A.T.; Niedzielska, A.; Biele, C.; Krejtz, I. Eye tracking cognitive load using pupil diameter and microsaccades with fixed gaze. *PLoS ONE* **2018**, *13*, e0203629. [CrossRef]
142. Yarosh, O. Neurobranding in territorial development: From traditional to innovative. In Proceedings of the International Scientific-Practical Conference "Business Cooperation as a Resource of Sustainable Economic Development and Investment Attraction" (ISPCBC 2019), Pskov, Russia, 21–23 May 2019.
143. Niehorster, D.C.; Andersson, R.; Nyström, M. Titta: A toolbox for creating PsychToolbox and Psychopy experiments with Tobii eye trackers. *Behav. Res. Methods* **2020**, *52*, 1970–1979. [CrossRef]
144. Ding, N. The effectiveness of evacuation signs in buildings based on eye tracking experiment. *Nat. Hazards* **2020**, *103*, 1201–1218. [CrossRef]
145. Banire, B.; Al Thani, D.; Qaraqe, M.; Mansoor, B.; Makki, M. Impact of mainstream classroom setting on attention of children with autism spectrum disorder: An eye-tracking study. *Univ. Access Inform. Soc.* **2020**. [CrossRef]

146. Katona, J.; Kovari, A.; Costescu, C.; Rosan, A.; Hathazi, A.; Heldal, I.; Helgesen, C.; Thill, S.; Demeter, R. The examination task of source-code debugging using GP3 eye tracker. In Proceedings of the 2019 10th IEEE International Conference on Cognitive Infocommunications (CogInfoCom), Naples, Italy, 23–25 October 2019; pp. 329–334.
147. Gibaldi, A.; Sabatini, S.P. The saccade main sequence revised: A fast and repeatable tool for oculomotor analysis. *Behav. Res. Methods* **2020**, *53*, 167–187. [CrossRef]
148. Van Renswoude, D.R.; Raijmakers, M.E.J.; Koornneef, A.; Johnson, S.P.; Hunnius, S.; Visser, I. Gazepath: An eye-tracking analysis tool that accounts for individual differences and data quality. *Behav. Res. Methods* **2018**, *50*, 834–852. [CrossRef] [PubMed]
149. Ghose, U.; Srinivasan, A.A.; Boyce, W.P.; Xu, H.; Chng, E.S. PyTrack: An end-to-end analysis toolkit for eye tracking. *Behav. Res. Methods* **2020**. [CrossRef]
150. Khandekar, R.; Al Harby, S.; Abdulmajeed, T.; Helmi, S.A.; Shuaili, I.S. Validity of vision screening by school nurses in seven regions of Oman. *East Mediterr. Health J* **2004**, *10*, 528–536.
151. Donaldson, L.A.; Karas, M.; O'Brien, D.; Woodhouse, J.M. Findings from an opt-in eye examination service in English special schools. Is vision screening effective for this population? *PLoS ONE* **2019**, *14*, e0212733. [CrossRef] [PubMed]
152. Yawn, B.P.; Lydick, E.G.; Epstein, R.; Jacobsen, S.J. Is school vision screening effective? *J. Sch. Health* **1996**, *66*, 171–175. [CrossRef] [PubMed]
153. Ethan, D.; Basch, C.E. Promoting healthy vision in students: Progress and challenges in policy, programs, and research. *J. Sch. Health* **2008**, *78*, 411–416. [CrossRef] [PubMed]
154. Campbell, L.R.; Charney, E. Factors associated with delay in diagnosis of childhood amblyopia. *Pediatrics* **1991**, *87*, 178–185.
155. Alvarez-Peregrina, C.; Sánchez-Tena, M.Á.; Andreu-Vázquez, C.; Villa-Collar, C. Visual health and academic performance in school-aged children. *Int. J. Environ. Res. Public Health* **2020**, *17*, 2346. [CrossRef] [PubMed]
156. Ali, Q.; Heldal, I.; Helgesen, C.G.; Krumina, G.; Tvedt, M.N. Technologies supporting vision screening: A protocol for a scoping review. *BMJ Open* **2021**, *11*, e050819. [CrossRef] [PubMed]

Article

Application Experiences Using IoT Devices in Education

Jan Francisti [1], Zoltán Balogh [1,*], Jaroslav Reichel [1], Martin Magdin [1], Štefan Koprda [1] and György Molnár [2]

[1] Department of Informatics, Constantine the Philosopher University, 94974 Nitra, Slovakia; jan.francisti@ukf.sk (J.F.); jreichel@ukf.sk (J.R.); mmagdin@ukf.sk (M.M.); skoprda@ukf.sk (Š.K.)
[2] Department of Technical Education, Budapest University of Technology and Economics, Műegyetem rkp.3, 1111 Budapest, Hungary; molnar.gy@eik.bme.hu
* Correspondence: zbalogh@ukf.sk

Received: 16 September 2020; Accepted: 16 October 2020; Published: 18 October 2020

Abstract: The Internet of Things (IoT) is becoming a regular part of our lives. The devices can be used in many sectors, such as education and in the learning process. The article describes the possibilities of using commonly available devices such as smart wristbands (watches) and eye tracking technology, i.e., using existing technical solutions and methods that rely on the application of sensors while maintaining non-invasiveness. By comparing the data from these devices, we observed how the students' attention affects their results. We looked for a correlation between eye tracking, heart rate, and student attention and how it all impacts their learning outcomes. We evaluate the obtained data in order to determine whether there is a degree of dependence between concentration and heart rate of students.

Keywords: internet of thing (IoT); eye tracking; heart rate (HR); measurements; data analysis

1. Introduction

The Internet of Things (IoT) phenomenon plays an important role in many areas today. Every year, the number of devices that we could consider as elements of the Internet of Things is growing. The Internet of Things can be understood as a large network with different connected types of objects, able to communicate with each other and exchange information, regardless of whether they belong to the same group. The network creation consisting of mutually communicating devices provides the user with the possibility of more effective management of all connected devices. The Internet of Things allows objects to be captured and controlled remotely over existing network infrastructures, creating opportunities for further direct integration of the physical world into computer systems, resulting in increased efficiency, accuracy and economic benefits [1]. IoT is augmented by sensors and actuators, making it a cyber-physical system that includes technologies such as smart grids, smart homes, intelligent transport, smart cities, and also smart education [2,3].

The choice of measurement methods and sensors is a complex process in which a large number of questions are asked. Physiological parameters can be measured in the same way as the physical principles of signal acquisition. The measurement technology in relation to the individual sensors creates a huge number of choices.

Common devices connected to the Internet provide us with a wealth of information and data. Such devices include, for example, ordinary smart wristbands, smart mobile devices, eye tracking cameras, etc. These devices were used to look for relationships between the physiological state of the student and the impact on his results.

The aim of the article is to point out the relationship between the concentration we measured with eye tracking technology and the excitement we measured with heart rate (HR) as a stress measurement tool and how it all affects the results of the study. We are looking for dependencies (connections) between physiological functions and students' knowledge. In this article, we try to find the relationships between the values of test results in students and their heart rate during testing. We assume that during testing, the level of stress may increase and thus the heart rate will also increase. The dependence of heart rate and stress has already been studied by many authors [4–9]. We also reveal the dependence of the student's heart rate before learning that they would be tested and during testing, and we monitored this change. We have formulated all these questions into research questions, which are defined in the article.

2. Related Work

Eye tracking has been used in the last decade, especially in the field of neuromarketing [10–16]. Recently, research has emerged pointing to the possibilities of its application and use in the educational process [17–20] or virtual reality [21]. In terms of using eye tracking, data about the so-called saccades and fixations, i.e., how long and where the person is looking, are important for researchers. In the case of neuromarketing, it is primarily about understanding human behavior through eye tracking studies related to recognizing the needs of consumers in marketing products [22].

In the case of the educational process, the interest of researchers is focused primarily on understanding the internal experience of a particular situation and its impact on learning outcomes through a set of learning styles [23].

The study of human behavior through eye tracking is a growing multidisciplinary field that combines electronics, psychology and cognitive sciences in connection with the study of human behavior, especially in problem solving or decision making and learning analytics [22]. Nguyen [24] describes that learning analytics (LA) involves the analysis of cognitive, social, and emotional processes in learning scenarios in order to make informed decisions about the design and provision of teaching.

According to Sharma [25], the success of acquiring a wealth of knowledge and skills during the educational process is influenced by many factors. For example, students' efforts to complete a task are often a manifestation not only of how they are assessed during the learning process but also of how actively they engage in individual learning activities based on different levels of motivation. It is possible (very difficult) for students' efforts to be observed directly, for example in the case of an online learning environment. Multimodal data from online learning environments can provide further insight into learning processes and can estimate their efforts. Sharma explored various approaches to the classification of efforts in the context of adaptive assessment. Using the eye tracking method, he evaluated the behavior of 32 students during the adaptive self-assessment activity. Using this method, the author predicted the students' effort to complete the upcoming task based on the discovered patterns of behavior using a combination of the hidden Markov model and the Viterbi algorithm. In his research, Sharma has shown that eye tracking and HMM can identify moments in providing preventive/normative feedback to students in real-time by building a relationship between patterns of behavior and the efforts of learners.

The use of eye tracking in the educational process is really broad-based. For example, Davis and Zhu [26] used eye tracking to gain an overview of how future programmers learn to read and process code. According to Davis and Zhu, there is a significant need to improve methodologies with active practical training techniques for programmers to learn practical strategies to mitigate software vulnerabilities; for the protection of private data; and finally write the secure code in the first place. They recorded the eye movements of future programmers as they read the code, thus removing problem areas in the source code. The study involved 29 students mitigating (eliminating) software bugs using manual source code analysis. Eye tracking data allowed them to objectively study the behavior of future programmers and at the same time gain an overview of their knowledge. The analysis of the study suggests that there is a difference in the learning phase for students who answered correctly compared to students who did not provide the right mitigation strategy. In particular, research suggests that it is possible to use the eye tracking method to understand students' behavior so that we are able to develop improved practical teaching materials.

Latini [27], by deploying the eye tracking method, investigated the effects of the medium on reading (printed vs. digital) in the context of processing and understanding the illustrated text. The participants were 100 university students enrolled in natural science post-graduate programs. The results showed that participants who read the illustrated text on paper showed more integrated processing during reading than participants who read exactly the same text on a computer. Integrative processing had a positive effect on comprehension, resulting in a mediated effect of the reading medium on comprehension through integrative processing. The reading medium did not have a major impact on integrated comprehension.

According to Emerson [28], a characteristic feature of online multimedia learning and game-based environments is the ability to create learning experiences that are effective and engaging. Advances in sensor-based technologies, such as facial expression analysis and vision tracking, have provided an opportunity over the last 10 years to use multimodal data streams to teach analytics. Teaching analytics works with multimodal data captured during students' interactions with playful learning environments and is an important promise for developing a deeper understanding of game-based teaching, detecting online student behavior, and supporting individualized learning.

Therefore, together with the eye tracking method, other types of sensors for sensing student activity (EEG, ECG, HR, temperature, GSR, etc.) are also used for detailed learning analytics. Most often, such tools must meet strict conditions. One of them is that they must not under any circumstances restrict the movement of the student in any way, so they must be a natural (technical and technological) part of the educational process. One way to ensure this is to use a wristband.

Zhang [29] explains that the rapid development of wristband has enabled the continuous recording of neurophysiological signals in natural environments, such as the classroom. The study conducted by Zhang aimed to examine the neurophysiological correlates of the academic results of high school students. The study used electrodermal signals (EDA) and heart rate (HR), which were collected through a wristband from 100 students during daily Chinese and math classes for 10 days over two weeks. Significant correlations were found between academic performance, reflected in students' final examination scores, and EDA responses. A study conducted by Zhang provides evidence to support the feasibility of predicting learning outcomes using wristbands through neurophysiological records.

The results of the Zhang study are also confirmed by Fortenbacher [30] which states that the latest advances in sensor technology make it possible to examine students' emotional and cognitive states. Obtaining data from sensors is a complex effort, all the more so when considering physiological data to support learning. Fortenbacher has developed a comprehensive solution for an adaptive learning system using data from GSR and HR sensors. To obtain data, the author used a wristband, which obtains physiological and environmental data, for which he completed a tablet application (SmartMonitor) for monitoring and visualizing data from the wristband. For learning analytics, he created a learning analysis backend that processes, and stores sensor data obtained from SmartMonitor and learning applications using these features. During his research, Fortenbacher found that while the possibilities of using physiological data for learning analytics are very promising, their use in a real learning environment requires interdisciplinary research implying combination of pedagogy, psychology, and several areas of computer science.

Motivation

Early detection of students' attention (concentration) can change the success (effectiveness) of studying and the way of teaching. For this reason, it is important to test the reliability (dependencies) of the procedures used to determine students' attention and to measure the influence of external factors, such as stressful situations (filling in quiz questions over time), on changes in data collected from sensory networks [31,32]. Students are an interesting group to study the effects of stressful situations because they are faced with a variable number of stressful situations in their daily activities in a short time.

The research was also motivated by previously conducted experiments focusing on the use of commonly available devices to measure the physiological function of the heart rate in order to identify stress or other emotions.

Jha, Prakash and Sagar [33] used a smart wristband and a sensory network to identify the emotion of anger. They used heart rate and body temperature sensors in the research. Prior to the research, both sensors were calibrated. Subsequently, they defined the limit values for heart rate and temperature, on the basis of which the anger scale could be determined. As a result of the testing, the authors found that the heart rate and also the skin temperature increased when the anger was manifested.

Several studies [34–37] used wristbands to measure heart rate during sports. Athletes wore wristbands and performed the assigned task. They created a comprehensive IoT system that was able to receive data from sensors and send it for evaluation.

These articles motivated us to test the sensory networks and to establish whether they are able to detect the presence of stress in any way and whether with the help of these devices it is possible to explore students' attention, thus affecting their results.

3. Material and Methods

The aim of the article is to find connections between the possibilities of individual devices connected to the Internet (i.e., IoT, eye tracking technology) and the student's concentration influenced by physiological conditions. The experiment was performed on a sample of students who had to answer as many questions as possible in 3 min. The time was set empirically so that the time quantum allocated to answer the questions evokes competition among students and thus indirectly the middle situation and the change (expected increase) of HR. The relationship between stress and HR has already been described by the authors in their research [3,4]. A smart wristband was used to monitor the participants' physiological conditions, measuring their heart rate. Correct answers and overall control were indicated by using eye tracking technology. The task of the research was to find the dependence between physiological functions (with which we can identify stress and calm condition) and the results of students' knowledge, i.e., how concentration and heart rate affect the overall result of students. We are looking for dependencies between the results of eye tracking technology, HR and their impact on students' knowledge.

The testing methodology consisted of several steps:

1. Create a quiz—LMS Canvas was used to create the quiz activity;
2. Create a question bank in the quiz—The created question bank contained 40 questions, consisting of four categories, which are described in the continuation of the chapter;
3. Choosing a wristband capable of measuring heart rates—The choice of smart wristband consisted of a comparison of several manufacturers with a licensed device for use in medicine;
4. Smart Watch Sensor Setup—each student had a wristband calibrated according to the manufacturer's recommendations;
5. Adjust the eye tracker sensor—calibration of the device on the Tobii EyeX Eye Tracker was performed for each student according to the manufacturer's instructions;
6. Start quiz with a predefined time—The quiz time was set at 3 min, in order to indirectly induce stress and competition for students to achieve the highest possible rating. For time synchronization, the same times were set for both sensors, and at the same time after starting the quiz, heart rate measurement was started, as well as eye tracking. At the same time as quiz activity was stopped, heart rate measurement and eye tracking were stopped;
7. Quiz evaluation—A log from Canvas was used to evaluate the quiz, where the steps of completing the quiz were analyzed in detail;

8. Export the wristband data—The data obtained from the watch was automatically sent to the mobile device and then to the cloud;
9. Evaluation of exported data—the obtained data were synchronized with each other based on the setting of the same time for both sensors, while the heart rate is a connected function and eye tracking is a discrete or separate function. Before synchronization, the times were adjusted to Unix time for all sensors;
10. Interpretation of results—the results of the experiment are described in more detail in the Results chapter.

This section describes how the eye tracker can be connected with smartwatch from Apple (Apple Inc., Cupertino, CA, USA) for the purpose of analyzing students' heart rate. Students from the University of Novi Sad did online quiz on Canvas.

Canvas is an online tool for teaching and learning. Instructors on Canvas can share course content, post assignments and grades, conduct quizzes and collect online submissions. It was developed by Instructure Inc. in order to support the continued development of a new learning management system (LMS).

The aim of this experiment was to measure and compare students' heart rate, which were gained from Apple Watch devices. Students attempted to answer the quiz, and they needed to answer as many questions as they could in a period of 3 min. The purpose of the time limit was to put pressure on students in order to see from heart rate results if there was any difference between questions at the start and questions that were done in the last 30 s or so.

The first step was to create the quiz for the participants to take. LMS Canvas was used because of using detail logs for every student attempt, measured in seconds. Logs were important because the Apple Watch can generate a report based on measures every 2–3 s. Also, we could use Quiz Statistics to get some useful information. The quiz has 40 questions that represent different scientific areas, like mathematics, geography, biology, and informatics, i.e., Halley's comet orbits the sun every 65, 75, or 85 years. Students answered different types of questions, such as multiple choice (one correct), true/false questions, and fill in the blank questions. For every correct answer, students earn one point. They needed to answer as many questions as they could within a time limit of 3 min. The test was set for three minutes in order to indirectly induce stress and competition in the students, as the goal was to get as many correct answers as possible. The time was also set to three minutes so that students would not go through all the questions even if they clicked through the quiz. The quiz had the shuffle answers option, which was enabled; thus, the questions were not repeated.

3.1. Eye Tracking Data

Eye tracking deals not only with the movement of the eye based on its physiology but also with a condition where the eye remains in one place for a certain period of time. Devices that detect the movements of the eye and its states monitor when the eye temporarily stops, e.g., on just a word while reading. This is called fixation: the eye is fixed at one point and stopped moving. Such a stop can last from 10s of milliseconds to several seconds or more. The length of fixation is influenced by many factors, including the size of the study area. Really fast eye movements between points are referred to as saccadic movements or saccades [38].

Students answered the questions while using the eye tracking device, Tobii EyeX. They needed to navigate with their eyes to answer questions. In the background, using Visual Studio, an application measured and converted the coordinates of student view into a red ellipse, which represents the students' point of view, shown in Figure 1.

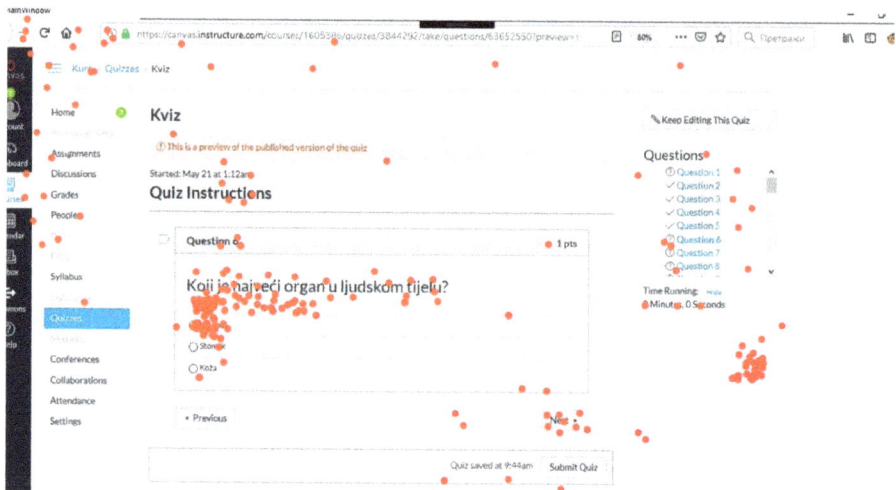

Figure 1. Student's eye activity during the quiz.

Eye movement was recorded at intervals of an average of 1.5 recordings per second, so that 270 recordings could be obtained during the quiz, which lasted three minutes. Segmentation of images using the nearest squares method was used to determine which places the student was looking at during the quiz.

$$J = \sum_{j=1}^{k} \sum_{i=1}^{n} \left\| x_i^{(j)} - c_j \right\|^2 \qquad (1)$$

where J is objective function, k is number of clusters, n is number of cases, x_i is i-th case and c_j is centroid for cluster j.

The method was created using the Python programming program using the OpenCV library. The individual records in the form of points were divided into several clusters based on the nearest squares' method. In the research, we divided the records into several parts, the place of the question, the place of the "next" button and other records. The same method was repeated for each student [39–41].

3.2. Heart Rate Measurement

As a sensor for measuring the heart rate of students, we have chosen a heart rate sensor built into the smartwatch from Apple. In order to get the most accurate data from the watch for setting the quiz time, we put the students a few minutes early so that they were not upset by the event. After a few minutes, the students could start the quiz and then we recorded the time when they started. On the watch, we have activated an activity application that allows them to measure heart activity at intervals of 3–6 s. The quiz itself was set to 3 min for each student. After the quiz time has expired, we stopped the app and recorded a quiz ending time according to the Canvas log to determine which final value of heart rate to include in the analysis for a particular student. We repeated the same procedure for each student.

3.3. Export Measured Data

The measured data was stored in the Health application on the mobile device paired with the watches. The Health application allows us to view measured data as well as export it in csv format. With the function export health data, we exported all data to a computer, where they were subsequently processed. The exported file was opened with the Excel tool from Microsoft, which contained all the data that the watch records (image).

The data we needed for research was written under the record type HK Quantity Type Identifier Heart Rate. As all the data were together in one column, the data had to be cleaned. We performed the cleaning function by changing the text to columns and we received the data in the table. Subsequently, we removed unnecessary listings, such as the date (as we did the experiment on the same day), as well as the end time, because heart rate measurements were made in one second. We also removed the word parts of the notation to leave only numeric values. The heart rate measurement time was formatted as time type.

3.4. Data Analysis

After cleaning the data, Table 1 contained the heart rate measurement time (Time in Sec) and heart rates per minute (Value).

Table 1. Example of Heart rate recording.

Time in Sec	Value
00:00:03	93
00:00:06	94
00:00:11	103
00:00:20	94
00:00:23	95
00:00:29	91
00:00:35	97
00:00:08	96
00:00:43	96
00:00:47	96
...	...

The quiz was set to 3 min, and time for each student was recorded when the quiz started and finished. Based on the time of the beginning and ending quiz, we were able to assign which values belong to a particular student. Since the time was written in the form when a particular value was measured, it had to be recalculated. The new time was calculated as the time difference obtained from the watch and the time that was written to the table. Based on time calculation we got the opportunity to compare the heart rate for a particular time. The LMS Canvas for the quiz tool was used for test, and the logs contained details of the quiz's progress. We could compare the watch data and the quiz that way. Since we recalculated the time set from zero to three minutes in the watch data table, we had the opportunity to analyze in more detail what the student was doing in the quiz when the heart rate value suddenly changed.

4. Results

During our experiment, we measured the heart rate of students before and during the quiz. Our goal was to verify whether a particular course activity would affect their heart rate. Fifteen students were tested. They dealt with a general knowledge quiz. We verified whether there was any dependence between the overall quiz result and

1. average heart rate during the quiz;
2. heart rate before testing;
3. heart rate at the start of testing.

Table 2 shows a sample of students who were tested. Table 2 shows the measured values of AVG HR, which is the average value of the student's heart rate during the test; Pretest HR, which is the heart rate before testing; Start HR, which is the heart rate at the start of testing, and Total score, which is the final score of the student from the test.

Table 2. Research data about students

User ID	AVG HR	Pretest HR	Start HR	Total Score
4	95	76	93	6
5	93	95	93	1.33
6	93	103	98	0.67
7	100	107	106	0.67
8	108	102	114	0.67
9	92	92	101	4
10	75	94	84	2
13	73	73	70	6
14	105	85	111	2
15	101	103	116	3
17	79	82	89	1
18	109	111	110	0
19	97	103	101	4
20	89	65	88	1
21	91	92	86	4

Indices numbered 1 and 2 have been omitted in order to test the correct operation of the wristband used to measure the heart rate, as well as the device for recording eye activity. Similarly, for index 12, the same device testing was performed.

The research aimed to verify whether there are significant dependencies between the results of these measurements (Table 2) and the score obtained by the students. The research questions, which we verified during the experiment, were also based on this aim.

The research questions were based on the dependence between these values in relation to the score obtained by the student.

Research questions:

Q (1): Does the student's test result depend on his/her average heart rate during the test?

Q (2): Does the student's test result depend on the heart rate the student has before he/she learns that he/she will be tested?

Q (3): Does the student's test result depend on the heart rate that the student has at the beginning of the test?

These research questions were converted into hypotheses, which were statistically verified in the analysis. We used correlation analysis to verify these dependencies. This dependence was verified by means of a correlation analysis that expresses the tightness of the statistical dependence between quantitative variables. It is the linear dependence rate of two variables. The correlation coefficient between two variables was calculated by the relation

$$\rho_{XY} = \frac{\sigma_{XY}}{\sqrt{\sigma_X^2 \sigma_Y^2}} \qquad (2)$$

where σ_{XY} is covariance. By the ratio of covariance and standard deviations, we get a correlation coefficient between −1 and 1. Where −1 is indirect proportionality, 1 is direct proportionality and 0 is the independence of X and Y [42].

By applying a correlation analysis using the Spearman correlation coefficient, we have obtained the results shown in Table 3. Spearman's rank correlation coefficient is calculated according to the formula:

$$r_s = 1 - \frac{6}{n(n^2-1)} \sum_{i=1}^{n} (R_i - Q_i)^2, \qquad (3)$$

where R_i is the rank in the arrangement and Q_i is the rank of the values in an ordered arrangement. This coefficient is calculated from the rank of the ordered values.

Table 3. Spearman correlation.

	Valid N	Spearman R	T(N-2)	p-Value
AVG HR and Score	15	−0.404349	−1.59402	0.134945
Pretest HR and Score	15	−0.519935	−2.19462	0.046959
Start HR new and Score	15	−0.374899	−1.45806	0.168556

This non-parametric statistical method was used because the variables *AVG HR*, *Pretest HR*, and *Start HR* contain relatively few values and do not have a normal distribution. We verified this based on the Lilliefors Test for Normality. In all three cases, the *p*-value was higher than 0.2. Based on these results, we can verify the null hypotheses, which were converted from research questions:

Hypothesis 1 (H1). *There is no statistically significant correlation between the student's average heart rate during the quiz and his/her score.*

Hypothesis 2 (H2). *There is no statistically significant correlation between the student's heart rate before testing and his/her score.*

Hypothesis 3 (H3). *There is no statistically significant correlation between the student's heart rate at the start of testing and his/her score.*

From the above results, we can see that (H1) and (H3) are not rejected. Thus, there is no statistically significant relationship between the result score and the average heart rate or heart rate at the start of testing. However, we also see that based on the *p*-value for the hypothesis 2 (H2), we can reject (H2), which means that there is a statistically significant relationship between total scores and heart rate before testing. The correlation coefficient has a value of −0.52, i.e., a relatively high negative dependence. This means that if a student had a high heart rate before the quiz, then their score was low compared to a student who had a lower heart rate before testing. The dependence in the case of the variable Pretest HR against the Score is not very significant and we would classify it as a large dependence. It is also statistically significant at the 5% significance level.

Higher heart rate in this case probably means higher student stress. Stress may be due to the very fact that the student will be tested or if the student fails the quiz. The results show that if the heart rate before the quiz is higher, i.e., the student is under stress, the quiz result will be worse.

The data obtained from observing students' eyes when solving quizzes can tell how much the students were focused on solving the given tasks. The system recorded the direction of the student's gaze three times in 2 s. This can identify quite accurately how long the student watched which part of the screen. For this purpose, we divided the screen into zones shown in Figure 2.

- *Question*—the area of the screen where the text of the question and answer was located,
- *Next button*—the area of the Next button that was used to switch the screen to the next question,
- *Other*—the area outside the *Question* and *Next button* areas,
- *Out of screen*—unrecorded points from eye tracking (since each student had 3 min to solve the question, which is 270 records, so this value was created by calculating 270—(*Question* + *Next button* + *Other*).

Figure 2. Areas of research (interest).

Table 4 contains the numbers of red dots (eye tracking records) that were included in the previously described zones. These values are listed for each student (*User ID*) separately.

Table 4. Eye tracking records.

User ID	Question	Next Button	Other	Out of Screen
4	85	8	86	91
5	122	9	123	16
6	122	17	103	28
7	51	0	187	32
8	87	9	70	104
9	79	47	91	53
10	79	25	141	25
13	109	8	132	21
14	71	26	126	47
15	75	6	34	155
17	78	8	123	61
18	104	4	113	49
19	55	25	172	18
20	88	5	131	46
21	136	12	75	47

If the student focused mainly on the *Question* area, we assume that this student was more focused on solving the problem compared to a student whose value was lower. The values from the variable *Other* indicate a lower concentration rate. We assume that the *Out of screen* value, which speaks of how much time the student took his gaze off the screen, means the lowest level of concentration. We also judge this on the basis that the students were under time pressure and therefore did not have too much time to look around. The variable *Next button* was created so that we could also identify how much time students spent switching between questions. This information is not very revealing, as it focuses on an activity that the student has to perform but does not indicate concentration or non-concentration. However, the *Next button* was needed, for example, to more accurately determine the *Out of screen* values.

Using the above data, we determined whether there is a relationship between concentration (based on the above interpretation of eye tracking data) and *Start HR*. Specifically, we determined whether the degree of correlation between the variables *Out of screen* and *Start HR* is statistically significant. Since the variable *Out of screen* does not have a normal distribution and we work with a relatively small sample, it was necessary to use a nonparametric statistical method to evaluate the correlation.

The Kendall coefficient tau expresses the difference between the probability that the values of two variables are in the same order compared to the probability that the values are not in the same order [42]:

$$\tau = \frac{n_c - n_d}{n(n-1)/2} \qquad (4)$$

where n_c is the number of matching pairs and n_d is the number of mismatched pairs. Pair (x_i, y_i) and (x_j, y_j), where $i, j = 1, ..., n$ and $i \neq j$ is concordant if $x_i > x_j$ and $y_i > y_j$ or $x_i < x_j$ and $y_i < y_j$. A pair is discordant if $x_i > x_j$ and $y_i < y_j$ or $x_i < x_j$ and $y_i > y_j$. A pair is indecisive if $x_i = x_j$ or $y_i = y_j$.

The variables *Out of screen* and *Start HR* have a statistically significant correlation, as Kendall tau came out to 0.39, which in this case is a statistically significant dependence on the level of statistical significance of 0.05.

Based on the above, we can say that the heart rate that the student had at the beginning of the quiz had a statistically significant effect on its concentration during the quiz. The correlation turned out to be positive, which means that if the student had a higher HR, he often looked around and was not focused on solving the assigned tasks.

5. Discussion

When evaluating the relationship of the student's final score to heart rate at different stages of testing, we found an indirect dependence. Correlation coefficients expressing the relationships between heart rate and *Score* were in the range from −0.37 to −0.52. In any case, the variables *AVG HR* and *Start HR* are moderately correlated according to Cohen [43]. For the variable Pretest HR, this correlation is −0.52, which we consider being a large correlation and, based on our results, also statistically significant. Therefore, we can say that the result of the test largely depends on the heart rate of the student, but mostly on the student's heart rate before the test. The student's heart rate is a consequence of other physiological phenomena in the student, such as stress. However, the heart rate as an indicator of these changes and processes helps us to identify the student who will be able to pass the test successfully or unsuccessfully.

The knowledge we gained from the analysis of eye tracking in connection with the heart rate could be used in teaching to identify students who do not sufficiently concentrate on the lesson. If such students can be identified and that information can be passed on to the teacher, the teacher can adapt his or her teaching to increase the concentration of the group of students. These systems can also be deployed in real time.

6. Conclusions

In this article, we focused on commonly available IoT devices, which can be used to collect valuable information about the student's condition in a non-invasive way. This information has some revelatory value, enabling us to set up the next steps and procedures. Our experiment was performed on respondents (students) who passed the test in the specified time span (interval).

The research was divided into several stages described in the Materials and Methods section. Prior to the research itself, it was necessary to set up devices that were used as sensors to measure heart rate and eye activity. It was also necessary to set up a quiz through which the research was carried out. Based on the analysis of the results, it was found that students who achieved higher heart rate values before the quiz did not fully concentrate during the quiz, which indicated that, in addition to looking at the screen, they were looking outside and therefore achieved weaker quiz results. On the other hand, students who were calmer before taking the quiz had a higher level of concentration, looked at the screen longer, and achieved higher quiz results.

Using the data obtained from eye tracking and HR, we found out whether there is a dependence between the concentration on solving the problem of students and their HR. Based on the performed experiment, we can conclude that the heart rate that the student had at the beginning of testing had a statistically significant effect on his concentration during the test. The correlation turned out to be positive, which means that if a student had a higher HR, he often looked around and was not focused on solving the assigned tasks. In the future, we plan to create a comprehensive sensory system (IoT), with which we will be able to obtain and evaluate results in real time. These results will help students to adapt their studies and thus streamline the learning process. It will be possible to use the methodology in other areas as well, such as industry, smart homes, the automotive industry, etc.

Author Contributions: Conceptualization, J.F. and Z.B.; methodology, J.F. and Z.B.; investigation, J.F., M.M. and Z.B.; resources, J.F. and M.M.; data curation, J.R. and J.F.; writing—original draft preparation, J.F., J.R. and Z.B; writing—review and editing, J.F., Z.B., J.R., M.M, Š.K. and G.M.; visualization, J.F and J.R.; supervision, Z.B; funding acquisition, Z.B. All authors have read and agreed to the published version of the manuscript.

Funding: This research was funded by the Cultural and educational agency of the Ministry of Education of the Slovak Republic, grant number: KEGA 036UKF-4/2019, "Adaptation of the learning process using sensor networks and the Internet of Things."

Acknowledgments: This research has been supported by the Bolyai János Research Stipend.

Conflicts of Interest: The authors declare no conflict of interest.

References

1. Francisti, J.; Balogh, Z. Wireless Sensor Network as a Part of Internet of Things. In Proceedings of the 12th Internation Science Conference Distance Learning in Applied Informatics (DIVAI), Sturovo, Slovakia, 2–4 May 2018.
2. Francisti, J.; Balogh, Z. Identification of Emotional States and Their Potential. In *Advances in Intelligent Systems and Computing*; Springer: Bangkok, Thailand, 2019; Volume 924, pp. 687–696. [CrossRef]
3. Orosz, B.; Kovács, C.; Karuovic, D.; Molnár, G.; Major, L.; Vass, V.; Szűts, Z.; Námesztovszki, Z. Digital education in digital cooperative environments. *J. Appl. Tech. Educ. Sci.* **2019**, *9*, 55–69.
4. Cubillos-Calvachi, J.; Piedrahita-Gonzalez, J.; Gutiérrez-Ardila, C.; Montenegro-Marín, C.; Gaona-García, P.; Burgos, D. Analysis of stress's effects on cardiac dynamics: A case study on undergraduate students. *Int. J. Med. Inform.* **2020**, *137*, 104104. [CrossRef] [PubMed]
5. Slavich, G.M.; Taylor, S.; Picard, R.W. Stress measurement using speech: Recent advancements, validation issues, and ethical and privacy considerations. *Stress* **2019**, *22*, 408–413. [CrossRef] [PubMed]
6. Novani, N.P.; Arief, L.; Anjasmara, R.; Prihatmanto, A.S. Heart Rate Variability Frequency Domain for Detection of Mental Stress Using Support Vector Machine. In Proceedings of the 2018 International Conference on Information Technology Systems and Innovation(ICITSI), Bandung–Padang, Indonesia, 22–26 October 2018; pp. 520–525. [CrossRef]
7. Taelman, J.; Vandeput, S.; Spaepen, A.; van Huffel, S. Influence of mental stress on heart rate and heart rate variability. In *IFMBE Proceedings*; Springer: Antwerp, Belgium, 2008; Volume 22, pp. 1366–1369. [CrossRef]
8. Vanitha, L.; Suresh, G.R. Hierarchical SVM to detect mental stress in human beings using Heart Rate Variability. In Proceedings of the IEEE International Caracas Conference on Devices, Circuits and Systems (ICDCS), Combiatore, India, 6–8 March 2014. [CrossRef]
9. Chen, N.; Zhao, M.; Gao, K.; Zhao, J. The physiological experimental study on the effect of different color of safety signs on a virtual subway fire escape—An exploratory case study of zijing mountain subway station. *Int. J. Environ. Res. Public Health* **2020**, *17*, 5905. [CrossRef] [PubMed]
10. Hamelin, N.; Thaichon, P.; Abraham, C.; Driver, N.; Lipscombe, J.; Naik, M.; Pillai, J.N. Storytelling, the scale of persuasion and retention: A neuromarketing approach. *J. Retail. Consum. Serv.* **2020**, *55*, 102099. [CrossRef]
11. Paulus, Y.T.; Hiramatsu, C.; Syn, Y.K.H.; Remijn, G.B. Measurement of viewing distances and angles for eye tracking under different lighting conditions. In Proceedings of the 2nd International Conference on Automation, Cognitive Science, Optics, Micro Electro-Mechanical System, and Information Technology (ICACOMIT), Jakarta, Indonesia, 23–24 October 2017; pp. 54–58. [CrossRef]

12. Funke, G.; Greenlee, E.; Carter, M.; Dukes, A.; Brown, R.; Menke, L. Which Eye Tracker Is Right for Your Research? Performance Evaluation of Several Cost Variant Eye Trackers. In Proceedings of the Human Factors and Ergonomics Society Annual Meeting, Washington DC, USA, 8 September 2016; Volume 60, pp. 1240–1244. [CrossRef]
13. Zhao, Y.; Siau, K. Cognitive neuroscience in information systems research. *J. Database Manag.* **2016**, *27*, 58–73. [CrossRef]
14. Jason, V.L.; Stachecki, L.P.; Magee, J. Eye-gaze with predictive link following improves accessibility as a mouse pointing interface. In Proceedings of the 18th International ACM SIGACCESS Conference on Computers and Accessibility, Reno, NV, USA, 23–26 October 2016; pp. 297–298. [CrossRef]
15. Hensel, D.; Wolter, L.C.; Znanewitz, J. A Guideline for ethical aspects in conducting neuromarketing studies. In *Ethics and Neuromarketing: Implications for Market Research and Business Practice*; Springer International Publishing: Cham, Switzerland, 2016; pp. 65–87.
16. Khushaba, R.N.; Wise, C.; Kodagoda, S.; Louviere, J.; Kahn, B.E.; Townsend, C. Consumer neuroscience: Assessing the brain response to marketing stimuli using electroencephalogram (EEG) and eye tracking. *Expert Syst. Appl.* **2013**, *40*, 3803–3812. [CrossRef]
17. Shotton, T.; Kim, J.H. Assessing Differences on Eye Fixations by Attention Levels in an Assembly Environment. In *Advances in Intelligent Systems and Computing*; Springer: San Diego, CA, USA, 2020; Volume 1201 AISC, pp. 417–423.
18. Rotariu, C.; Costin, H.; Bozomitu, R.G.; Petroiu-Andruseac, G.; Ursache, T.I.; Cojocaru, C.D. New assistive technology for communicating with disabled people based on gaze interaction. In Proceedings of the 2019 7th E-Health and Bioengineering Conference (EHB), Iasi, Romania, 21–23 November 2019. [CrossRef]
19. Zhang, X.; Sugano, Y.; Bulling, A. Evaluation of appearance-based methods and implications for gaze-based applications. In Proceedings of the 2019 CHI Conference on Human Factors in Computing Systems, Glasgow, UK, 4–9 May 2019; pp. 1–13. [CrossRef]
20. Jones, P.R. Myex: A MATLAB interface for the Tobii Eyex eye-tracker. *J. Open Res. Softw.* **2018**, *6*. [CrossRef]
21. Komoriya, K.; Sakurai, T.; Seki, Y.; Takimoto, M.; Kambayashi, Y. User interface in virtual space using vr device with eye tracking. In *Advances in Intelligent Systems and Computing*; Springer: Paris, France, 2020; Volume 1253 AISC, pp. 316–321. [CrossRef]
22. Zamani, H.; Abas, A.M.; Amin, M. Eye Tracking Application on Emotion Analysis for Marketing Strategy. *J. Telecommun. Electron. Comput. Eng.* **2016**, *8*, 87–91.
23. Crescenzi-Lanna, L. Multimodal Learning Analytics research with young children: A systematic review. *Br. J. Educ. Technol.* **2020**, *51*, 1485–1504. [CrossRef]
24. Nguyen, H.; Ahn, J.; Young, W.; Campos, F. Where's the Learning in Education Crowdsourcing? In Proceedings of the Seventh ACM Conference on Learning@ Scale, Virtual Event, USA, 12–14 August 2020; Association for Computing Machinery: New York, NY, USA, 2020; pp. 305–308. [CrossRef]
25. Sharma, K.; Papamitsiou, Z.; Olsen, J.K.; Giannakos, M. Predicting learners' effortful behaviour in adaptive assessment using multimodal data. In Proceedings of the Tenth International Conference on Learning Analytics & Knowledge, Frankfurt, Germany, 23–27 March 2020; pp. 480–489. [CrossRef]
26. Davis, D.; Zhu, F. Understanding and improving secure coding behavior with eye tracking methodologies. In Proceedings of the 2020 ACM Southeast Conference (ACMSE), Tampa, FL, USA, 2–4 April 2020; pp. 107–114. [CrossRef]
27. Latini, N.; Bråten, I.; Salmerón, L. Does reading medium affect processing and integration of textual and pictorial information? A multimedia eye-tracking study. *Contemp. Educ. Psychol.* **2020**, *62*, 101870. [CrossRef]
28. Emerson, A.; Cloude, E.B.; Azevedo, R.; Lester, J. Multimodal learning analytics for game-based learning. *Br. J. Educ. Technol.* **2020**, *51*, 1505–1526. [CrossRef]
29. Zhang, Y.; Qin, F.; Liu, B.; Qi, X.; Zhao, Y.; Zhang, D. Wearable neurophysiological recordings in middle-school classroom correlate with students' academic performance. *Front. Hum. Neurosci.* **2018**, *12*. [CrossRef] [PubMed]
30. Fortenbacher, A.; Ninaus, M.; Yun, H.; Helbig, R.; Moeller, K. Sensor based adaptive learning—Lessons learned. In *Lecture Notes in Informatics (LNI), Proceedings*; Series of the Gesellschaft fur Informatik (GI); Gesellschaft fur Informatik (GI): Bonn, Germany, 2019; Volume P-297, pp. 193–198. [CrossRef]

31. Katona, J.; Kovari, A. Examining the Learning Efficiency by a Brain-Computer Interface System. *Acta Polytech. Hung.* **2018**, *15*, 251–280.
32. Katona, J.; Kovari, A. The Evaluation of BCI and PEBL-based Attention Tests. *Acta Polytech. Hung.* **2018**, *15*, 225–249.
33. Jha, V.; Prakash, N.; Sagar, S. Wearable anger-monitoring system. *ICT Express* **2017**. [CrossRef]
34. Cvetković, B.; Szeklicki, R.; Janko, V.; Lutomski, P.; Luštrek, M. Article in press real-time activity monitoring with a wristband and a smartphone. *Inf. Fusion* **2017**. [CrossRef]
35. De Zambotti, M.; Baker, F.C.; Willoughby, A.R.; Godino, J.G.; Wing, D.; Patrick, K.; Colrain, I.M. Measures of sleep and cardiac functioning during sleep using a multi-sensory commercially-available wristband in adolescents. *Physiol. Behav.* **2016**. [CrossRef]
36. Kwak, Y.H.; Kim, W.; Park, K.B.; Kim, K.; Seo, S. Flexible heartbeat sensor for wearable device. *Biosens. Bioelectron.* **2017**. [CrossRef]
37. Yong, B.; Xu, Z.; Wang, X.; Cheng, L.; Li, X.; Wu, X.; Zhou, Q. IoT-based intelligent fitness system. *Comput. J. Parallel Distrib. Comput* **2017**. [CrossRef]
38. Holmqvist, K.; Nyström, N.; Andersson, R.; Dewhurst, R.; Jarodzka, H.; van de Weijer, J. *Eye Tracking: A Comprehensive Guide to Methods and Measures*; Oxford University Press: Oxford, UK, 2011.
39. Tian, K.; Li, J.; Zeng, J.; Evans, A.; Zhang, L. Segmentation of tomato leaf images based on adaptive clustering number of K-means algorithm. *Comput. Electron. Agric.* **2019**, *165*, 104962. [CrossRef]
40. Khalid, M.; Pal, N.; Arora, K. Clustering of Image Data Using K-Means and Fuzzy K-Means. *Int. J. Adv. Comput. Sci. Appl.* **2014**, *5*. [CrossRef]
41. Dhanachandra, N.; Manglem, K.; Chanu, Y.J. Image Segmentation Using K-means Clustering Algorithm and Subtractive Clustering Algorithm. *Procedia Comput. Sci.* **2015**, *54*, 764–771. [CrossRef]
42. Munk, M. *Počítačová Analýza Dát*; Univerzita Konštantína Filozofa v Nitre: Nitra, Slovakia, 2011.
43. Cohen, J. *Statistical Power Analysis for the Behavioral Sciences*; Routledge: New York, NY, USA, 2013.

Publisher's Note: MDPI stays neutral with regard to jurisdictional claims in published maps and institutional affiliations.

© 2020 by the authors. Licensee MDPI, Basel, Switzerland. This article is an open access article distributed under the terms and conditions of the Creative Commons Attribution (CC BY) license (http://creativecommons.org/licenses/by/4.0/).

Article

Case Study: Students' Code-Tracing Skills and Calibration of Questions for Computer Adaptive Tests

Robert Pinter [1,*], Sanja Maravić Čisar [1], Attila Kovari [2], Lenke Major [3], Petar Čisar [4] and Jozsef Katona [5]

1. Department of Informatics, Subotica Tech—College of Applied Science, 24000 Subotica, Serbia; sanjam@vts.su.ac.rs
2. Department of Natural Sciences and Environmental Protection, Institute of Engineering, University of Dunaujvaros, 2400 Dunaújváros, Hungary; kovari@uniduna.hu
3. University of Novi Sad Hungarian Language Teacher Training Faculty, 24000 Subotica, Serbia; major.lenke@magister.uns.ac.rs
4. Department of Informatics, University of Criminal Investigation and Police Studies, 11080 Belgrade-Zemun, Serbia; petar.cisar@kpu.edu.rs
5. CogInfoCom based LearnAbility Reseacrh Team, University of Dunaujvaros, 2400 Dunaújváros, Hungary; katonaj@uniduna.hu
* Correspondence: pinter.robert@vts.su.ac.rs

Received: 31 August 2020; Accepted: 8 October 2020; Published: 11 October 2020

Featured Application: Authors are encouraged to provide a concise description of the specific application or a potential application of the work. This section is not mandatory.

Abstract: Computer adaptive testing (CAT) enables an individualization of tests and better accuracy of knowledge level determination. In CAT, all test participants receive a uniquely tailored set of questions. The number and the difficulty of the next question depend on whether the respondent's previous answer was correct or incorrect. In order for CAT to work properly, it needs questions with suitably defined levels of difficulty. In this work, the authors compare the results of questions' difficulty determination given by experts (teachers) and students. Bachelor students of informatics in their first, second, and third year of studies at Subotica Tech—College of Applied Sciences had to answer 44 programming questions in a test and estimate the difficulty for each of those questions. Analyzing the correct answers shows that the basic programming knowledge, taught in the first year of study, evolves very slowly among senior students. The comparison of estimations on questions difficulty highlights that the senior students have a better understanding of basic programming tasks; thus, their estimation of difficulty approximates to that given by the experts.

Keywords: computer adaptive testing; code tracing; basic programming skills

1. Introduction

Modern technologies offer numerous possibilities for improving knowledge assessment and the process of education. In higher education, testing is one of the most commonly used methods of measuring student knowledge. The main goal of testing is to determine the level of students' knowledge in one or more areas of the course material on which the test is based. In a class held in an online environment, self-testing via computer-based tests (CBT) provides feedback that shows the students how well they are progressing in the acquisition of knowledge or skills.

The most common form of computer tests are linear fixed-length tests. All students are given the same test; in fact, it "imitates" traditional pen-and-paper tests, representing their digital version. This type of test does not take into account the abilities of each individual student. While doing such a one-size-fits-all test, respondents may feel discouraged if the questions are too difficult, or, on the other hand, they may lose interest if the tasks are too easy for their level of knowledge. The solution to this problem may be the application of computer adaptive tests, which have the ability to change the level of question difficulty on the basis of the respondents' abilities similar to in an oral exam.

A computer adaptive test offers the ability to create a test based on the respondent's actual abilities. Questions are selected from a database, and the individual capabilities of the candidates are taken into account during the test. By applying the Item Response Theory (IRT) in CAT, different versions of the test are possible, e.g., if the respondent answers the question correctly, the following question is selected from a group of questions that are one level higher than the previous question. On the other hand, if the respondent's answer is incorrect, the next question is selected from a group of questions that are one level easier than the previous questions the respondent failed to answer correctly. Many different possibilities (algorithms) for selecting the next question can be applied. With this adaptive selection, the respondents with less knowledge will be given easier questions, while those better prepared for the test will receive a set of more difficult questions. Therefore, the end result of the test for each respondent also depends on the level of question difficulty that they answered correctly. Thus, individual students may have the same percentage of correct answers, but those who answered the more difficult questions will be given a higher grade [1].

There are several advantages of computer adaptive tests worth considering, when compared to traditional linear tests, such as

- individualization of tests,
- promoting a positive attitude toward testing,
- determining the level of knowledge with greater accuracy.

To develop a system with adaptive tests requires a database with many questions. Every question needs calibration, i.e., objective values indicating the question's difficulty. With the difficulty values determined properly, adaptive algorithms can then select the "right" next question in the test. One way of calibration is when the difficulty of the question is determined by the expert(s). Another way of calibration relies on a given population, e.g., a set of students. The large number of answers from participants for whom the adaptive tests will be created can provide information about the question's difficulty [2].

The primary goal of this study is to compare results of how experts (teachers) determined the difficulty of the questions with the values that were determined by the students. The authors aim to use the results and the experience acquired in this research in the further development of CS1 courses (introductory programming courses) and CAT systems. The secondary goal is to analyze what happens with the basic programming knowledge students are taught in their first year of study throughout the course of their education. In the case of students of informatics, does such knowledge develop, or will it be mainly forgotten in later years of study?

Writing a computer program is a difficult cognitive skill to master and it is also difficult to measure it [3,4]. Despite the best efforts of teachers, many students are still challenged by programming. In order to write a simple program, they need to possess a basic knowledge of variables, input/output of data, control structures, and other areas. Even given all necessary theoretical knowledge, they face a problem when having to apply their knowledge as a whole and actually write a programming code [5].

The authors used a code-tracing type of task and questions to assess students' programming skills. It has been confirmed that students cannot learn to write code without having previously learned to read/trace code as a precursory skill [6–8]. It must also be pointed out that being aware of how code functions is not the same as being able to use it in problem solutions [9].

The characteristics of code tracing can be summarized as follows:

- Tracing refers to following the flow and data progression of a given program.
- In many instances, tracing represents a single user's journey through a program or program snippet.
- Its purpose is not reactive, but instead, it focuses on optimization.

- By tracing, developers can identify bottlenecks and focus on improving performance.
- When a problem does occur, tracing allows the user to see how it came to be: which function, duration of a function, which parameters passed, and how deep into the function the user could delve.

2. Related Works

Problems that arise in novices when learning programming is a field that has been the subject of numerous studies, as have the ways of adopting those new concepts [10]. Xie et al. [11] proposed a theory that identified four distinct skills that novices learned incrementally. These skills were tracing, writing syntax, comprehending templates, and writing code with templates. They assumed that the explicit instruction of these skills decreased cognitive demand. The authors conducted an exploratory mixed-methods study and compared students' exercise completion rates, error rates, ability to explain code, and engagement when learning to program. They compared the learning material that reflected this theory to more traditional material that did not distinguish between the skills. The findings of their study were as follows: teaching skills incrementally resulted in an improved completion rate on practice exercises and decreased error rate and improved understanding on the post-test.

The report of a 2001 ITiCSE (Innovation and Technology in Computer Science Education) working group [12] assessed the programming ability of 216 post-CS1 students from eight tertiary institutions in various countries. The "McCracken group" used a common set of programming problems. The majority of students performed much worse than their teachers had expected. The average score was 22.89 out of 110 points. While such a report by an author at a single institution might be dismissed as a consequence of poor teaching at that particular institution, dismissing a multinational study is not done so lightly. Given the scale and the multinational nature of the collaboration, these results were widely viewed as significant and compelling. The McCracken study did not isolate the causes of the problem. A popular explanation for the students' poor performance was that they lacked the ability to problem-solve. In fact, students lack the ability to take a problem description, decompose it into sub-problems, implement the necessary steps, and then reassemble the pieces to create a complete solution. Based on the McCracken group research, an ITiCSE 2004 working group (the "Leeds Group") tested students from seven countries in two ways. First, students were tested on their ability to predict the outcome of executing a short piece of code. Next, the students were given the desired function of a short piece of near-complete code and tested on their ability to select the correct completion of the code out of a small set of possibilities. An alternative explanation is that many students have a weak grasp of the basic programming principles and were missing the ability to systematically carry out routine programming tasks, such as tracing through code [13].

This working group established that many students lacked knowledge and skills that are a precursor to problem-solving. These missing elements were more associated with the students' ability to read code than to write it. Many showed weakness in systematically analyzing a short piece of code. The working group did not argue that all students who manifested weakness in problem-solving were doing so due to reading-related factors. They accepted that a student who scored high on the type of tests used in this study, but was unable to write a novel code of similar complexity, was most likely suffering from a weakness in problem-solving. The working group merely stated that any research project aiming to study problem-solving skills in novice programmers had to include a mechanism to screen for subjects weak in the precursor, code reading-related skills.

Kopec et al. [14] analyzed programmers' examination errors but focused on intermediate programmers (i.e., with some programming experience and understanding of basic programming concepts). They concluded that educators had to pay careful attention to their problem description and presentation, since novices were easily confused by nested loops and recursion, which differentiated between intermediate programmers' and novice programmers' errors.

Code-tracing problems (e.g., debugging a program, identifying the output of a program) are attractive in that they can be solved in shorter stints of time, using formats such as multiple choice that are easier to grade [13]. In this context, Computer Science researchers were keen to see whether there was a correlation between students' performance on code-tracing problems and their ability to write code. Lopez et al. [6] studied the responses of novices in an examination and found strong support for a relation between their skills for code tracing and code writing, and between explaining code and writing code. They showed that there were strong correlations between code tracing and code writing.

The hypothesis of Kumar's [7] study was that tracing code would lead to an improvement in code-writing skills. The results of this study indicated that code-tracing activities helped students learn to write both syntactic and semantic components of code. However, the extend of the effect was found to be small to medium. Therefore, code-tracing exercises can be used as a supplement rather than a substitute for code-writing exercises.

Moreover, some other research areas present alternative technologies such as virtual reality (VR) [15,16], alternative approaches for teaching programming [17], brain–computer interface (BCI) systems [18,19], and serious games [20], which could support learning efficiency and help complement learning difficulties as well.

3. Research

The authors of this paper conducted a study among all three student cohorts of bachelor Informatics students at Subotica Tech—College of Applied Sciences. The research included 182 student participants who were taking an introductory course called 'Algorithms and data structures'. The course is composed of lectures, practices, and laboratory practices. The course runs in the spring semester over 15 weeks in the first year of study. It consists of one 90 min lecture per week as well as a 45 min practice and another 90 min lab practice.

The aim of the course is to introduce the students to the basic concepts of algorithms and data structures in C/C++ programming language. The course covers the basic principles of programming, variables, control statements (*if, while, do-while, break* and *continue* statements), arrays, functions, pointers, and some basic algorithms (sort and search).

Based on historical data, the grade average in this given course is 7.02 (5 is the minimum grade and means fail, and the maximum grade is 10), while the pass rate is about 50%.

During their studies, informatics students learn new developing methods and developing environments as well as new programming languages. The novel techniques help them solve more complex information and communications technology (ICT) problems. Although the problems are more complex, and the developing environment and the programming language are new to them, when solving a particular problem, students frequently rely on those basic algorithms and data structures that they learned in their first year of study, e.g., iterations, arrays, or conditional statements. The programming languages that students use in their second and third years are so-called C-like programming languages, so the syntax and logic of the language are highly similar to those covered in the course 'Algorithms and data structures' during the first year of study.

As stated earlier, the aim of this research is to analyze what happens to those basic programming skills acquired during the first year. The following questions arise:

1. Do computer science students forget the basic algorithms and elements of language, or because of using them actively, do they understand them even better?
2. What is the impact of the constantly developing programming skills when estimating the difficulty of basic tasks?

The authors formulated three hypotheses:

Hypotheses 1 (H1): *Students in later years of study have a better understanding of basic algorithms and data structures.*

Hypotheses 2 (H2): *Subjective difficulty estimation of one task at later years of study will be lower.*

Hypotheses 3 (H3): *The difference between the students' and teachers' subjective estimation decreases with the years of study.*

4. Data Collection and Analysis

The primary data collected were the students' answers in a test. A total of 182 students contributed data to this part of the study, but only 117 of those gave answers to all 44 questions.

The students in the second and third year of a study were tested in January 2020. The test was conducted as a paper-and-pencil test with limited time. During the test, students were not allowed to use computers, smart phones, or any other type of help in answering the questions. The test consisted of multiple-choice questions, so the students had to mark one or more of the given answers or write the output of the code snippet. The students were given a time limit of 90 min for completing the test.

The participants in Year 1 were given the same test in May 2020, but they had to complete it in an online environment due to the COVID-19 pandemic. Students took the test in the Moodle system. In all aspects, including questions themselves, their order, and the time limit, the online test was identical to the paper-and-pencil test.

Altogether, 182 records were collected. With the relatively short time limit and the large number of questions, even for students with an average programming language knowledge and tracing skills, completing this test correctly was not an easy task. The underlying reason for compiling a challenging test was to highlight the differences in individual skills. As a result of such a test design, there were students who did not answer all the questions either in the tracing part or the difficulty estimation of questions part.

4.1. Test Design

The test was composed of 44 tasks or questions. Out of these, 24 were multiple-choice questions with four possible answers, while 22 questions required the students to write down the output of the program or program snippet. The test questions were categorized based on two criteria: the subject matter of the question and the subjective estimation of the task's difficulty. Students filling in the test essentially had to do two things for each question: first, to solve the question as required by the task, then, to give their subjective estimate regarding the difficulty of the given question. In short, the authors asked them to do the task and then to indicate how easy or difficult (or medium) they found the task.

4.2. The Categories of the Questions.

Apart from skills tracing, the authors were also keen to explore students' programming knowledge; therefore, the test contained six types of tasks (i.e., six categories). The task categories were the following:

1. **Statements.** Through these questions, the authors aimed to explore the students' knowledge of various basic concepts of the C programming language. For example, students had to calculate an arithmetic problem with different operators, determine the correct names of variables, find syntax errors, etc. The "statement" category contained 12 questions.
2. **Conditional statements.** This category consisted of questions that used the *if, if-else, if-else-if:* (ternary operator), and switch structures. The conditions that the students had to examine/conclude ranged from simple to complex. Altogether, there were six conditional statements in this category.
3. **Iterations.** With the help of these iteration questions, the authors studied how the students would solve questions with *for, while,* and *do-while* cycles. The variation in task difficulty was achieved by adjusting the conditions of the cycle: they ranged from simple to complex. There were ten iteration-type questions in this category.

4. **Structures.** This category of questions focused on the use of arrays and two-dimensional arrays with integers and characters, although there were no tasks of data sorting. This category contained six questions.
5. **Functions.** Questions in this category were related to the knowledge of function arguments (call by value, call by reference) and the use of local, static, and global variables. There was a total of six function-related questions in the category.
6. **Recursion.** Questions in this category were consciously created as "difficult", so that they could reveal whether senior students still knew the principles of recursion functioning or would use other methods. There were four recursion questions in the category, three of which were basic, while the fourth question had double recursion.

The following list contains a few examples of categories, with the boxes for their level of difficulty that had to be ticked:

- Conditional statements

```
What is the output of the following code? Hard □ Medium □ Easy □
void main(){
int i = 0;
if((13 + 1)/2) {
if(13 > 5)
i+=5;
else
i++;
} else
i+=2;
cout<<i;
}
```

- Iteration statements

```
What is the output of the following code? Hard □ Medium □ Easy □

main (){
int i = 0;
int s = 9;
for (; i < s; i++)
s -= i;
    cout<<s;
}
```

- Data structures

```
What is the output of the following code? Hard □ Medium □ Easy □

void main(){
int j, i;
int ar[] = {1, 2, 3, 4, 5, 6, 11};
for(j = 5, i = 0 ; i < j ; j--,i++)
ar[i] = ar[j];
    i = ar[i];
    cout<<i;
}
```

- Functions

```
What is the output of the following code?  Hard □ Medium □ Easy □
void swap(int&, int&);
int main(void) {
int a = 10, b=20;
    swap (a, b);
    cout<<a<<"\t"<<b<<"\n";
    return 0;
}
void swap(int& x, int& y) {
    x+=2;
    y+=3;
```

```
}
    a.  14, 24
    b.  11, 21
    c.  10, 20
    d.  Nothing, there is an error in the code
    e.  12, 23
```

4.3. Estimation of the Question Difficulty

Students' subjective estimation of question difficulty is very useful feedback. Based on the students' answers, teachers gain information about which part of the study material the students understand the least. Furthermore, the teachers can rethink and revise their own estimation of the questions' difficulty. For example, a given question thought to be 'easy' by the teacher may be perceived as an unsolvable problem by the students themselves. This feedback can be used by teachers when preparing questions for the exam in order to only include questions with approximately the same difficulty. As mentioned above, a subjective estimation of question difficulty can also serve in computerized adaptive testing that adapts to the examinee's level of ability. This is advantageous from the examinee's perspective, as the difficulty of the exam seems to tailor itself to their level of ability. For instance, if an examinee performs well on an item of intermediate difficulty, they will then be presented with more difficult questions. Or, if they performed poorly, they will be presented with a simpler question.

The evaluation of task difficulty was conducted on two levels, by the participant teachers and students. The teachers involved in this study gave their subjective estimation both of the difficulty (how challenging a given task was) and complexity (how complicated the question was, how many steps it involved) of each question, resulting in two separate marks for every question. The underlying reason was to increase the objectivity and accuracy of difficulty estimation. This step was followed by a detailed professional discussion of the given marks, which were then transformed into the difficulty values of "easy", "medium" and "hard". While the teachers performed this double estimation, primarily to make the overall estimation more objective, such double estimation was not required from the students, because the authors believed it would prove too distracting for students, given that they had to pay close attention to actually solving the question and then evaluating its difficulty. In the test itself, the students were able to give a subjective estimation of the question difficulty by selecting one of the three offered options: "easy", "medium" and "hard", as seen above, next to the sample questions.

4.4. Participants

As stated before, the total number was 182 bachelor students of informatics. Table 1. shows the structure of the participating students by year of study.

Table 1. The number of students per year of study.

	n	%
First year	78	42.9
Second year	58	31.9
Third year	46	25.3
Total	182	100.0

Two professors from Subotica Tech, with more than two decades of experience in teaching computer science courses, have designed the test and conducted the data acquisition.

4.5. Comparison of the Performance in Each Year of Study

Comparing the test results for each year of study, the students from the first year gave the largest number of correct answers, averaging 25.73 out of 44. The students in Year 2 had an average of 21.21 points, while the weakest result came from the third-year students, who only achieved a point average of 17.39 (see Table 2.).

Table 2. Average of points.

Year of Study	Average Point	Standard Deviation (SD)
1st	25.73	6.1
2nd	21.21	7.2
3rd	17.39	6.0

The authors also examined the results achieved by the students on a question-by-question basis. The percentage of correct answers for each question is illustrated in Figure 1. The questioned revealed for which items the students' correct answer reached a level of 51% (the lower limit of the points to "just pass"). Those questions where the value did not surpass the 51% limit are marked (e.g., Q2, Q6, Q11, etc.)

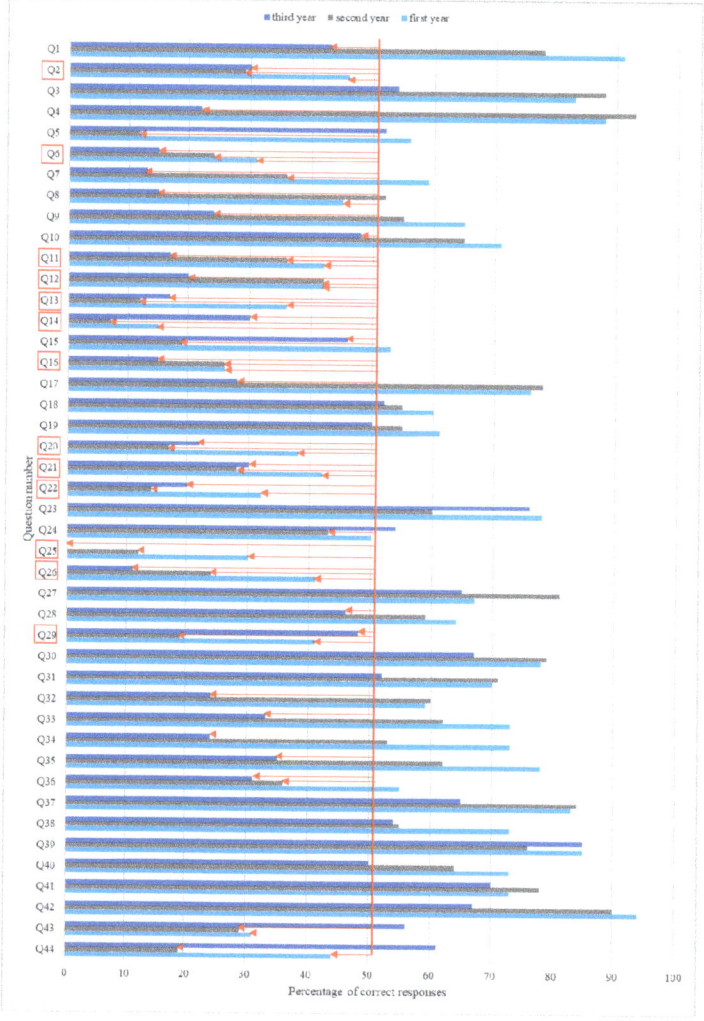

Figure 1. Percentage of correct answers for each question.

The first step of data analysis was the comparison of performance for each year of study separately. The one-way ANOVA test was used in order to determine the differences in grade results. Based on the Tukey B test, which followed the ANOVA, the following relationship can be recorded (Table 3):

Table 3. Results comparison by ANOVA.

	First Year		Second Year		Third Year		One-Way ANOVA	
	Mean	SD	Mean	SD	Mean	SD	F	p
Points	25.73	6.1	21.21	7.2	17.39	6.0	25.2	0.001

[first grade] > [second grade] > [third grade] $F = 25.2\ p = 0.001$.

This was followed by a two-sample t-test so that the authors could examine the ratio of deviations in the light of the above results (Tables 4–6).

Table 4. Comparing results for the first and the second-year students.

	Results		Two-Sample t-test	
	Mean	SD	t	p
First year	25.73	6.1	3.97	0.001
Second year	21.21	7.2		

Table 5. Comparing results for the first and the third-year students.

	Results		Two-Sample t-test	
	Mean	SD	t	p
First year	25.73	6.1	7.44	0.001
Third year	17.39	6.0		

Table 6. Comparing results for the second and the third-year students.

	Results		Two-Sample t-test	
	Mean	SD	t	p
Second year	21.21	7.2	2.92	0.004
Third year	17.39	6.0		

The results highlighted that students in Year 1 achieved significantly better results than those in Year 2 and Year 3. The results of those in Year 2 and Year 3 also show a statistical difference in favor of the second year. Studies in Year 3 performed considerably poorer than the students in Years 1 and 2.

Therefore, these results do not support the H1 hypothesis: the students in Year 1 performed significantly better than the second- and third-year students.

4.6. Analysis Based on the Subjective Difficulty Estimation of the Test Questions

4.6.1. Comparing Students' Estimations

The following step in examining the results was to compare how students evaluated the difficulty of each test questions.

Given that the test contained a large number of questions, the analysis was not conducted on the whole set of questions (full scale). Five questions were selected for which the students had given the highest percentage of correct answers, and another five questions were selected for which students had given the lowest percentage of correct answers.

The responses were ranked on a three-point Likert scale, where the easy questions were rated as 1, medium questions were rated as 2, and the hard questions were rated as 3. Although the Likert scale is strictly a ranking scale, it is generally accepted that in order to take advantage of the possibilities offered by complex statistical procedures, it is treated as an interval scale. According to Selltiz et al. [21], the scores can be added and averaged.

The following analysis presents the degree of estimated difficulty for the top five correctly answered questions and top five questions with the lowest correct-answer rate, for all students from all years of study. The results show estimations on the full scale.

As already indicated, the students could rate every single question as easy, medium, or hard.

In terms of the average of the responses for students from all three years of study, the top five correctly answered questions were Questions 3, 30, 37, 39, and 42. For these specific questions, most of the estimations fell into the easy category, which is in accordance with the number of correct answers to those questions (Figure 2).

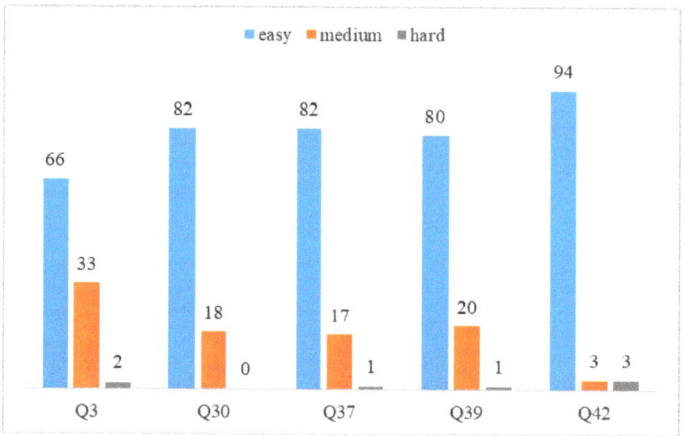

Figure 2. Estimation of the top five highest score questions (percentage of correct answers).

The descriptive results were analyzed with a one-way ANOVA test. Three out of the top five highest score questions (Q30, Q39, Q42) show a significant difference between students' estimation (Table 7). Based on the results, we can state that the students from the third year estimated the questions as significantly harder than the Year 1 and Year 2 students. The students from Year 2 estimated question Q39 as harder than the Year 1 and Year 3 students. The results for each question are less significant than the data for the whole scale, and this result partially supports the H2 hypothesis.

Table 7. The top five highest score questions.

	Year 1		Year 2		Year 3		One-Way ANOVA	
	Avg.	Sd.	Avg.	Sd.	Avg.	Sd.	F	p
Q3	1.31	0.5	1.36	0.4	1.48	0.5	1.52	0.02
Q30	1.12	0.3	1.17	0.3	1.30	0.4	3.18	0.04
Q37	1.22	0.4	1.24	0.4	1.00	0.0	2.92	0.05
Q39	1.12	0.3	1.33	0.4	1.20	0.4	4.49	0.01
Q42	1.00	0.00	1.10	0.2	1.24	0.6	6.1	0.003

The five worst-performing questions were Questions 6, 13, 14, 20, and 25 on the average of the responses for students from all three years of study. For these questions, most of the estimations fell into the medium or hard category (Figure 3). Based on the results, a review of question Q6 is definitely recommended, as the majority of students rated it as an easy question; however, they received one of the lowest scores (the lowest rate of correct answers) for this question.

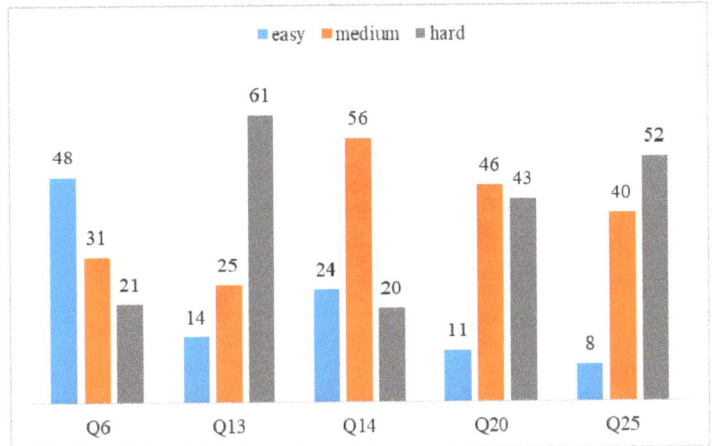

Figure 3. Estimation of the top five lowest-score questions (percentage of correct answers).

Four from the top five lowest-score questions (Q6, Q14, Q20, Q25) show a significant difference between students' estimation (Table 8). The Q6 question was estimated as significantly harder by the Year 3 students than the Year 1 and Year 2 students. In the case of Q14, opposite estimation was done: this question was "harder" for the Year 1 students. The result for Q14 supports the H2 hypothesis. The estimations for Q20 and Q25 questions differ from the previous three.

Table 8. The top five lowest-score questions.

	Year 1		Year 2		Year 3		One-Way ANOVA	
	Avg.	Sd.	Avg.	Sd.	Avg.	Sd.	F	p
Q6	1.47	0.6	1.60	0.6	2.37	0.7	23.2	0.001
Q13	2.48	0.7	2.44	0.7	2.48	0.6	0.02	0.9
Q14	2.12	0.5	1.96	0.6	1.70	0.8	5.93	0.003
Q20	2.21	0.6	2.62	0.5	2.11	0.6	6.86	0.001
Q25	2.22	0.6	2.63	0.4	2.59	0.7	7.23	0.001

These results do not support hypothesis H2 and show only that the students' estimation was close to the real difficulty of the questions: when a question was estimated as hard, then they achieved a lower score.

This section presents the results of the analysis answering the following questions:

- Is the lowest-score question estimated on the same level for all students?
- Is the highest-score question estimated on the same level for all students?
- What score did students achieve on those questions?

Score in this case refers to the rate of correct answers.

It was found that question Q25 received minimum correct answers (low score), and question Q42 was given the highest correct answers rate (high score) from students from all three years of study. The question with the highest score (Q42) was also rated as the easiest question by students of Year 1 and Year 2. Without exception, all students from Year 1 estimated its difficulty as easy, only 12% of students from Year 2, and 10% from Year 3 estimated the question's difficulty as medium difficult (Figure 4).

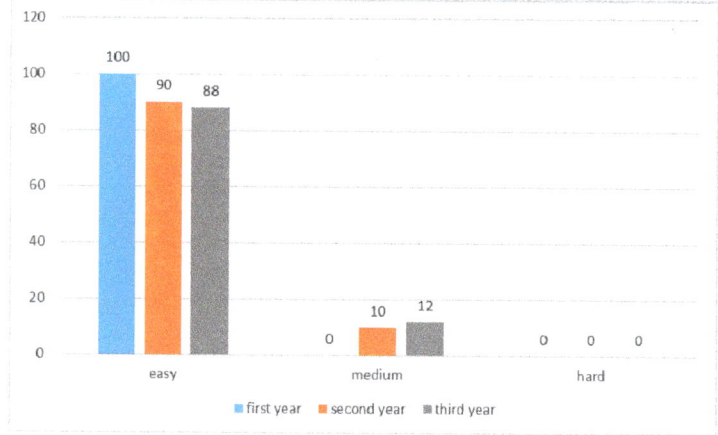

Figure 4. Estimation of difficulty for the highest-score (Q42) question.

As for the question with the lowest score (Q25), the students' estimations were quite different. The majority of students in Year 1 found the question moderately difficult (medium), while the majority of students in Years 2 and 3 found its difficulty hard (Figure 5). The same trend can be observed with the points achieved on these two questions: the Year 1 students received the highest amount of points, while the Year 2 and Year 3 students received the lowest points. At the same time, the Year 2 and Year 3 students also felt that it was more difficult for them to solve the task, which supports hypothesis H2.

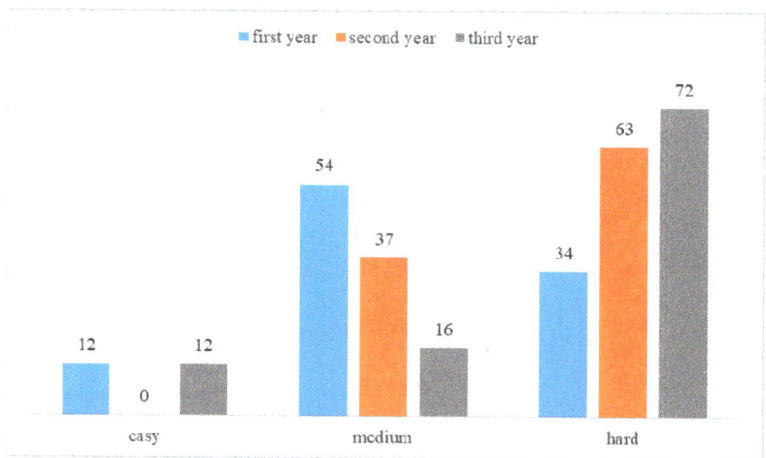

Figure 5. Estimation of difficulty for the lowest-score (Q25) question.

When estimating the degree of difficulty of the full scale (the whole test), the majority of students identified it as being of medium difficulty (Table 9).

Table 9. Estimating the degree of difficulty for the whole test.

	n	%
easy	30	16
medium	147	81
hard	5	3
Total	**182**	**100**

The differences in students' estimation for each year of study are illustrated in Figure 6.

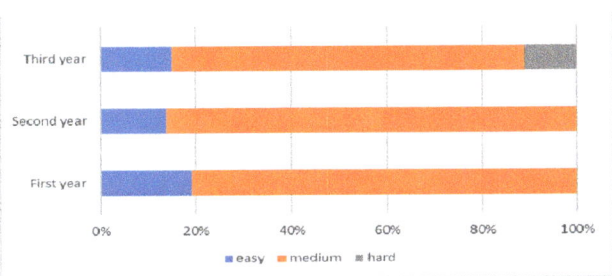

Figure 6. Estimating the difficulty of questions by year of study.

The results were tested with the chi square probe (Table 10).

Table 10. Correlations between year of study and estimated difficulty of the questions.

	Easy	Medium	Hard	Total
Year 1	12	66	0	78
Year 2	5	53	0	58
Year 3	7	36	3	46
Total	24	155	3	182

The data are consistent with the values shown in the descriptive statistics: the majority of students from all three years of the study estimated the test with medium difficulty value. The results show a correlation between the two variables (performance and task difficulty estimation) ($\chi^2 = 10.73$ $p = 0.03$). When students from a higher year of the study estimated a question as medium or hard, then, they achieved a lower score there. This result supports hypothesis H2.

Based on year of study, the authors separately examined the five best- and worst-performing questions and the estimation associated with them. For these questions, they also compared the estimated value of the questions given by students with those defined by the instructors.

In addition to Questions 42 and 25 already analyzed above, several questions received similar estimations from the students of all three years of study. Question 37 received a high score from the students in their first and second year, and Q39 received a high score from the students in their second and third year. Students in Year 1 and Year 2 achieved low scores with Q14. Questions 5 and 16 were those where students in Year 1 and Year 3 gave fewer correct answers.

In the estimation of the difficulty of the five highest-scoring questions, the opinions voiced by students in Year 1 and the instructor are very similar (Figure 7). Estimations were also compared with a nonparametric statistical analysis, where the result of the Mann–Whitney test also supported the agreement of the estimations.

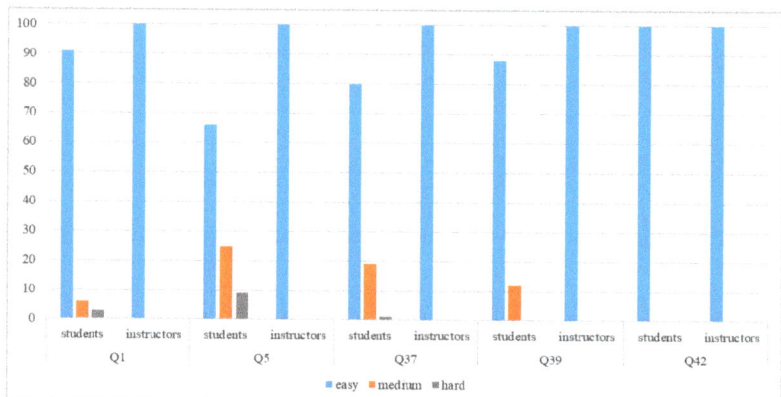

Figure 7. Comparing estimation difficulty of the five highest-scoring questions defined by the students in Year 1 and the instructors.

For the lowest-scoring questions, the estimations of the students and the instructor were no longer so consistent (Figure 8). For Questions 6 and 16, the majority of students and the instructor were of the same opinion. Question 25 was marked as medium by the students and hard by the instructor. Question 43 was rated as hard by the students as opposed to medium by the instructor. Statistically, these differences are not significant. However, for Question 14, where students estimated the question as medium and the instructor estimated the question as easy, there was a significant difference in estimation based on the Mann–Whitney test ($Z = -1.97$, $p = 0.04$). This result supports hypothesis H3: it indicates that the Year 1 students gave a different estimation of difficulty than the instructor.

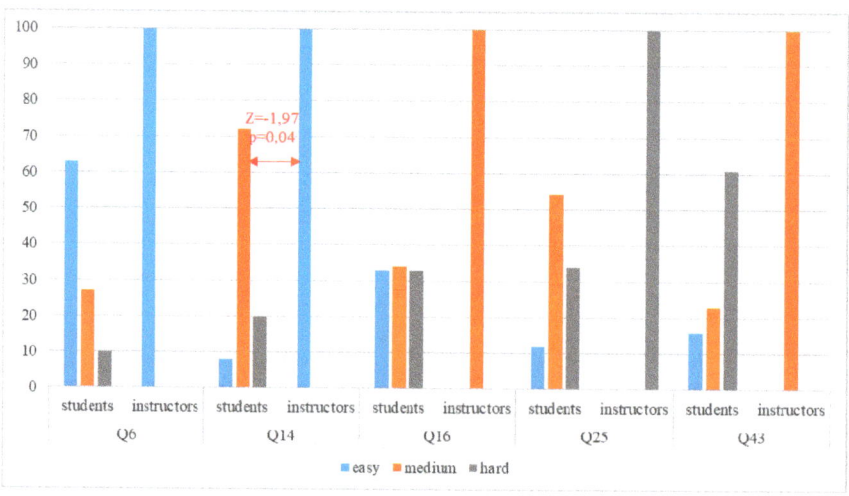

Figure 8. Comparing estimation difficulty of the five lowest-scoring questions defined by the students in Year 1 and the instructors.

The same two analyses were conducted for the estimation of students in Year 2. In the case of the questions with the highest scores, four questions (Q3, Q4, Q37, Q42) were estimated to be on the same level by both students and instructors (Figure 9). In the case of Question 27, the instructor rated the question as hard, while most of the students rated it as medium. The difference is only shown in the descriptive statistics; the Mann–Whitney test does not show a statistically significant difference. This result supports the H3 hypothesis. The questions with lowest scores showing in Figure 10.

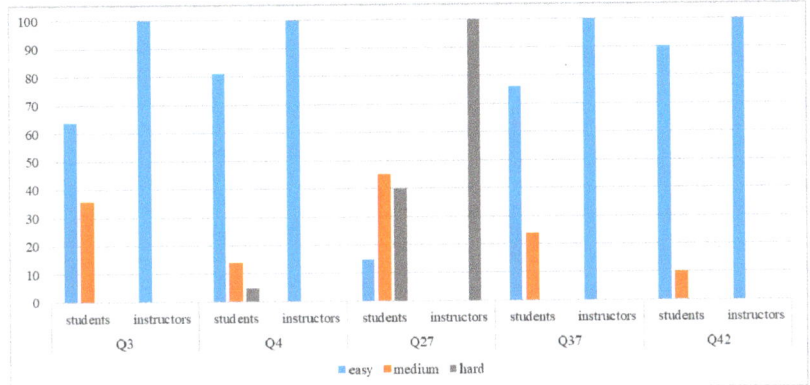

Figure 9. Comparing estimation difficulty of the five highest-scoring questions defined by the students in Year 2 and the instructors.

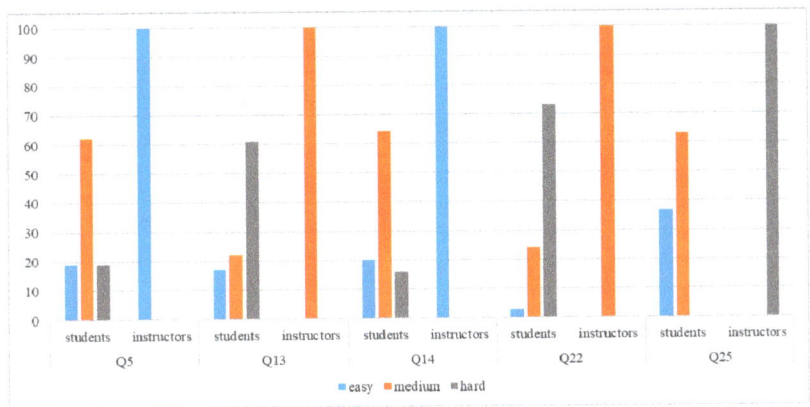

Figure 10. Comparing estimation difficulty of the five lowest-scoring questions defined by the students in Year 2 and the instructors.

With the students in Year 3, regarding the questions that received the most points, there was a difference in the estimation for the students and the teachers for Question 23 (Figure 11). The students rated the question as easy, whereas the teacher saw it as medium. The other questions' estimation did not show much diversion: the majority of students as well as the teachers found the questions easy.

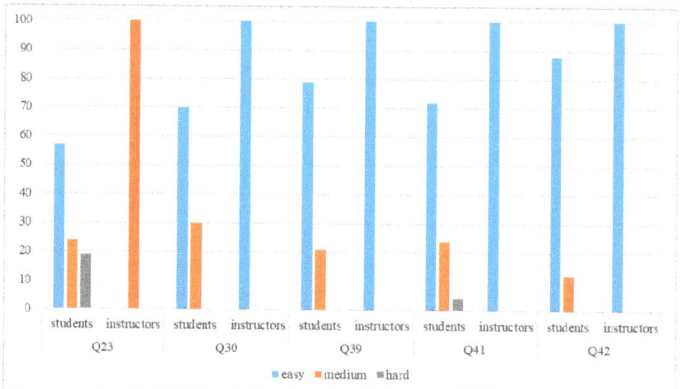

Figure 11. Comparing estimation difficulty of the five highest-scoring questions defined by the students in Year 3 and the instructors.

In the case of the questions with the lowest score (Figure 12), the estimations of the students and the instructor were the same for Questions 25 and 26, as both parties saw their difficulty as hard. Question 6 was considered hard by the students, as opposed to easy by the teacher. Question 7 was considered easy by the majority of students and of medium difficulty by the teacher. Question 16 was estimated as hard by the students and of medium difficulty by the teacher. In all cases, the Mann–Whitney test performed with these results showed no relevant difference between the estimation of the students from the third year and the teacher. This result also supports hypothesis H3 for the students of Year 3.

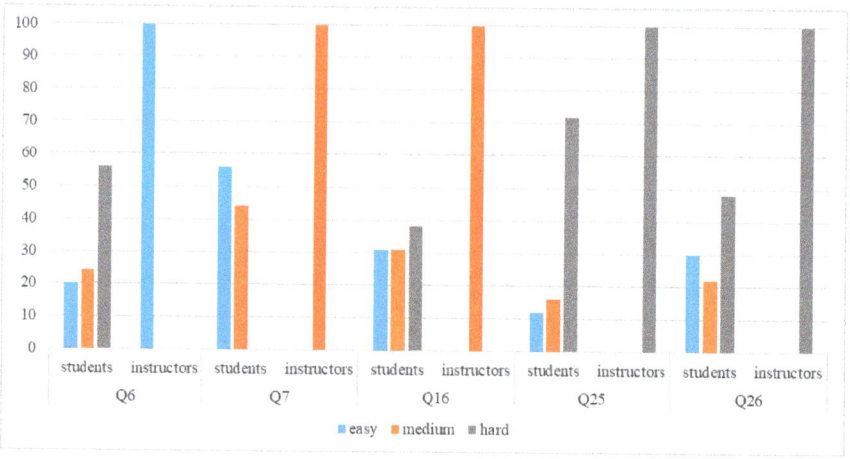

Figure 12. Comparing estimation difficulty of the five lowest-scoring questions defined by the students in Year 3 and the instructors.

4.6.2. Relationship between Student and Teacher Estimations

Apart from the students, all questions used in this research were also evaluated by the teachers. The teachers estimated the average difficulty of the questions' full scale as "medium". The results for the full scale are derived from the average score of the answers to each question. The difference between the teachers' and the students' evaluation was examined in a one-sample t-test, where the teachers' evaluation was viewed as an external standard average (Table 11).

Table 11. Comparing teacher and student estimation of test difficulty.

	Average of Marks by Students	SD	Average of Marks by Teachers	One-Sample *t*-test	
				t	p
First year	1.85	0.3	2	−3.74	0.001
Second year	1.91	0.2		−2.31	0.02
Third year	1.91	0.5		−1.27	0.2

Based on the obtained results, students in their first and second year estimated the questions as significantly easier than the medium difficulty level specified by the teachers. The third-year students' estimation was the same as that of the teachers.

The results support the H3 hypothesis, namely, there is no difference between the students' and the teachers' estimation in the third year of study on question difficulty. However, the discrepancy in the estimation of students in their first and second year suggests that the students perceived the tasks as easier than the teachers did.

5. Discussion

Three hypotheses were formulated in this research. The first was to prove that students were perfecting their knowledge and skills during their studies. The assumption was that if a given material was taught in the first year and subsequently used in the later years of education, that knowledge or skill would be improved. As a result, the students in the later years of study would solve the basic programming tasks more easily than their peers in the first year. The hypothesis was also supported by the assumption that during computer science education, students were also using the process of debugging to identify and remove errors from application in many other courses. Code tracing is one of the main techniques in debugging. To summarize these assumptions, the authors find that in the later years of study, students have more experience in understanding how the whole program, or just a segment of the code, works.

- The test results of this hypothesis show a reverse tendency. Possible reasons for hypothesis H1 not being confirmed may be as follows. First, programming requires many skills. Some of those programming skills are taught, while others are not. Those skills that are developed during the years of study tend to be independent from each other. For instance, improving algorithmic thinking or making different abstraction layers, which are the basis for developing a quality solution, does not depend on, e.g., code-writing, debugging, or code-tracing skills. This supports the view that the profession of a software developer is a complex one, and it is possible to define important skills for specific areas. Some research works confirm this [10,22].
- Students generally do not, or only to a very slight extent, associate knowledge from one course (subject) with the knowledge from another course with similar topics. This is possibly caused by the structure of the curriculum, which fails to sufficiently emphasize the connection between topics of different courses. There is another possible explanation: many students believe that a successfully passed exams means they are somehow "finished with it" and there will be no continuation. They erroneously assume that each new course, even with similar topics, is something completely new, for which they would not need to use previously learned topics. The fact that first-year students obtained better results can be explained by them not yet having completed the course 'Algorithms and data structures' and thus had not yet taken the final exam. For them, the first date to take the exam was three weeks away, counting from the date of doing the test for this research. Cognitive flexibility theory focuses on this behavior. The theory is largely concerned with the transfer of knowledge and skills beyond their initial learning situation. The theory asserts that effective learning is context-dependent, as argued by Spiro and Jehng. [4].

- The reason for poorer performance in later years of study may also lie in the fact that students with a greater amount of experience more often use libraries and advanced objects when solving problems. Functions from those libraries or objects' methods with one instruction solve a (complex) task, such as inserting a new element at the beginning of an array, doing binary search in array, sorting an array, etc. This may be the underlying reason for why senior students make less use of the basic elements of programming language such as statements, conditionals, and iterations to create a solution. Conversely, this also means that some of their vital software developing skills, such as code tracing, fail to develop and evolve.
- In their second and third years of the study, students frequently use ready-made solutions that can be found online. By querying the expression "how do you break string to words in C?", the Google search engine offers more than 170 million hits. One of those hits is sure to contain a suitable solution for the given situation. Students apply these ready-made solutions without too much thinking, analyzing or, in fact, understanding the actual code.

The second and third hypotheses are based on the assumption that during their studies, students will enhance their programming skills, because they have to solve an increasing number of programming tasks and develop projects with growing complexity. More hours spent solving programming tasks should ideally contribute to the creation of better abstraction layers, improved task comprehension, and finding the optimal solution. The authors' assumption was that due to the lack of experience in first-year students, their perception of task difficulty would mostly be medium or hard. The assumption was that the difference between subjective evaluation of the task difficulty given by students and teachers would be smaller, since students are improving their skills in solving programming tasks.

Most of the first-year students lack experience in code writing, algorithmic thinking, and debugging, thus making it more challenging for them to objectively evaluate the difficulty of test questions.

Since they do not have enough experience solving programming tasks, they do not have any basis of comparison for a given task's difficulty. This is precisely why they perceived the tasks as more difficult than they actually were and also more difficult than how the second- and third-year students evaluated them. Through participating in various projects and exercises, senior students gain experience in solving tasks and knowledge of how to solve similar problems, so that they can faster recognize the elements needed for creating the solution, thus enabling them to more objectively evaluate the difficulty of the task at hand. The results of the conducted research confirm this assumption, and by this, the second hypothesis is also confirmed.

The same trend can be observed regarding hypothesis H3: the results of estimating the task difficulty given by the third-year students approximate the evaluation given by the teachers. Since the teachers defined the level of every task by taking into account both task complexity and difficulty, it can be stated that objective estimation skills can be acquired with more and more years of education. This is why senior students were able to objectively evaluate the complexity and difficulty of the task.

The results of comparing the answers on questions on a full scale can be used in the so-called computerized adaptive testing (CAT). These tests adapt to the given student's level of ability. CAT selects questions so as to maximize the exam accuracy based on what is known about the student from previous questions [23]. For the student, this means that he or she will always be given a "tailored" test suited to their level of ability. For instance, if the student does well on a medium question (i.e., medium in difficulty), they will then receive a more difficult next question. However, a wrong answer to a given question means the student performed poorly and will be given simpler questions. For tailoring the optimal item (subsequent question) in CAT, the system uses IRT and requires a database in which all the levels of difficulty for all questions (item) are precisely determined (calibrated). Figure 13 shows the probability of a correct response depending on the skill level. The left-hand curve indicates an easy item, the middle-dotted line refers to a medium item, while the broken line on the right side signifies an item whose difficulty level is hard. The horizontal axis shows skill levels between −4 and 4.

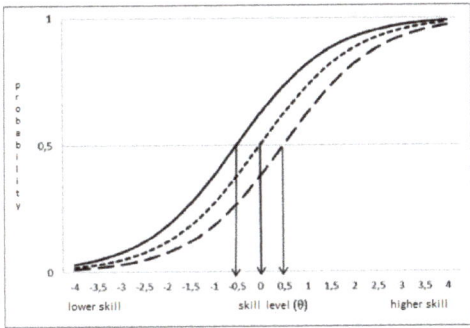

Figure 13. Probability for the correct answer.

The results of this research can help instructors in creating items for such a database. The following passage offers some ways to apply the results of the research to CAT.

Based on the score (correct answers) results (Figure 1), groups of difficulties can be formed. For example, questions for which less than 20% of the students supplied a correct answer would be classified as hard. Questions with scores between 21% and 40% would be medium, while the rest would be categorized as easy. This is a categorization of the items and not the calibration. However, this categorization can be used to build different variations of tests with approximately the same difficulty by selecting tasks from low, medium, or hard categories. The here-presented results can also be used in the process of item calibration for CAT.

By combining the score and the estimated difficulty, it is possible to pinpoint, for example, which are the easiest or most difficult questions in the set. Figure 2 presents those questions that were rated as easy; thus, they had the most correct answers, whereas Figure 3 contains those questions that were difficult for all students. Furthermore, instructors can identify the problematic questions in the set, such as Q6, which was rated as easy, but the majority of the students failed to answer it correctly. In this study, the analysis was limited to the five best- and five worst-performing questions. Extending the analysis to the whole set of questions may produce a subset of questions that can be used in CAT—and, due to very different scores and estimations, those that cannot be used in CAT.

The results, as summarized in Figures 7–12, show the level of difference between the estimation of difficulty given by the students and those given by the instructors. This comparison can help determine the real level of difficulty for a given question, because:

- It excludes an estimation that is based only on one teacher opinion or several of them.
- It excludes the possibility of defining the rate based on estimations given by students from a single year of study: the level of knowledge and skills may vary from one cohort to the next. Using questions with estimations from a so-called academically speaking "strong" cohort, i.e., students with good results, will give faulty results in the not-so-stellar, academically "weaker" cohort.

Significant differences between the estimation of difficulty given by students from different years of study and the teachers may indicate specific questions where the probability of giving the correct answer also depends on certain other factors. There may be situations when giving the correct answer depends on knowledge in some other topic, but then such items should be excluded from the database of questions.

As for future plans, the next phase of this particular research will include a more detailed analysis of the results. Comparing the results by the categories (sub-scale) can offer a better understanding of the efficiency of the implemented teaching methodology. The results of this comparison can aid teachers in pinpointing the specific parts of the material in the CS1 course that need to be taught or explained in a different way.

Author Contributions: Conceptualization, A.K.; methodology, R.P. and S.M.Č.; software, R.P. and S.M.Č.; validation, L.M., R.P. and S.M.Č.; formal analysis, A.K. and J.K.; investigation, R.P.; resources, S.M.Č.; data curation, S.M.Č.; writing—R.P. and A.K.; writing—review and editing, S.M,Č and J.K.; visualization, L.M.; supervision, P.Č.; project administration, A.K.; funding acquisition, A.K. All authors have read and agreed to the published version of the manuscript.

Funding: The project is sponsored by EFOP-3.6.2-16-2017-00018 "Produce together with the nature" project.

Conflicts of Interest: The authors declare no conflict of interest.

References

1. Reckase, M.D.; Ju, U.; Kim, S. How Adaptive Is an Adaptive Test: Are All Adaptive Tests Adaptive? *J. Comput. Adapt. Test.* **2019**, *7*, 1–14. [CrossRef]
2. Luecht, R.M.; Nungester, R.J. Some Practical Examples of Computer-Adaptive Sequential Testing. *J. Educ. Meas.* **2005**, *35*, 229–249. [CrossRef]
3. Lister, R.; Clear, T.; Simon, B.; Bouvier, D.J.; Carter, P.; Eckerdal, A.; Jacková, J.; Lopez, M.; McCartney, R.; Robbins, P.; et al. Naturally occurring data as research instrument: Analyzing examination responses to study the novice programmer. *ACM SIGCSE Bull.* **2010**, *41*, 156–173. [CrossRef]
4. Spiro, R.J.; Jehng, J.C. Cognitive flexibility and hypertext: Theory and technology for the nonlinear and multidimensional traversal of complex subject matter. In *Cognition, Education, and Multimedia: Exploring Ideas in High Technology*; Nix, D., Spiro, R.J., Eds.; Lawrence Erlbaum Associates, Inc.: Mahwah, NJ, USA, 1990; pp. 163–205.
5. Maravić Čisar, S.; Radosav, D.; Pinter, R.; Čisar, P. Effectiveness of Program Visualization in Learning Java: A Case Study with Jeliot 3. *Int. J. Comput. Commun. Control* **2011**, *6*, 668–680. [CrossRef]
6. Lopez, M.; Whalley, J.; Robbins, P.; Lister, R. Relationships between reading, tracing and writing skills in introductory programming. In Proceedings of the Fourth International Workshop on Computing Education Research (ICER '08), Sydney, Australia, 6–7 September 2008; pp. 101–112.
7. Kumar, A. Solving Code-tracing Problems and its Effect on Code-writing Skills Pertaining to Program Semantics. In Proceedings of the 2015 ACM Conference on Innovation and Technology in Computer Science Education (ITiCSE'15), Vilnius, Lithuania, 4–8 July 2015; pp. 314–319.
8. Cunningham, K.; Blanchard, S.; Ericson, B.; Guzdial, M. Using tracing and sketching to solve programming problems: Replicating and extending an analysis of what students draw. In Proceedings of the 2017 ACM Conference on International Computing Education Research, Tacoma, WA, USA, 18–20 August 2017; pp. 164–172.
9. Nelson, G.L.; Xie, B.; Ko, A.J. Comprehension First: Evaluating a Novel Pedagogy and Tutoring System for Program Tracing in CS1. In Proceedings of the 2017 ACM Conference on International Computing Education Research, Tacoma, WA, USA, 18–20 August 2017; pp. 2–11.
10. Sheard, J.; Souza, D.D.; Klemperer, P.; Porter, L.; Zingaro, D. Benchmarking Introductory Programming Exams: Some Preliminary Results. In Proceedings of the 2016 ACM Conference on International Computing Education Research, Melbourne, Australia, 8–12 September 2016; pp. 103–111.
11. Xie, B.; Loksa, D.; Nelson, G.L.; Davidson, M.J.; Dong, D.; Kwik, H.; Tan, A.H.; Hwa, L.; Li, M.; Ko, A.J. A theory of instruction for introductory programming skills. *Comput. Sci. Educ.* **2019**, *29*, 205–253. [CrossRef]
12. McCracken, M.; Almstrum, V.; Diaz, D.; Guzdial, M.; Hagen, D.; Kolikant, Y.; Laxer, C.; Thomas, L.; Utting, I.; Wilusz, T. A Multi-National, Multi-Institutional Study of Assessment of Programming Skills of First-year CS Students. *ACM SIGCSE Bull.* **2001**, *33*, 125–180. [CrossRef]
13. Lister, R.; Adams, E.S.; Fitzgerald, S.; Fone, W.; Hamer, J.; Lindholm, M.; McCartney, R.; Mostrom, J.E.; Sanders, K.; Seppala, O.; et al. A Multi-National Study of Reading and Tracing Skills in Novice Programmers. *ACM SIGCSE Bull.* **2004**, *36*, 119–150. [CrossRef]
14. Kopec, D.; Yarmish, G.; Cheung, P. A description and study of intermediate student programmer errors'. *ACM SIGCSE Bull.* **2007**, *39*, 146–156. [CrossRef]
15. Lampert, B.; Pongracz, A.; Sipos, J.; Vehrer, A.; Horvath, I. MaxWhere VR-learning improves effectiveness over clasiccal tools of e-learning. *Acta Polytech. Hung.* **2018**, *15*, 125–147.
16. Budai, T.; Kuczmann, M. Towards a modern, integrated virtual laboratory system. *Acta Polytech. Hung.* **2018**, *15*, 191–204.

17. Csapó, G. Placing event-action-based visual programming in the process of computer science education. *Acta Polytech. Hung.* **2019**, *16*, 35–57.
18. Katona, J.; Kovari, A. Examining the learning efficiency by a brain-computer interface system. *Acta Polytech. Hung.* **2018**, *15*, 251–280.
19. Katona, J.; Kovari, A. The evaluation of bci and pebl-based attention tests. *Acta Polytech. Hung.* **2018**, *15*, 225–249.
20. Sik-Lanyi, C.; Shirmohammmadi, S.; Guzsvinecz, T.; Abersek, B.; Szucs, V.; Van Isacker, K.; Lazarov, A.; Grudeva, P.; Boru, B. How to develop serious games for social and cognitive competence of children with learning difficulties. In Proceedings of the 2017 8th IEEE International Conference on Cognitive Infocommunications (CogInfoCom), Debrecen, Hungary, 11–14 September 2017; pp. 321–326.
21. Selltiz, C.; Jahoda, M.; Deutsch, M.; Cook, S.W.; Chein, I.; Proshansky, H.M. *Research Methods in Social Relations*; Methuen & Co. LTD.: London, UK, 1964.
22. Cooper, S.; Wang, K.; Israni, M.; Sorby, S. Spatial skills training in introductory computing. In Proceedings of the Eleventh Annual International Conference on International Computing Education Research, Omaha, NE, USA, 9–13 August 2015; pp. 13–20.
23. Maravić Čisar, S.; Pinter, R.; Čisar, P. Evaluation of knowledge in Object Oriented Programming course with computer adaptive tests. *Comput. Educ.* **2016**, *92–93*, 142–160. [CrossRef]

© 2020 by the authors. Licensee MDPI, Basel, Switzerland. This article is an open access article distributed under the terms and conditions of the Creative Commons Attribution (CC BY) license (http://creativecommons.org/licenses/by/4.0/).

Article

Analysis of the Learning Process through Eye Tracking Technology and Feature Selection Techniques

María Consuelo Sáiz-Manzanares [1,*], Ismael Ramos Pérez [2], Adrián Arnaiz Rodríguez [2], Sandra Rodríguez Arribas [3], Leandro Almeida [4] and Caroline Françoise Martin [5]

1 Departamento de Ciencias de la Salud, Facultad de Ciencias de la Salud, Universidad de Burgos, Research Group DATAHES, P° Comendadores s/n, 09001 Burgos, Spain
2 Departamento de Ingeniería Informática, Escuela Politécnica Superior, Universidad de Burgos, Research Group ADMIRABLE, Escuela Politécnica Superior, Avda. de Cantabria s/n, 09006 Burgos, Spain; ismaelrp@ubu.es (I.R.P.); adrianar@ubu.es (A.A.R.)
3 Departamento de Ingeniería Informática, Escuela Politécnica Superior, Universidad de Burgos, Research Group DATAHES, Escuela Politécnica Superior, Avda. de Cantabria s/n, 09006 Burgos, Spain; srarribas@ubu.es
4 Instituto de Educação, Universidade do Minho, Research Group CIEd, Campus de Gualtar, 4710-057 Braga, Portugal; leandro@ie.uminho.pt
5 Departamento de Filología Inglesa, Universidad de Burgos, P° Comendadores s/n, 09001 Burgos, Spain; caroline.martin@ubu.es
* Correspondence: mcsmanzanares@ubu.es

Citation: Sáiz-Manzanares, M.C.; Pérez, I.R.; Rodríguez, A.A.; Arribas, S.R.; Almeida, L.; Martin, C.F. Analysis of the Learning Process through Eye Tracking Technology and Feature Selection Techniques. *Appl. Sci.* **2021**, *11*, 6157. https://doi.org/10.3390/app11136157

Academic Editors: Attila Kovari and Cristina Costescu

Received: 25 May 2021
Accepted: 26 June 2021
Published: 2 July 2021

Publisher's Note: MDPI stays neutral with regard to jurisdictional claims in published maps and institutional affiliations.

Copyright: © 2021 by the authors. Licensee MDPI, Basel, Switzerland. This article is an open access article distributed under the terms and conditions of the Creative Commons Attribution (CC BY) license (https://creativecommons.org/licenses/by/4.0/).

Abstract: In recent decades, the use of technological resources such as the eye tracking methodology is providing cognitive researchers with important tools to better understand the learning process. However, the interpretation of the metrics requires the use of supervised and unsupervised learning techniques. The main goal of this study was to analyse the results obtained with the eye tracking methodology by applying statistical tests and supervised and unsupervised machine learning techniques, and to contrast the effectiveness of each one. The parameters of fixations, saccades, blinks and scan path, and the results in a puzzle task were found. The statistical study concluded that no significant differences were found between participants in solving the crossword puzzle task; significant differences were only detected in the parameters saccade amplitude minimum and saccade velocity minimum. On the other hand, this study, with supervised machine learning techniques, provided possible features for analysis, some of them different from those used in the statistical study. Regarding the clustering techniques, a good fit was found between the algorithms used (k-means ++, fuzzy k-means and DBSCAN). These algorithms provided the learning profile of the participants in three types (students over 50 years old; and students and teachers under 50 years of age). Therefore, the use of both types of data analysis is considered complementary.

Keywords: machine learning; cognition; eye tracking; instance selection; clustering; information processing

1. Introduction

The eye tracking technique has represented an important advance in research in different fields, for example, cognitive psychology, as it records evidence on the cognitive processes related to attention during the resolution of different types of tasks. In particular, this technology provides the researcher with knowledge of the eye movements that the learner performs to solve different tasks [1]. This implies an important advance in the study of information processing, as this technique will allow us to obtain empirical indicators in different metrics, all of which offers a guarantee of precision to the psychology professional for the interpretation of each user's information processing. However, the measurements are complex and, above all, lengthy in time, which often means that the ratios of participants are not very large. In summary, technological advances are improving the study of information processing in different learning tasks. The use of these resources is an opportunity for cognitive and instructional psychology to delve into the analysis of the variables that facilitate deep learning in different tasks. In addition, these tools allow the visualisation of the learning patterns of apprentices during the resolution of different activities. Initial research in this field [2] indicated that readers with prior knowledge showed little interest in the images embedded in the learning material. Furthermore, recent research [3] has found significant differences in eye tracking behaviour between experts vs. novices. It seems that experts allocate their attention more efficiently and learn more easily if automated monitoring processes are applied in learning proposals. Similarly, other studies [2] have indicated that the use of multimedia resources that incorporate zoom effects makes it easier for information to remain longer in short-term memory (STM). Likewise, if this information is accompanied by a narrating voice, attention levels and semantic comprehension increase [4,5]. A number of methods are used to analyse the effectiveness of the learning process including eye tracking-based methods. This technique offers an evaluation of eye movement in different metrics [1–5]. The eye tracking technique can use different algorithms [6–9]. They can be used to extract different metrics (more detailed explanations are given below). Specifically, eye tracking technology allows the analysis of the relationship between the level of visual attention and the eye–hand coordination processes during the resolution of different tasks within the executive attention processes [7,8]. Clearly, rapid eye movement has also been associated with the learner's fixation on the most relevant elements of the material being learned [2].

In this context, attention is considered to be the beginning of information processing and the starting point for the use of higher-order executive functions. In the same way, observational skills relate to eye tracking, which is directly related to the level of arousal and the transmission of information first to the STM and then its processing in working memory [6]. This development is influenced by learner-specific variables such as age, level of prior knowledge, cognitive ability and learning style [7]. However, some studies show that prior knowledge can compensate for the effects of age [8]. On the other hand, eye tracking technology is one of the resources that is supporting this new way of analysing the learning process. This technology is centred on evidence-based software engineering (EBSE) [9]. This technological resource makes it possible to study attentional levels and relate them to the cognitive processes that the learner uses in the course of solving a task [10,11]. Thus, eye tracking technology provides different metrics based on the recording of the frequency of gaze on certain parts of a stimulus. These metrics can be previously defined by the researcher and are called areas of interest (AOI), which can be relevant or irrelevant. This information will allow the practitioner to determine which learners are field-dependent or field-independent, based on their access to irrelevant vs. relevant information [12]. Likewise, the use of multimedia resources, such as videos, which include Self-Regulation Learning (SRL) aids through the teacher's voiceover or the figure of an avatar seems to be an effective resource for maintaining attention and comprehension of the task and even compensating for the lack of prior knowledge of the learners. One possible explanation is that they enhance self-regulation in the learning process [13–15]. However, the design of learning materials seems to be a key factor in maintaining attention during task performance. Therefore, it is necessary to know which elements are relevant vs. irrelevant, not only for the teacher but also for the learners' perception [16]. This is why the knowledge of measurement metrics in eye tracking technology, together with their interpretation, is a relevant component for the design of learning activities for different types of users.

1.1. Measurement Parameters in Eye Tracking Technology

As mentioned above, eye tracking technology facilitates the collection of different metrics. First, it enables the recording of the learner's eye movement or eye tracker while performing an activity. In addition, the use of eye tracking technology allows the definition of relevant vs. non-relevant areas (AOI) in the information being learned [17]. Within these metrics, different parameters can be studied, such as the fixation time of the eye on the part of the stimulus (interval between 200 and 300 ms). In this line, recent studies [18] indicate that the acquisition of information is related to the number of eye fixations of the learner. Similarly, another important metric is the saccade, which is defined as the sudden and rapid movement of a fixation (the interval is 40–50 ms). Sharafi et al. [18,19] found differences in the type of saccade depending on the phase of information encoding the learner was at. Another relevant parameter is the scan path or tracking path. This metric collects, in chronological order, the steps that the learner performs in the resolution of the learning task within the AOI marked by the teacher [18,19]. Likewise, eye tracking technology allows the use of supervised machine learning techniques to predict the level of learners' understanding, as this seems to be related to the number of fixations [20]. Recent studies indicate that variability in gaze behaviour is determined by image properties (position, intensity, colour and orientation), task instructions, semantic information and the type of information processing of the learner. These differences are detected using AOIs that are set by the experimenter [21].

In summary, eye tracking technology records diverse types of parameters that provide different interpretations of the underlying cognitive processes during the execution of a task. These parameters fall into three categories: fixations, saccades and scan path. The first one, fixations, refers to the stabilisation of the eye on a part of the stimulus during a time interval between 200 and 300 ms. In addition, eye tracking technology provides information about the start and the end time in x and y coordinates. The meaning of the cognitive interpretation is related to the perception, encoding and processing of the stimulus. The second ones, saccades, refers to the movement from one fixation to another, which is very fast and in the range of 40–50 ms. The third ones, scan path, refers to a series of fixations in the AOIs in chronological order of execution. This cognitive metric is useful for understanding the behavioural patterns of different participants in the same activity. Furthermore, each of these metrics has its own measurement specifications. Table 1 below shows the most significant ones and, where appropriate, their relationship with information processing.

Table 1. Most representative parameters that can be obtained with the eye tracking technique and their significance in information processing.

Metric	Acronym	Metric Meaning	Learning Implications
Fixation Count	FC	Counts the number of specific bindings on AOIs in all stimuli	A greater number and frequency of fixations on a stimulus may indicate that the learner has less knowledge about the task or difficulty in discriminating relevant vs. non-relevant information. These are measures of global search performance [22].
Fixation Frequency Count	FFC		
Fixation Duration	FD	Duration of fixation	It gives an indication of the degree of interest and reaction times of the learner. Longer duration is usually associated with deeper cognitive processing and greater effort. For more complicated texts, the user has a longer average fixation duration. Fixation duration provides information about the search process [22].
Fixation Duration Average	AFD	Average duration of fixation	Longer fixations refer to the learner spending more time analysing and interpreting the information content within the different areas of interest (AOIs). The average duration is considered to be between 200 and 260 ms.
Fixation Duration Maximum	FDMa	Maximum duration of fixation	They refer to reaction times.
Fixation Duration Minimum	FDMi	Minimum duration of fixation	

Table 1. Cont.

Metric	Acronym	Metric Meaning	Learning Implications
Fixation Dispersion Total	FDT	Sum of all dispersions of fixations in X and Y	It refers to the perception of information in different components of the task.
Fixation Dispersion Average	FDA	Sum of all fixation dispersions in X and Y divided by the number of fixations in the test	It analyses the dispersions in each of the fixations in the different stimuli.
Saccades Count	SC	Total number of saccades in each of the stimuli	A greater number of saccades implies greater search strategies. The greater the breadth of the saccade, the lower the cognitive effort. It may also refer to problems in understanding information.
Saccade Frequency Count	SFC	Sum of all saccades	They refer to the frequency of use of saccades that are related to search strategies.
Saccade Duration Total	SDT	Sum of the duration of all saccades	
Saccades Duration Average	SDA	Average duration of saccades in each of the AOIs	It allows discriminating field-dependent vs. non-dependent trainees.
Saccade Duration Maximum	SDMa	Maximum saccade duration.	They refer to the perception of information in different components of the task.
Saccade Duration Minimum	SDMi	Minimum saccade duration.	
Saccade Amplitude Total	SAT	Sum of the amplitude of all saccades	Newcomers tend to have shorter saccades.
Saccade Amplitude Maximum	SAMa	Maximum of saccade amplitude	
Saccade Amplitude Minimum	SAMi	Minimum of the saccade amplitude	
Saccade Velocity Total	SVT	Sum of the velocity of all saccades	They are directly related to the speed of information processing in moving from one element to another within a stimulus.
Saccade Velocity Maximum	SVMa	Maximum value of the saccade velocity	
Saccade Velocity Minimum	SVMi	Minimum value of saccade speed	
Saccade Latency Average	SLA	It is equal to the time between the end of one saccade and the start of the next saccade	It is directly related to reaction times in information processing. The initial saccade latency provides detailed temporal information about the search process [22].
Blink Count	BC	Number of flashes in the test	It is related to the speed of information processing. Novice learners report a higher frequency.
Blink Frequency Count	BFC	Number of blinks of all selected trials per second divided by number of selected trials	Blinks are related to information processing during exposure to a stimulus to generate the next action. Learners with faster information processing may have shorter blinks of shorter duration. However, this action may also occur when attention deficit problems are present. These results will have to be compared with those obtained in the other metrics in order to adjust the explanation of these results within the analysis of a learning pattern.
Blink Duration Total	BDT	Sum of the duration of all blinks of the selected trials divided by the number of trials selected	
Blink Duration Average	BDA	The sum of the duration of all blinks of all selected trials divided by the number of selected trials	
Blink Duration Maximum	BDMa	Longest duration of recorded blinks	
Blink Duration Minimum	BDMi	The shortest duration of recorded blinks	
Scan Path Length	SPL	It provides a pattern of learning behaviours for each user	The study of the behavioural patterns of learning will facilitate the teacher's orientations in relation to the way of learning. The length of the scan path provides information about reaction times in tasks with no predetermined duration.

In summary, the use of eye tracking technology for the analysis of information processing during the resolution of tasks in virtual learning environments has been shown to be a very effective tool for understanding how each student learns [23]. Moreover, recent studies conclude the need to integrate this technology in the usual learning spaces such as classrooms, although its use is still conditioned to important technical and interpretation knowledge on the part of the teacher [24]. Therefore, more research studies are needed to find out which of the presentation conditions of a learning task are more or less effective in learning depending on the characteristics of each learner (age, previous knowledge, learning style, etc.) [25].

1.2. Use of Data Mining and Pattern Mining Techniques in the Interpretation of the Results Obtained with the Eye Tracking Methodology

There are many studies on the application of eye tracking technology that address the model of understanding the results obtained in the different metrics. To do so, they analyse the differences in results between experts vs. novices. Experts use additional information and solve a task faster and in less time. These studies also analyse behavioural patterns by comparing the type of participant, the type of pattern and the efficiency in solving the task. Cluster analysis metrics on frequency, time and effort are used to perform these analyses. Experts vs. novices use the additional information, e.g., colour and layout, in order to use the most efficient way of navigating the platform [11]. Additionally, experts seem to be faster, meaning they will solve tasks faster and more accurately. However, novice students seem to have a greater ability to understand the tasks [13]. Nevertheless, a comparative analysis of the performance of either the same learner in their learning process or between different types of learners (e.g., novices vs. experts) [26,27] requires the use of different data mining techniques [21,28]. These can be supervised learning (related to prediction or classification) [21] or unsupervised learning (related to the use of clustering techniques) [29]. Such techniques applied to the analysis of user learning have been called educational data mining (EDM) techniques [30]. Likewise, especially in the field of analysing student behaviour during task solving, the importance of using pattern analysis techniques within what has been called educational process mining (EPM) [31] stands out. EPM is a process that focuses on detecting among the possible variables of a study those that have a greater predictive capacity. These variables may be unknown or partially known. In short, EPM thus focuses on assuming a different type of data called events. Each event belongs to a single instance of the process, and these events are related to the activities. EPM is interested in end-to-end processes and not in local patterns [31]. The general objective of instance selection techniques (e.g., prototype selection) is to "try to eliminate from the training set those instances that are misclassified and, at the same time, to reduce possible overlaps between regions of different classes, i.e., their main goal is to achieve compact and homogeneous groupings" [32] (p. 2). These analyses would belong to the supervised machine learning techniques of classification and also to the statistical techniques related to knowing which possible independent variable or variables are the ones that have a significant weight on the dependent variable or variables. The common aim of these techniques would be the elimination of noise [33], which in experimental psychology would be related to the development of pre-experimental descriptive studies [34].

In summary, feature selection techniques are a very important part of machine learning and very useful in the field of education, as they will make it possible to eliminate those attributes that contribute little or nothing to the understanding of the results in an educational learning process. Knowledge of these aspects will be essential for the proposal of new research and in the design of educational programmes [8,35]. In brief, the use of sequence mining techniques [36] and the selection of instances used in studies on the analysis of the metacognitive strategies used during task resolution processes will be very useful for the development of personalised educational intervention proposals.

1.3. Application of the Use of Eye Tracking Technology

The cognitive procedure in the process of visual tracking of images, texts or situations in natural contexts is based on the stimulus–processing–response structure. Information enters via the visual pathway (retina-fovea) and is processed at the level of the subcortical and cortical regions within the central nervous system. This processing results in a sensory stimulation response. Specifically, saccades are a form of sensory-to-motor transformation from a stimulus that has been found to be significant. Saccadic eye movements are used to redirect the fovea from one point of interest to another. Fixation is also used to keep the fovea aligned on the target during subsequent analysis of the stimulus. This alternative saccade–fixation behaviour is repeated several hundred thousand times a day and is essential in complex behaviours such as reading and driving. Saccades can be triggered by the appearance of a visual stimulus that is motivating to the subject or initiated voluntarily by the person's interest in an object. Saccades can be suppressed during periods of visual fixation. In these situations, the brain must inhibit the automatic saccade response [37]. Eye tracking technology collects, among others, metrics related to fixations, saccades and blinks. This technology is also used in studies on information processing in certain learning processes (reading, driving machines or vehicles, marketing, etc.) in people without impairments [38–45] or in groups with different impairments such as attention deficit hyperactivity disorder or autism spectrum disorder [45]. In these cases, the objective is to analyse the users' difficulties in order to make proposals for therapeutic intervention. This technology is also being used as an accident avoidance strategy [43]. Similarly, this technology can be used to study the behavioural patterns of subjects and to analyse the differences or similarities between different groups [44–46]. Eye tracking is also currently being used to test the human–machine interface based on monitoring the control of smart homes through the Internet of Things [47]. In addition, this technology is being incorporated into mobile devices. This will soon facilitate its use by users in natural contexts [48]. Similarly, eye tracking technology is being incorporated into virtual and augmented reality scenarios as the software for registration is included within the glasses [49–51]. Similarly, eye tracking technology is being incorporated into the control of industrial robots [52,53]. Finally, systems are being implemented to improve the calibration and tracking of gaze tracking for users who were previously unable to use it, due to various neurological conditions (stroke paralysis or amputations, spinal cord injuries, Parkinson's disease, multiple sclerosis, muscular dystrophy, etc.) [53]. However, these applications are still very novel and require very specific knowledge of application, and processing and interpretation of the metrics. However, progress is being made in this aspect with the implementation of interpretation algorithms in software, such as machine learning techniques for supervised learning of classification, including algorithms such as *k*-nn and random forests [54].

Based on the above theoretical foundation, a study was carried out on the analysis of the behaviour of novice vs. expert learners during the performance of a self-regulated learning task. This task was carried out in a virtual environment with multimedia resources (self-regulated video) and was monitored using eye tracking technology.

In this study, two types of analysis were used. On the one hand, statistical techniques based on analysis of covariance (ANCOVA) were used on two fixed effects factors which have been shown to be relevant in the research literature, the type of participant (novice vs. expert) and age (in this study, over 50 years old vs. under 50 years old). In addition, whether the participant is a student vs. a teacher was considered as a covariate on the dependent variables learning outcomes in solving a crossword puzzle task and eye tracking metrics (fixations, saccades, blinks and scan path length).

The hypotheses were as follows:

RQ1. Will there be significant differences in the results of solving a crossword puzzle depending on whether the participants are novices vs. experts, taking into account the covariate student vs. teacher?

RQ2. Will there be significant differences in fixations, saccades, blinks and scan path length metrics depending on the age of the participant (over 50 vs. under 50), taking into account the covariate student vs. teacher?

RQ3. Will there be significant differences in the metrics of fixations, saccades, blinks and scan path length depending on whether the participants are novices vs. experts, taking into account the covariate student vs. teacher?

On the other hand, this study applied a data analysis procedure using different supervised learning algorithms for feature selection. The objective was to find out the most significant attributes with respect to all the variables (characteristics of the participants and metrics obtained with eye tracking technology).

2. Materials and Methods

2.1. Participants

A disaggregated description of the sample with respect to the variables age, gender and type of participant (prior knowledge vs. no prior knowledge; teacher vs. student) can be found in Table 2.

Table 2. Descriptive statistics of the sample.

Participant Type	N	With Prior Knowledge (n = 17)						Without Prior Knowledge (n = 21)					
		N	n	Men		n	Woman		N	n	Men		n	Woman
				M_{age}	SD_{age}		M_{age}	SD_{age}			M_{age}	SD_{age}		M_{age}	SD_{age}
Students	14	9	5	49.00	23.40	4	45.25	23.47	5	4	30.25	7.93	1	22.00	-
Teachers	24	8	5	47.40	8.62	3	42.67	11.85	16	9	43.00	11.79	7	52.29	4.79

Note. M_{age} = mean age; SD_{age} = standard deviation age.

2.2. Instruments

The following resources were used:

1. Eye tracking equipment iView XTM, SMI Experimenter Center 3.0 and SMI BeGazeTM. These tools record eye movements, their coordinates and the pupillary diameters of each eye. In this study, 60 Hz, static scan path metrics (fixations, saccades, blinks and scan path) were used. In addition, participants viewed the performance of the learning task on a monitor with a resolution of 1680 × 1050.

2. Ad hoc questionnaire on the characteristics of each participant (age, gender, level of studies, branch of knowledge, current employment situation and level of previous knowledge).

The questions were related to the following:

(a) Age;
(b) Gender;
(c) Level of education;
(d) Field of knowledge;
(e) Employment status (active, retired, student);
(f) Knowledge about the origin of monasteries in Europe.

3. Ad hoc crossword puzzle on the knowledge of the information in 5 questions related to the content of the video seen and referring to the origin of monasteries in Europe.

4. Learning task that consisted of a self-regulated video through the figure and voice of an avatar that narrated the task about the origins of monasteries in Europe. The duration of the activity was 120 s.

The questions were related to the following:

(a) Monks belonging to the order of St. Benedict;
(b) Powerful Benedictine monastic centre founded in the 10th century, whose influence spread throughout Europe;
(c) Space around which the organisation of the monastery revolves;
(d) Set of rules that govern monastic life;
(e) Each of the bays or sides of a cloister.

2.3. Procedure

An authorisation was obtained from the Bioethics Committee of the University of Burgos before starting the research. In addition, convenience sampling was used to select the sample. The participants did not receive any financial compensation. They were previously informed of the objectives of the research, and a written informed consent was obtained from all of them. The first phase of the study consisted of collecting personal data and testing the level of prior knowledge. Subsequently, the calibration test was prepared for each participant, using the standard deviation of 0.1–0.9, for both eyes, with a percentage adjustment of between 86.5% and 100%. Subsequently, a test was applied, which consisted of watching a 120-s video about the characteristics of a medieval monastery. The video was designed by a specialist teacher in art history, and the voiceover was provided by a specialist in SRL. After watching the video, each participant completed a crossword puzzle with five questions about the concepts explained in the video. The evaluation sessions were always conducted by the same people: a psychologist with expertise in SRL and a computer engineer, both with experience in the operation of eye tracking technology. Figure 1 shows an image of the calibration procedure and Figure 2 shows the viewing of the video and the completion of the crossword puzzle.

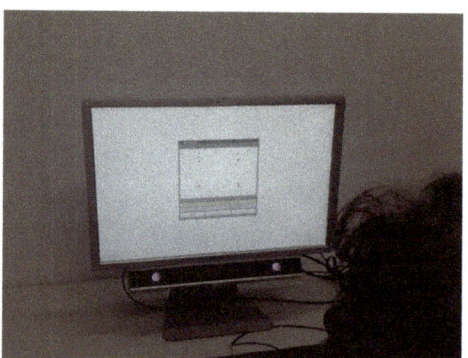
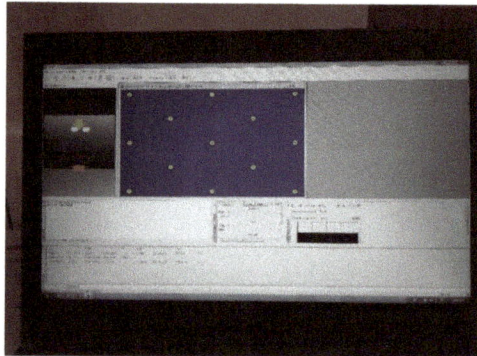

Figure 1. Calibration with eye tracking.

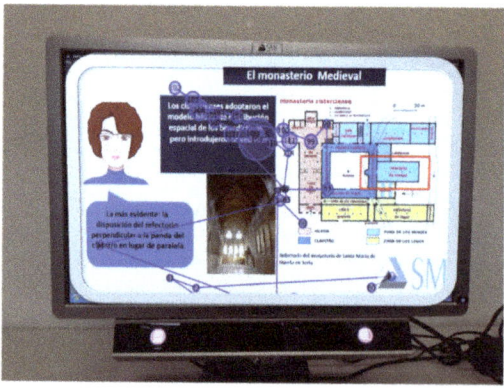
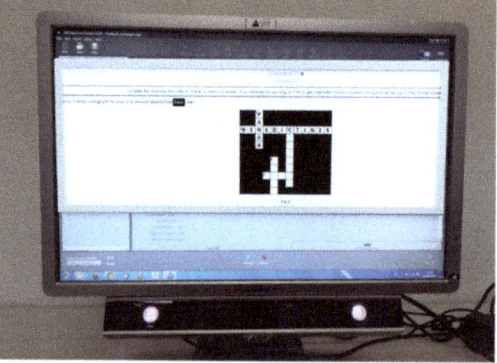

Figure 2. Watching the video and carrying out the crossword puzzle. Note. The circles on the image on the right indicate each point of fixation of the learner on each element in the visual tracking sequence of the image.

2.4. Data Analysis

2.4.1. Statistical Study

A study was conducted using three-factor fixed effects analysis of variance (ANOVA) statistical techniques (type of participant, i.e., student vs. teacher, age (over 50 years old vs. under 50 years old) and knowledge (expert vs. novices)) and eta squared effect value analysis (η^2). Analyses were performed with the SPSS v.24 statistical package [55].

A 2 × 2 × 2 factorial design (experts vs. non-experts, students vs. teachers, age (over 50 years old vs. under 50 years old)) was used [34]. The independent variables were type of participant (experts vs. novice), age (over 50 years old vs. under 50 years old) and participant type (students vs. teachers). The dependent variables were as follows:

- Solving crossword puzzle results;
- Fixations (fixation count, fixation frequency count, fixation duration total, fixation duration average, fixation duration maximum, fixation duration minimum, fixation dispersion total, fixation dispersion average, fixation dispersion maximum, fixation dispersion minimum);
- Saccades (saccade count, saccade frequency count, saccade duration total, saccade duration average, saccade duration maximum, saccade duration minimum, saccade amplitude total, saccade amplitude average, saccade amplitude maximum, saccade amplitude minimum, saccade velocity total, saccade velocity average, saccade velocity maximum, saccade velocity minimum, saccade latency average);
- Blinks (blink count, blink frequency count, blink duration total, blink duration average, blink duration maximum, blink duration minimum) and scan path length.

These metrics are related to the analysis of the cognitive procedure during visual tracking. This procedure is based on the stimulus–processing–response structure. Information enters via the visual pathway (retina-fovea) and is processed at the level of subcortical and cortical regions within the central nervous system. This processing results in a sensory stimulation response. Specifically, saccades constitute a form of sensory-to-motor transformation in response to a stimulus that has been found to be significant and a sensorimotor control of the processing. Saccadic eye movements are used to redirect the fovea from one point of interest to another. Likewise, fixation is used to keep the fovea aligned on the target during subsequent image analysis. This alternating saccade–fixation behaviour is repeated several hundred thousand times a day in humans and is central to complex behaviours such as reading. Saccades can be triggered by the appearance of a visual stimulus that is motivating to the subject or initiated voluntarily by the person's interest in a particular object. Saccades can be suppressed during periods of visual fixation, in which case the brain must inhibit the automatic saccade response [37]. The whole process is summarised in Figure 3. In addition, a video (https://youtu.be/DlRK21afGgo access on 28 June 2021) on the process of performing the task applied in this study can be consulted. In this video, the fixation and saccade points can be seen.

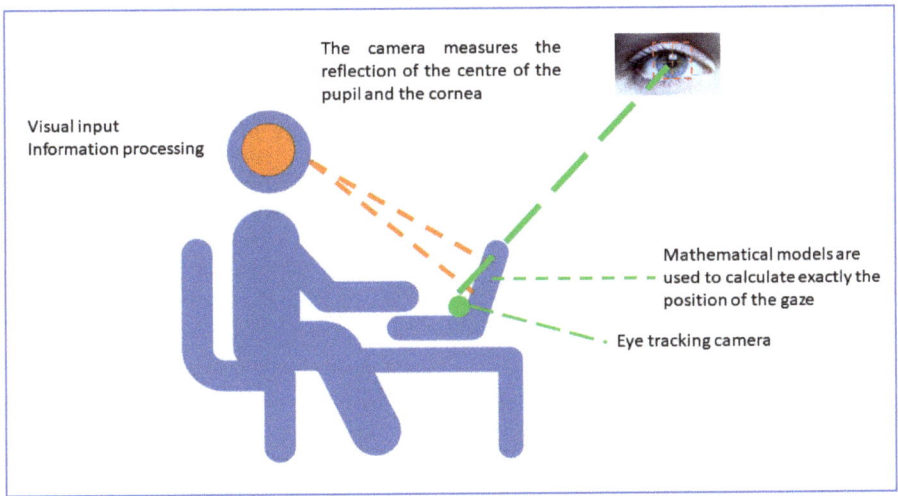

Figure 3. Visual tracking process during the resolution of a task.

2.4.2. Study Using Machine Learning Techniques

As stated in the introduction, machine learning techniques can be divided into supervised learning techniques, which in turn can be subdivided into classification and prediction techniques [21], and unsupervised learning, which refers to the use of clustering techniques [29]. Specifically, supervised learning techniques of pattern analysis are used for human behavioural analysis. These would fall within the supervised learning techniques of clustering [31,32,36]. Concretely, in this study, we used supervised automatic learning techniques for classification (the gain ratio, symmetrical uncertainty and chi-square algorithms were applied) and unsupervised clustering (the k-means ++, fuzzy k-means and DBSCAN algorithms were applied). The analyses were performed with the R programming language [56].

In the study with machine learning techniques, a descriptive correlational design was applied [34]. A supervised learning analysis of classification and non-supervised clustering was applied on all features.

3. Results
3.1. Statistical Study
3.1.1. Previous Analyses

Before starting the testing of the hypotheses, it was checked whether the sample followed a normal distribution, for which a study was conducted on the values of skewness (values below |2.00| are considered accepted values, and a value of skewness = −0.22 was found) and kurtosis (values below |8.00| are considered accepted values, and a value of kurtosis = −2.06 was found). Therefore, the results indicate that the distribution follows the assumptions of normality, which is why parametric statistics were used to test the hypotheses.

3.1.2. Hypothesis Testing Analysis

To test RQ1, a one-factor fixed effects ANCOVA was applied for the participant type "expert vs. novice" considering the covariate (participant type "student vs. teacher") with respect to the dependent variable crossword result. No significant differences were found, but a mean effect value was found (F = 1.91, $p = 0.40$, $\eta^2 = 0.66$). Additionally, no effect of the covariate was found (F = 0.03, $p = 0.90$, $\eta^2 = 0.03$), and in this case, the effect value was low.

To test RQ2, a one-factor fixed effects ANCOVA (participant type "over 50 vs. under 50") was applied considering the covariate (participant type "student vs. teacher"). No significant differences were found in the metrics of fixations, saccades, blinks and scan path length. A covariate effect was only found in the metrics of saccade amplitude minimum (F = 5.19, $p = 0.03$, $\eta^2 = 0.13$) and saccade velocity minimum (F = 5.18, $p = 0.03$, $\eta^2 = 0.13$), in both cases with a low effect value. All results can be found in Table A1 in Appendix A.

Regarding test RQ3, a one-factor fixed effects ANCOVA (participant type "novice vs. expert") was applied considering the covariate (participant type "student vs. teacher"). No significant differences were found in the metrics of fixations, saccades, blinks and scan path length. The effect of the covariate was only found in the metrics of saccade amplitude minimum (F = 6.90, $p = 0.01$, $\eta^2 = 0.16$) and saccade velocity minimum (F = 7.67, $p = 0.01$, $\eta^2 = 0.18$), and in both cases, the effect value was medium. All results can be found in Table A2 in Appendix A.

3.2. Study with Supervised Learning Machine Learning Techniques: Feature Selection

A feature selection analysis was performed with the R programming package mclust, selecting from all possible variables those that received a positive ranking. The gain ratio, symmetrical uncertainty and chi-square algorithms were used for feature selection. Table 3 shows the best values found with each of them for feature selection.

Table 3. Best performing features in the gain ratio, symmetrical uncertainty and chi-square feature selection algorithms.

Features	Gain Ratio	Symmetrical Uncertainty	Chi-Square
Previous Knowledge	0.199	0.199	0.453
Group Type	0.238	0.171	0.421
Employment Status	0.238	0.171	0.421
Gender	0.108	0.067	0.372
Level Degree	0.100	0.082	0.263
Knowledge Branch	0.084	0.057	0.251

(a) The gain ratio is a feature selection method that belongs to the filtering methods. It relies on entropy to assign weights to discrete attributes based on their correlation between the attribute and a target variable (in this study, the results in solving the crossword puzzle). The gain ratio focuses on the information gain metric [57], traditionally used to choose the attribute at a node of a decision tree with the ID3 method. This is the one that generates a partition in which the examples are distributed less randomly among the classes. This method was improved by Quinlan in 1993 [58], as he detected that the information gain was calculated with an unfair favouritism towards attributes with many results. To correct this, he added a value correction based on standardisation by the entropy of that attribute. If Y is the variable to be predicted, then the gain ratio standardises the gain by dividing by the entropy of X. Thus, the C4.5 decision tree construction method uses this measure. From a data mining point of view, this attribute selection could be understood as the selection of attributes as best candidates for the root of a decision tree, which in this study will predict the solving crossword puzzle variable. With H being the entropy, the gain ratio equation is as follows:

$$gain\ ratio = \frac{H(Class) + H(Attribute) - H(Class, Attribute)}{H(Attribute)}$$

Figure 4 shows the correlation matrix found with the gain ratio algorithm in the selection of best features.

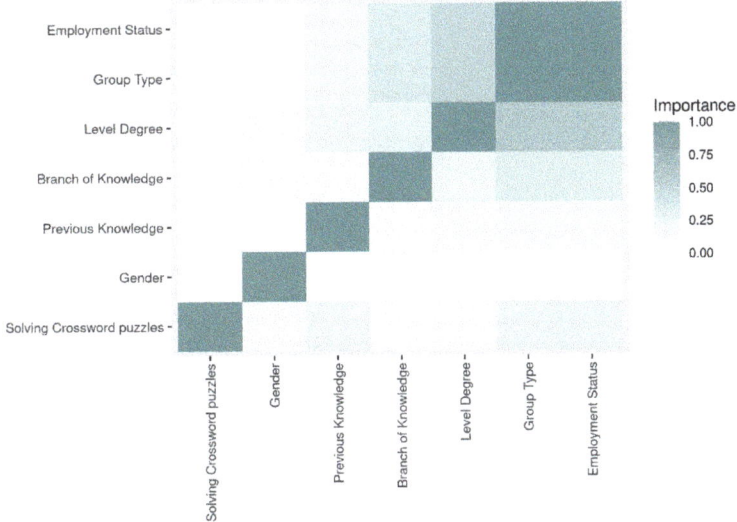

Figure 4. Relationship matrix on the selected features performed with the gain ratio algorithm.

(b) Symmetrical uncertainty is a feature selection method which, as with the gain ratio, belongs to the filter methods and is also based on entropy. Symmetrical uncertainty normalises the values in the range [0, 1]. It also normalises the gain by dividing by the sum of the attribute and class entropies, where H is the entropy.

$$symmetrical\ uncertainty = 2 \times \frac{H(Class) + H(Attribute) - H(Class,\ Attribute)}{H(Attribute) + H(Class)}$$

Figure 5 shows the correlation matrix found with the symmetrical uncertainty algorithm on the best features.

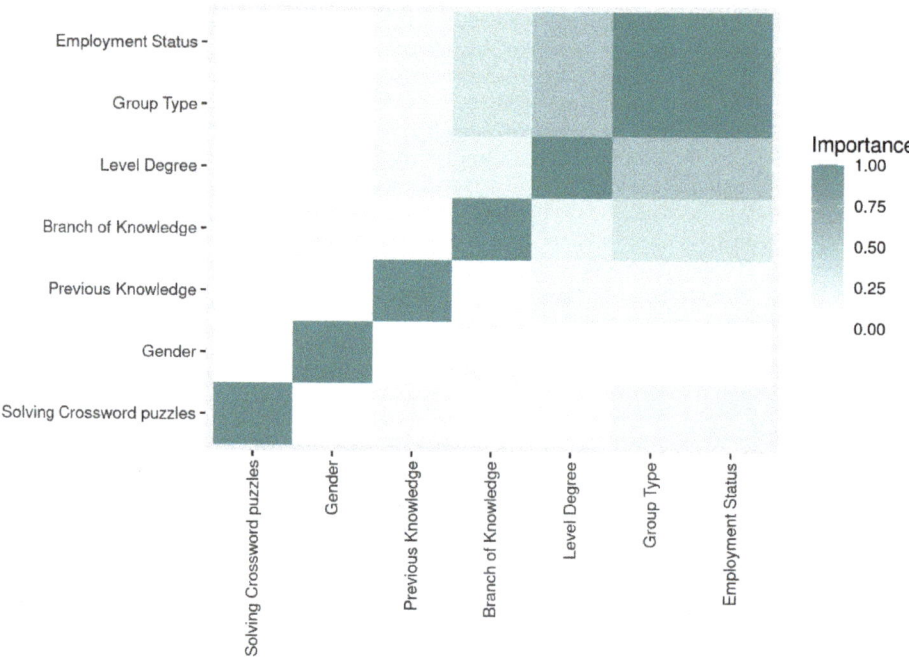

Figure 5. Relationship matrix on the selected features performed with the symmetrical uncertainty algorithm.

(c) Chi-square is a feature selection algorithm that belongs to the filter type and tries to obtain the weights of each feature by using the chi-square test (in case the features are not nominal, it discretises them). The selection result is the same as Cramer's V coefficient. The chi-square equation is as follows:

$$\chi^2 = \sum_{i=1}^{k} \frac{(Oi + Ei)^2}{Ei}$$

where Oi is the observed or empirical absolute frequency and Ei is the expected frequency. Figure 6 shows the correlation matrix found with chi-square (χ^2) [59].

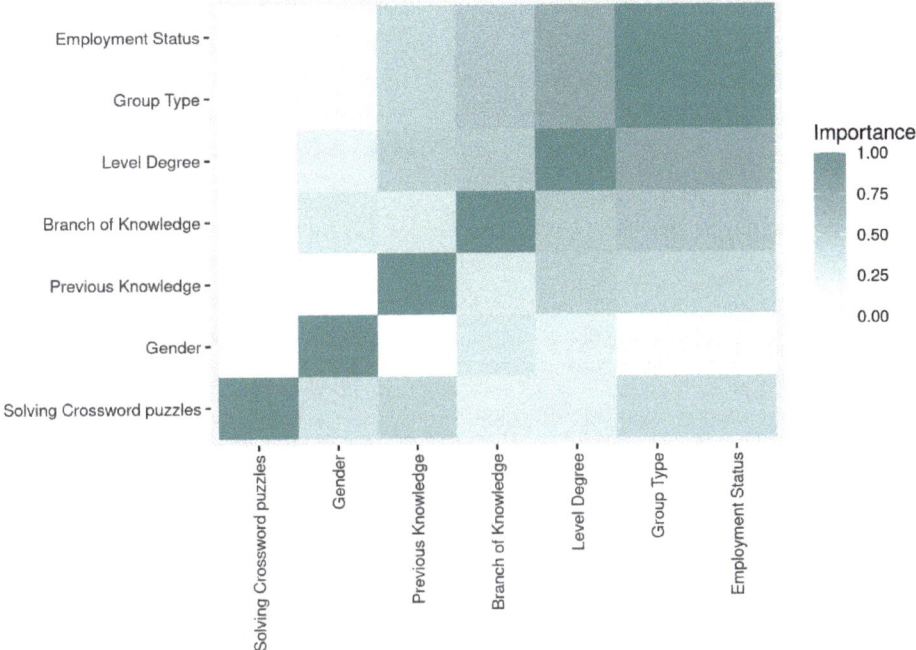

Figure 6. Relationship matrix on the selected characteristics performed with the chi-square algorithm.

3.3. Study with Unsupervised Learning Machine Learning Techniques: Clustering

Finally, cluster detection was performed on the data with unsupervised learning techniques, ignoring the solving crossword puzzles parameter in order to detect patterns in the instances. Nominal variables were transformed into dummy variables in such a way that a variable with n possible different values was divided into n-1 new binary variables, meaning that each of them indicated belonging to one of the previous values. The data were normalised by normalising the mean of the attributes to 0 and the standard deviation to 1. The following clustering algorithms were used:

(a) *k*-means++ is an algorithm for choosing the initial values of the centroids for the *k*-means clustering algorithm. It was proposed in 2007 by Arthur and Vassilvitskii [60] as an approximation algorithm for solving the NP-hard *k*-means problem. That is, a way to avoid the sometimes poor clustering encountered by the standard *k*-means clustering algorithm.

$$D^2(\mu_0) \leq 2D^2(\mu_i) + 2||\mu_i - \mu_0||^2$$

where μ_0 is the initial point selected and D is the distance between point μ_i and the nearest centre of the cluster. Once the centroids are chosen, the process is like the classical *k*-means.

(b) The fuzzy *k*-means algorithm combines the methods based on the optimisation of the objective function with those of fuzzy logic [61,62]. This algorithm performs cluster formation through a soft partitioning of the data. That is, a piece of data would not belong exclusively to a single group but could have different degrees of belonging to several groups. This procedure calculates initial means (m_1, m_2, ..., m_k) to find the degree of membership of data in a cluster. As long as there are no changes in these means, the degree of membership of each data item x_j in cluster i is calculated.

$$u(j, i) = \frac{e^{-(||x_j - m_i||^2)}}{\sum_j e^{-(||x_j - m_i||^2)}}$$

where m_i is the fuzzy mean of all the examples in cluster i.

$$m_i = \frac{\sum_j^i u(j,i)^2 x_j}{\sum_j u(j,i)^2}$$

(c) DBSCAN (density-based spatial clustering of applications with noise) [63] is understood as an algorithm that identifies clusters describing regions with a high density of observations and regions of low density. DBSCAN avoids the problem that other clustering algorithms have by following the idea that, for an observation to be part of a cluster, there must be a minimum number of neighbouring observations (minPts) within a proximity radius (epsilon) and that clusters are separated by empty regions or regions with few observations.

As all remaining variables were nominal after feature selection, after pre-processing the data, only clustering with binary variables was used, which complicated the processing of the k-means++ algorithm by placing the centroids at different locations in the space when the number of centroids was bigger than three. For this reason, the parameter value k in the k-means++ and fuzzy k-means algorithms was equal to 3.

The value of the DBSCAN algorithm parameters was 5 for the minPts variable as it is the default value in the library [64]. To choose the epsilon value, the elbow method was applied. Figure 7 shows the average distance of each point to its nearest neighbouring minPts, and the value 2.97 was chosen for this parameter.

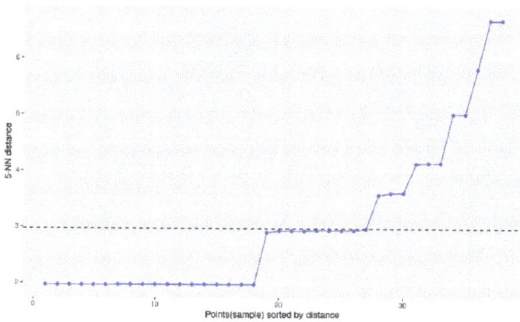

Figure 7. Elbow method in the DBSCAN algorithm.

The visualisation of the clustering results can be seen in Figure 8, which shows the data after applying dimensionality reduction with the principal component analysis (PCA) method. The clusters selected by the k-means++ and fuzzy k-means algorithms are identical, while DBSCAN only found two clusters, leaving instances out of them. These instances labelled as noise in this study are assigned to an additional cluster.

Finally, it has to be considered that when applying an unsupervised learning method, such as clustering, there is no objective variable to evaluate the goodness of the distribution of instances in clusters. However, the goodness of clustering can be tested using the adjusted Rand index (ARI), in order to compare how similar the clustering algorithms are to each other. Thus, if many algorithms perform similar partitions, the conclusion will be consistent [65]. That is, if a pair of instances is in the same cluster in both partitions, this fact will represent similarity between these partitions. In the opposite case, where a pair of instances is in the same cluster in one partition and in different clusters in the other category, it will represent a difference. With n being the number of instances, a being the number of pairs of instances grouped in the same cluster in both partitions and b being the number of pairs of instances grouped in different clusters in different partitions, the Rand index (without adjustment and correction) would be as follows:

$$a = |S_{eq}|, \text{ where } S_{eq} = \{(o_i, o_j) | o_i, o_j \in X_k, o_i, o_j \in Y_l, \}$$
$$b = |S_{eq}|, \text{ where } S_{eq} = \{(o_i, o_j) | o_i \in X_{k1}, o_j \in X_{k2}, o_i \in X_{l1}, o_j \in Y_{l2}, \}$$
$$\text{Rand index} = \frac{a+b}{\binom{n}{2}}$$

A correction is made to the original intuition of the Rand index, since the expected similarity between two partitions established with random models can have pairs of instances that coincide, and this fact would cause the Rand index to never be 0. To make the correction, the adjusted Rand index algorithm, ARI, was applied, in which negative values can be found if the similarity is less than expected, being equal to

$$\text{Adjusted rand index} = \frac{Index - Expected\ Index}{Maximun\ Index - Expected\ Index}$$

The applied ARI formula is therefore

$$ARI = \frac{\sum_{ij} \binom{n_{ij}}{2} - \left[\sum_i \binom{a_i}{2} \sum_j \binom{b_j}{2}\right] / \binom{n}{2}}{\frac{1}{2}\left[\sum_i \binom{a_i}{2} \sum_j \binom{b_j}{2}\right] / \binom{n}{2}}$$

where if $X = \{X_1, X_2, ..., X_r\}$ and $Y = \{Y_1, Y_2, ..., Y_s\}$, then $n_{ij} = Xi \cap Yj$, $ai = \sum_{jnij}$ and $b_i = \sum_{inij}$.

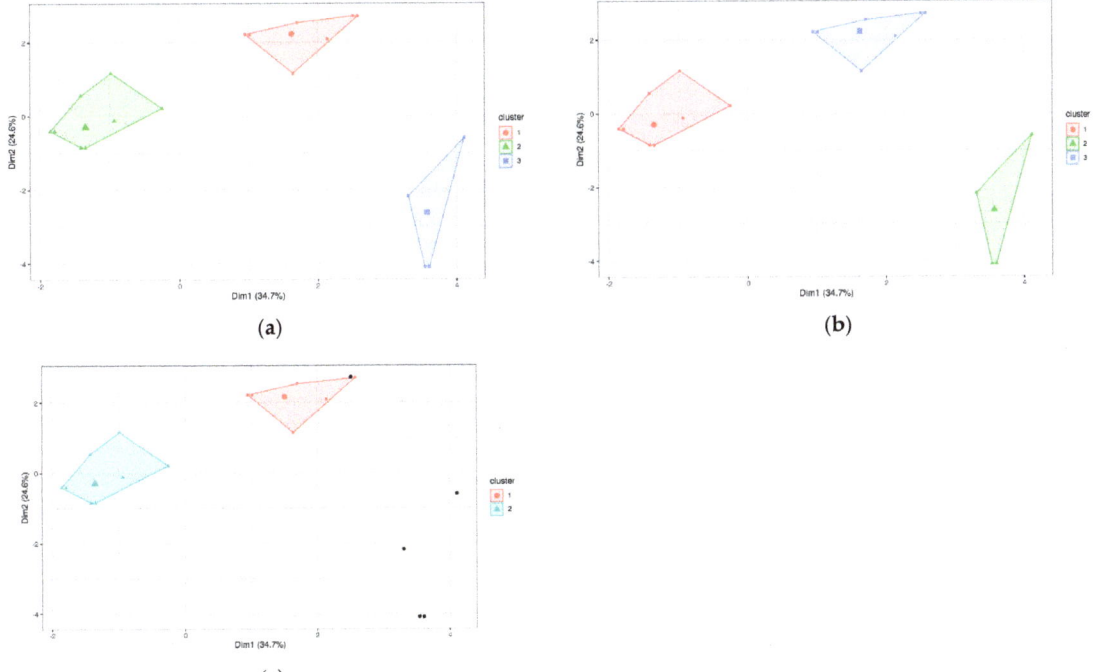

Figure 8. (a) Clustering with the *k*-means ++ algorithm; (b) clustering with the fuzzy means algorithm; (c) clustering with the DBSCAN algorithm.

Thus, the ARI can have a value between −1 and 1, where 1 indicates that the two data clusters match exactly in every pair of points, 0 is the expected value for randomly created clusters and −1 is the worst fit. The results indicate that the algorithms that provide the best fit are *k*-means ++ and fuzzy *k*-means (ARI = 1), *k*-means ++ and DBSCAN (ARI = 0.96) and fuzzy *k*-means and DBSCAN (ARI = 0.9), where the higher the intensity, the higher the relationship. It can therefore be concluded that the degree of fit between the algorithms applied in this study is good for all possible associations (show Figure 9).

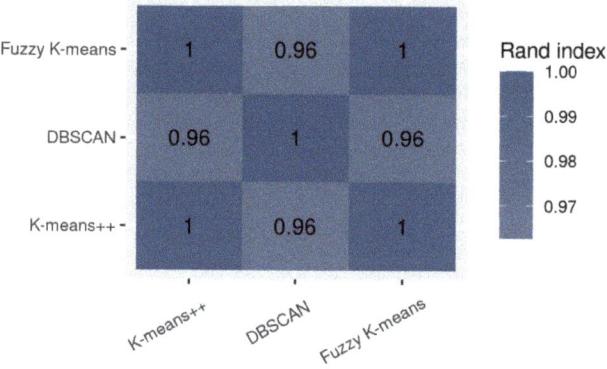

Figure 9. Adjusted Rand index (ARI).

4. Discussion

Regarding the results found in the RQ1 check, it was not confirmed that participants with prior knowledge performed better on the crossword puzzle solving test than non-experts. In line with studies by Eberhard et al. [2], Takacs and Bus [4] and Verhallen and Bus [5], this may be explained by the fact that the task was presented in a video that included self-regulated speech. This technique has been shown to be very effective in mitigating the differences between novice vs. experienced learners [12–14]. However, although no significant differences were found with respect to the independent variable, a mean effect value was found. This suggests that the participant type variable "novice vs. expert" is an important variable in task resolution processes. However, in this study, this effect may have been mitigated by the way the task was presented (self-regulated procedure). This result coincides with the findings of studies that conclude that the lack of prior knowledge in novice learners can be compensated by the proposal of self-regulated multi-measure tasks [12–15,35,36]. The explanation is that self-regulated video may facilitate homogeneity in the encoding of information, attention to relevant vs. non-relevant information and in the route taken in the scan path [18,19].

Regarding RQ2, no effect of age was found on the metrics of fixations, saccades, blinks and scan path length. This may be explained by the way the task was presented (self-regulated video), or by the participants' prior knowledge. In this line, research [8] supports that prior knowledge compensates for the effects of age on cognitive functioning, for example, on long-term memory processes or reaction times. In addition, it has been found that the covariate participant type "student vs. teacher" does weigh on task performance. Specifically, differences were found in the saccade amplitude minimum and saccade velocity minimum parameters. These data can be related to the findings of studies indicating that age effects can be mitigated by learners' prior knowledge of the task [8] and also by self-regulated presentation of the task [18,19]. In fact, the significant differences found in the covariate focused on saccade amplitude and minimum saccade velocity, which is consistent with studies that found differences in saccade type depending on the phase of information encoding the learner was at [18,19]. This result is important for future research proposals. The reason is that the way students vs. teachers process information might be influencing the way they learn. For example, teachers might develop more systematic processing that would compensate for their lack of knowledge in a task. Alternatively, younger students might implement more effective learning, and thus processing, strategies even though they are novices [13]. These hypotheses will be explored in future studies.

Regarding RQ3, no significant differences were found in the metrics of fixations, saccades, blinks, and scan path length depending on whether the participant was a novice or an expert. This may be explained by the way the task was presented (a self-regulated video with a set time duration). However, future studies could test the results on videos that did not include self-regulation and/or that could be viewed more than once. We also found that there is an effect of the covariate participant type "student vs. teacher" on the saccade amplitude minimum and saccade velocity minimum parameters. As indicated in RQ2, this is an important fact to consider in future research, as the way students vs. teachers process information could be influencing the type of information processing. Similarly, future studies could test whether the form of task presentation (self-regulated vs. non-self-regulated; timed vs. untimed, etc.) could be influencing the form of processing (fixations, saccades, blinks, scan path length). Similarly, processing patterns could be found for different participant types (novice vs. expert, with different age intervals, etc.), and the types of patrons could be tested according to the type of participant.

According to the analysis performed with supervised learning methods of feature selection, it was found that the different algorithms applied (gain ratio, symmetrical uncertainty, chi-square) provided valuable information regarding the most significant attributes in the study. In this case, the following attributes were considered as important: previous knowledge, group type, employment status, gender, level degree and knowledge branch. This result is very interesting for future research, as it provides information on the possible effects of characteristics that were not considered as independent variables in the statistical study (employment status, gender, level degree and knowledge branch).

Regarding the study with unsupervised learning techniques (clustering), it allowed us to know the grouping, i.e., the similar interaction patterns of the participants in the selected characteristics. The three algorithms applied had a good ARI. This result is important for future studies, as a learning style profile can be extracted for each group and its relationship with the outcome of the learning tasks and with the reaction times for the execution of the tasks can be checked.

5. Conclusions

The use of the eye tracking technique provides evidence on the processing of information in different types of participants during the resolution of different tasks [9–11]. This fact facilitates research in behavioural sciences [37]. Working with this technology opens up many fields of research applied to numerous environments (learning to read and write, logical-mathematical reasoning, physics, driving vehicles, driving dangerous machines, marketing, etc.) [38–42]. It can also be used to find out how people with different learning disabilities [45] (ADHD, ASD, etc.) learn. Therefore, it could improve their learning style and make proposals for personalised intervention according to the needs observed in each of them. In addition, this technology can be used to improve driving practices and accident prevention with regard to the handling of dangerous machinery. This training is being carried out in virtual and/or augmented reality scenarios [49–51] that apply eye tracking technology. All these possibilities open an important field to be addressed in future research.

Another relevant aspect to take into account is the way tasks are presented. This study has shown that the use of self-regulated tasks facilitates the processing of information and homogenises learning responses between novice and expert learners [12–15,35,36]. Therefore, in future studies, we will study participants' processing in different types of tasks (self-regulated designs with avatars, zooming in on the most relevant information, etc.). Likewise, the results will be tested in different educational stages (early childhood education, primary education, secondary education, university education and non-formal education) and in different subjects (experimental vs. non-experimental).

Subsequently, this study has shown that the use of different automatic learning techniques such as feature selection facilitates the knowledge of attributes that may be more significant for the research. This functionality is very useful in research that works with a large volume of features or instances. Moreover, if this technique is combined with the use of machine learning techniques and traditional statistics, the results can provide more information, especially related to future lines of research. In fact, in this study, it has been found that some of the variables considered as independent in the statistical study were also selected as relevant features in the study that applied supervised learning techniques of instance selection (e.g., prior knowledge, type of participant (student vs. teacher)). However, the feature selection techniques have also provided clues to be taken into account in future studies on the influence of other variables (e.g., gender, employment status, level of education and field of knowledge). In this line, the use of different algorithms to test both feature selection and clustering in unsupervised learning provides the researcher with a repertoire of results whose fit can be contrasted with the ARI. This will make it possible to know the groupings among the learners and to isolate the patterns of the types of learners in order to be able to offer educational responses based on personalised learning. On the other hand, the use of statistical analysis methods makes it possible to ascertain whether the variables indicated as independent have an effect on the dependent variables. In summary, perhaps the most useful procedure is, first, to apply the techniques of supervised learning of characteristics and then, depending on the variables detected, to pose the research questions and apply the relevant statistical analyses to test them.

Finally, the results of this study must be taken with caution, as this study has a series of limitations. These are mainly related to the size of the sample, which is small, and the selection of the sample, which was conducted convenience sampling. However, it must be considered that the use of the eye tracking methodology requires a very exhaustive control of the development of tasks in laboratory spaces, an aspect that makes it difficult for the samples to be large and randomised. Another of the limiting elements of this work is that a very specific task (acquisition of the concepts of the origins of monasteries in Europe and verification of this acquisition through the resolution of a crossword puzzle) was used in a specific learning environment (history of art). For this reason, possible future studies have been indicated in the Discussion and Conclusions sections.

Author Contributions: Conceptualisation, M.C.S.-M., I.R.P., L.A. and A.A.R.; methodology, M.C.S.-M., I.R.P. and A.A.R.; software, I.R.P.; validation, M.C.S.-M., I.R.P. and A.A.R.; formal analysis, M.C.S.-M., I.R.P. and A.A.R.; investigation, M.C.S.-M.; resources, M.C.S.-M.; data curation, M.C.S.-M.; writing—original draft preparation, M.C.S.-M.; writing—review and editing, M.C.S.-M.; I.R.P., C.F.M., A.A.R., L.A. and S.R.A.; visualisation, M.C.S.-M., I.R.P. and A.A.R.; supervision, M.C.S.-M.; project administration, M.C.S.-M.; funding acquisition, M.C.S.-M., L.A. and S.R.A. All authors have read and agreed to the published version of the manuscript.

Funding: This work was funded through the European Project "Self-Regulated Learning in SmartArt" 2019-1-ES01-KA204-065615.

Institutional Review Board Statement: The Ethics Committee of the University of Burgos approved this study No. IR27/2019.

Informed Consent Statement: Written informed consent was obtained from all participants in this study in accordance with the Declaration of Helsinki guidelines.

Data Availability Statement: The datasets generated for this study are available on request to the corresponding author.

Acknowledgments: The authors would like to thank all the people and entities that have collaborated in this study, especially the research group ADIR of Universidad de Oviedo for loaning the eye tracking equipment iView XTM, SMI Experimenter Center 3.0 and SMI BeGazeTM, and the Director of the Experience University of the University of Burgos.

Conflicts of Interest: The authors declare no conflict of interest.

Appendix A

Table A1. One-factor ANCOVA with fixed effects (age over 50 vs. under 50) and covariate (student vs. teacher).

Type of Access	N	n	G1 M (SD)	n	G2 M (SD)	df	F	p	η²
Independent Variable (novel vs. expert)									
Fixation Count	38	17	654.18 (138.56)	21	625.19 (189.87)	1,35	0.09	0.76	0.003
Fixation Frequency Count	38	17	3.01 (0.67)	21	2.96 (0.91)	1,35	0.09	0.76	0.003
Fixation Duration Total	38	17	166,132.18 (44,244.00)	21	152,531.78 (48,472.19)	1,35	0.49	0.49	0.01
Fixation Duration Average	38	17	255.93 (72.55)	21	254.36 (88.36)	1,35	0.004	0.95	0.000
Fixation Duration Maximum	38	17	1189.10 (484.92)	21	1286.40 (623.33)	1,35	0.42	0.52	0.01
Fixation Duration Minimum	38	17	83.21 (0.05)	21	83.21 (0.04)	1,35	0.03	0.86	0.001
Fixation Dispersion Total	38	17	47,498.23 (11,528.53)	21	46,202.81 (15,068.91)	1,35	0.01	0.93	0.000
Fixation Dispersion Average	38	17	72.50 (5.00)	21	73.58 (5.00)	1,35	0.42	0.52	0.01
Fixation Dispersion Maximum	38	17	99.98 (0.04)	21	98.89 (0.39)	1,35	0.79	0.38	0.02
Fixation Dispersion Minimum	38	17	11.54 (4.65)	21	9.94 (5.01)	1,35	0.85	0.36	0.02
Saccade Count	38	17	664.29 (136.78)	21	632.24 (195.64)	1,35	0.12	0.73	0.003
Saccade Frequency Count	38	17	3.15 (0.65)	21	3.00 (0.93)	1,35	0.12	0.73	0.003
Saccade Duration Total	38	17	32,282.09 (27,891.81)	21	31,241.80 (21,906.09)	1,35	0.03	0.86	0.001
Saccade Duration Average	38	17	59.47 (89.43)	21	52.58 (44.21)	1,35	0.23	0.63	0.02
Saccade Duration Maximum	38	17	629.38 (1332.76)	21	467.00 (414.22)	1,35	0.49	0.49	0.01
Saccade Duration Minimum	38	17	16.57 (0.05)	21	16.49 (0.30)	1,35	2.00	0.17	0.05
Saccade Amplitude Total	38	17	4825.23 (6815.38)	21	4740.95 (4780.52)	1,35	0.02	0.88	0.001
Saccade Amplitude Average	38	17	10.74 (25.55)	21	8.69 (10.82)	1,35	0.26	0.61	0.02
Saccade Amplitude Maximum	38	17	156.15 (300.65)	21	119.67 (96.25)	1,35	0.47	0.50	0.013
Saccade Amplitude Minimum	38	17	0.03 (0.05)	21	0.05 (0.07)	1,35	0.35	0.56	0.010
Saccade Velocity Total	38	17	61,828.35 (17,396.33)	21	66,554.31 (26,550.69)	1,35	0.50	0.49	0.014
Saccade Velocity Average	38	17	96.85 (36.43)	21	113.01 (45.91)	1,35	0.95	0.34	0.03
Saccade Velocity Maximum	38	17	878.00 (190.65)	21	844.58 (173.95)	1,35	0.25	0.62	0.01
Saccade Velocity Minimum	38	17	2.81 (1.43)	21	3.62 (2.47)	1,35	0.75	0.39	0.02
Saccade Latency Average	38	17	279.93 (64.36)	21	295.73 (106.93)	1,35	0.12	0.73	0.003
Blink Count	38	17	33.12 (25.59)	21	45.00 (37.00)	1,35	1.14	0.29	0.03
Blink Frequency Count	38	17	0.15 (0.12)	21	0.21 (0.18)	1,35	1.14	0.29	0.03
Blink Duration Total	38	17	6777.12 (9174.98)	21	20,619.05 (41,403.88)	1,35	1.36	0.25	0.04
Blink Duration Average	38	17	202.48 (244.52)	21	545.61 (1352.20)	1,35	0.77	0.39	0.02
Blink Duration Maximum	38	17	898.75 (2087.86)	21	5951.36 (19,080.82)	1,35	0.88	0.36	0.02
Blink Duration Minimum	38	17	85.19 (5.57)	21	84.80 (5.05)	1,35	0.07	0.79	0.002
Scan Path Length	38	17	122,506.94 (21157.24)	21	117,620.71 (36,042.24)	1,35	0.16	0.69	0.01

Table A1. Cont.

Type of Access	N	n	G1 M (SD)	n	G2 M (SD)	df	F	p	η^2
Covariable (type of participant student vs. professor)									
Fixation Count	38	17		21		1,35	1.61	0.21	0.04
Fixation Frequency Count	38	17		21		1,35	1.53	0.23	0.04
Fixation Duration Total	38	17		21		1,35	1.12	0.30	0.03
Fixation Duration Average	38	17		21		1,35	0.001	0.98	0.000
Fixation Duration Maximum	38	17		21		1,35	0.60	0.44	0.02
Fixation Duration Minimum	38	17		21		1,35	0.04	0.84	0.001
Fixation Dispersion Total	38	17		21		1,35	1.36	0.25	0.04
Fixation Dispersion Average	38	17		21		1,35	0.002	0.97	0.000
Fixation Dispersion Maximum	38	17		21		1,35	0.08	0.78	0.002
Fixation Dispersion Minimum	38	17		21		1,35	0.12	0.73	0.004
Saccade Count	38	17		21		1,35	1.73	0.20	0.047
Saccade Frequency Count	38	17		21		1,35	1.65	0.21	0.045
Saccade Duration Total	38	17		21		1,35	0.11	0.74	0.003
Saccade Duration Average	38	17		21		1,35	1.05	0.31	0.03
Saccade Duration Maximum	38	17		21		1,35	1.09	0.30	0.03
Saccade Duration Minimum	38	17		21		1,35	2.41	0.13	0.06
Saccade Amplitude Total	38	17		21		1,35	0.44	0.51	0.01
Saccade Amplitude Average	38	17		21		1,35	1.18	0.28	0.03
Saccade Amplitude Maximum	38	17		21		1,35	1.01	0.32	0.03
Saccade Amplitude Minimum	38	17		21		1,35	5.19	0.03 *	0.13
Saccade Velocity Total	38	17		21		1,35	0.28	0.60	0.01
Saccade Velocity Average	38	17		21		1,35	1.27	0.27	0.04
Saccade Velocity Maximum	38	17		21		1,35	0.08	0.77	0.002
Saccade Velocity Minimum	38	17		21		1,35	5.18	0.03 *	0.13
Saccade Latency Average	38	17		21		1,35	1.19	0.28	0.03
Blink Count	38	17		21		1,35	0.02	0.81	0.001
Blink Frequency Count	38	17		21		1,35	0.000	0.98	0.000
Blink Duration Total	38	17		21		1,35	0.93	0.34	0.03
Blink Duration Average	38	17		21		1,35	0.58	0.45	0.02
Blink Duration Maximum	38	17		21		1,35	0.53	0.47	0.02
Blink Duration Minimum	38	17		21		1,35	0.09	0.77	0.003
Scan Path Length	38	17		21		1,35	0.21	0.65	0.01

Note. G1 = participants younger than 50 years; G2 = participants older than 50 years; M = mean; SD = standard deviation; df = degrees of freedom; η^2 = eta squared effect value; * $p < 0.05$.

Table A2. One-factor ANCOVA with fixed effects (age over 50 vs. under 50) and covariate (student vs. teacher).

Type of Access	N	n	G1 M (SD)	n	G2 M (SD)	df	F	p	η^2
Independent Variable (novel vs. expert)									
Fixation Count	38	25	628.92 (183.40)	13	655.92 (136.21)	1,35	0.55	0.46	0.02
Fixation Frequency Count	38	25	2.98 (0.88)	13	3.10 (0.65)	1,35	0.51	0.48	0.01
Fixation Duration Total	38	25	152,469.04 (54,256.14)	13	170,437.56 (23,520.29)	1,35	1.98	0.17	0.05
Fixation Duration Average	38	25	243.64 (69.26)	13	277.03 (98.19)	1,35	1.49	0.23	0.04
Fixation Duration Maximum	38	25	1184.55 (512.27)	13	1355.00 (650.35)	1,35	1.06	0.31	0.03
Fixation Duration Minimum	38	25	83.22 (0.06)	13	83.20 (0.00)	1,35	1.16	0.29	0.03
Fixation Dispersion Total	38	25	46,170.27 (14,279.23)	13	47,959.40 (12,120.32)	1,35	0.39	0.54	0.01

Table A2. Cont.

Type of Access	N	n	G1 M (SD)	n	G2 M (SD)	df	F	p	η²
Fixation Dispersion Average	38	25	73.19 (4.65)	13	72.89 (5.72)	1,35	0.04	0.84	0.001
Fixation Dispersion Maximum	38	25	99.90 (0.36)	13	99.99 (0.03)	1,35	1.03	0.32	0.03
Fixation Dispersion Minimum	38	25	11.02 (4.59)	13	9.95 (5.44)	1,35	0.30	0.59	0.01
Saccade Count	38	25	638.44 (186.76)	13	662.23 (139.21)	1,35	0.47	0.50	0.01
Saccade Frequency Count	38	25	3.03 (0.89)	13	3.12 (0.67)	1,35	0.36	0.55	0.01
Saccade Duration Total	38	25	35,070.90 (28,103.57)	13	25,238.52 (13,761.84)	1,35	1.57	0.22	0.04
Saccade Duration Average	38	25	65.17 (81.24)	13	37.38 (14.49)	1,35	2.06	0.16	0.06
Saccade Duration Maximum	38	25	633.59 (1130.76)	13	358.96 (252.82)	1,35	1.12	0.30	0.03
Saccade Duration Minimum	38	25	16.52 (0.26)	13	16.52 (0.16)	1,35	0.07	0.80	0.002
Saccade Amplitude Total	38	25	5679.51 (6777.34)	13	3046.24 (1794.51)	1,35	2.31	0.14	0.06
Saccade Amplitude Average	38	25	12.20 (22.60)	13	4.62 (2.49)	1,35	2.05	0.16	0.06
Saccade Amplitude Maximum	38	25	161.21 (254.94)	13	87.50 (56.00)	1,35	1.49	0.23	0.04
Saccade Amplitude Minimum	38	25	0.04 (0.07)	13	0.03 (0.05)	1,35	1.39	0.25	0.04
Saccade Velocity Total	38	25	66,146.90 (24,164.73)	13	61,157.70 (20,255.49)	1,35	0.31	0.58	0.01
Saccade Velocity Average	38	25	112.39 (47.64)	13	93.07 (26.12)	1,35	2.78	0.10	0.074
Saccade Velocity Maximum	38	25	882.20 (193.02)	13	815.95 (148.76)	1,35	1.02	0.32	0.03
Saccade Velocity Minimum	38	25	3.50 (2.41)	13	2.79 (1.20)	1,35	2.49	0.12	0.07
Saccade Latency Average	38	25	282.75 (76.56)	13	300.02 (113.31)	1,35	0.12	0.73	0.003
Blink Count	38	25	39.24 (29.03)	13	40.54 (39.73)	1,35	0.003	0.96	0.000
Blink Frequency Count	38	25	0.18 (0.14)	13	0.18 (0.20)	1,35	0.000	0.98	0.000
Blink Duration Total	38	25	15,683.28 (38,413.22)	13	12,009.94 (12,594.05)	1,35	0.32	0.58	0.01
Blink Duration Average	38	25	418.80 (1251.58)	13	340.78 (286.44)	1,35	0.16	0.69	0.01
Blink Duration Maximum	38	25	4263.06 (17603.20)	13	2590.82 (3294.93)	1,35	0.27	0.61	0.01
Blink Duration Minimum	38	25	85.22 (5.57)	13	84.51 (4.66)	1,35	0.20	0.66	0.01
Scan Path Length	38	25	122,693.40 (31,212.68)	13	114,255.23 (27,953.39)	1,35	0.52	0.48	0.02
Covariable (type of participant student vs. professor)									
Fixation Count	38	25		13		1,35	2.14	0.15	0.06
Fixation Frequency Count	38	25		13		1,35	2.03	0.16	0.06
Fixation Duration Total	38	25		13		1,35	2.13	0.15	0.06
Fixation Duration Average	38	25		13		1,35	0.04	0.84	0.001
Fixation Duration Maximum	38	25		13		1,35	0.74	0.40	0.02
Fixation Duration Minimum	38	25		13		1,35	0.14	0.71	0.004
Fixation Dispersion Total	38	25		13		1,35	1.69	0.20	0.05
Fixation Dispersion Average	38	25		13		1,35	0.04	0.85	0.001
Fixation Dispersion Maximum	38	25		13		1,35	0.39	0.54	0.01
Fixation Dispersion Minimum	38	25		13		1,35	0.16	0.69	0.01
Saccade Count	38	25		13		1,35	2.27	0.14	0.06
Saccade Frequency Count	38	25		13		1,35	2.12	0.15	0.06
Saccade Duration Total	38	25		13		1,35	0.29	0.59	0.01
Saccade Duration Average	38	25		13		1,35	1.53	0.23	0.04
Saccade Duration Maximum	38	25		13		1,35	1.28	0.27	0.04
Saccade Duration Minimum	38	25		13		1,35	1.75	0.20	0.05

Table A2. Cont.

Type of Access	N	n	G1 M (SD)	n	G2 M (SD)	df	F	p	η^2
Saccade Amplitude Total	38	25		13		1,35	0.89	0.35	0.03
Saccade Amplitude Average	38	25		13		1,35	1.67	0.21	0.05
Saccade Amplitude Maximum	38	25		13		1,35	1.27	0.27	0.04
Saccade Amplitude Minimum	38	25		13		1,35	6.90	0.01 *	0.16
Saccade Velocity Total	38	25		13		1,35	0.09	0.77	0.003
Saccade Velocity Average	38	25		13		1,35	2.67	0.11	0.07
Saccade Velocity Maximum	38	25		13		1,35	0.04	0.85	0.001
Saccade Velocity Minimum	38	25		13		1,35	7.67	0.01 *	0.18
Saccade Latency Average	38	25		13		1,35	1.17	0.29	0.032
Blink Count	38	25		13		1,35	0.10	0.75	0.003
Blink Frequency Count	38	25		13		1,35	0.03	0.87	0.001
Blink Duration Total	38	25		13		1,35	1.55	0.22	0.04
Blink Duration Average	38	25		13		1,35	0.95	0.34	0.03
Blink Duration Maximum	38	25		13		1,35	0.95	0.34	0.03
Blink Duration Minimum	38	25		13		1,35	0.12	0.74	0.003
Scan Path Length	38	25		13		1,35	0.15	0.70	0.004

Note. G1 = novice participants; G2 = expert participants; M = mean; SD = standard deviation; df = degrees of freedom; η^2 = eta squared effect value; * $p < 0.05$.

References

1. van Marlen, T.; van Wermeskerken, M.; Jarodzka, H.; van Gog, T. Effectiveness of eye movement modeling examples in problem solving: The role of verbal ambiguity and prior knowledge. *Learn. Instr.* **2018**, *58*, 274–283. [CrossRef]
2. Eberhard, K.M.; Tanenhaus, M.K.; Sciences, C.; Sedivy, J.C.; Sciences, C.; Hall, M. Eye movements as a window into real-time spoken language comprehension in natural contexts. *J. Psycholinguist. Res.* **1995**, *24*, 409–436. [CrossRef]
3. Bruder, C.; Hasse, C. Differences between experts and novices in the monitoring of automated systems. *Int. J. Ind. Ergon.* **2019**, *72*, 1–11. [CrossRef]
4. Takacs, Z.K.; Bus, A.G. How pictures in picture storybooks support young children's story comprehension: An eye-tracking experiment. *J. Exp. Child. Psychol.* **2018**, *174*, 1–12. [CrossRef]
5. Verhallen, M.J.A.J.; Bus, A.G. Young second language learners' visual attention to illustrations in storybooks. *J. Early Child. Lit.* **2011**, *11*, 480–500. [CrossRef]
6. Ooms, K.; de Maeyer, P.; Fack, V.; van Assche, E.; Witlox, F. Interpreting maps through the eyes of expert and novice users. *Int. J. Geogr. Inf. Sci.* **2012**, *26*, 1773–1788. [CrossRef]
7. Hilton, C.; Miellet, S.; Slattery, T.J.; Wiener, J. Are age-related deficits in route learning related to control of visual attention? *Psychol. Res.* **2020**, *84*, 1473–1484. [CrossRef]
8. Sáiz Manzanares, M.C.; Rodríguez-Díez, J.J.; Marticorena-Sánchez, R.; Zaparaín-Yáñez, M.J.; Cerezo-Menéndez, R. Lifelong learning from sustainable education: An analysis with eye tracking and data mining techniques. *Sustainability* **2020**, *12*, 1970. [CrossRef]
9. Kitchenham, B.A.; Dybå, T.; Jørgensen, M. Evidence-based software engineering. In Proceedings of the 26th International Conference on Software Engineering, Edinburgh, UK, 28 May 2004; pp. 273–281. [CrossRef]
10. Joe, L.P.I.; Sasirekha, S.; Uma Maheswari, S.; Ajith, K.A.M.; Arjun, S.M.; Athesh, K.S. Eye Gaze Tracking-Based Adaptive E-learning for Enhancing Teaching and Learning in Virtual Classrooms. In *Information and Communication Technology for Competitive Strategies*; Fong, S., Akashe, S., Mahalle, P.N., Eds.; Springer: Singapore, 2019; pp. 165–176. [CrossRef]
11. Rayner, K. Eye Movements in Reading and Information Processing: 20 Years of Research. *Psychol. Bull.* **1998**, *124*, 372–422. [CrossRef]
12. Taub, M.; Azevedo, R.; Bradbury, A.E.; Millar, G.C.; Lester, J. Using sequence mining to reveal the efficiency in scientific reasoning during STEM learning with a game-based learning environment. *Learn. Instr.* **2018**, *54*, 93–103. [CrossRef]
13. Taub, M.; Azevedo, R. Using Sequence Mining to Analyze Metacognitive Monitoring and Scientific Inquiry Based on Levels of Efficiency and Emotions during Game-Based Learning. *JEDM* **2018**, *10*, 1–26. [CrossRef]
14. Cloude, E.B.; Taub, M.; Lester, J.; Azevedo, R. The Role of Achievement Goal Orientation on Metacognitive Process Use in Game-Based Learning. In *Artificial Intelligence in Education*; Isotani, S., Millán, E., Ogan, A., Hastings, P., McLaren, B., Luckin, R., Eds.; Springer International Publishing: Cham, Switzerland, 2019; pp. 36–40. [CrossRef]
15. Azevedo, R.; Gašević, D. Analyzing Multimodal Multichannel Data about Self-Regulated Learning with Advanced Learning Technologies: Issues and Challenges. *Comput. Hum. Behav.* **2019**, *96*, 207–210. [CrossRef]
16. Liu, H.-C.; Chuang, H.-H. An examination of cognitive processing of multimedia information based on viewers' eye movements. *Interact. Learn. Environ.* **2011**, *19*, 503–517. [CrossRef]

17. Privitera, C.M.; Stark, L.W. Algorithms for defining visual regions-of-Interest: Comparison with eye fixations. *IEEE Trans. Pattern Anal. Mach. Intell.* **2000**, *22*, 970–982. [CrossRef]
18. Sharafi, Z.; Soh, Z.; Guéhéneuc, Y.G. A systematic literature review on the usage of eye-tracking in software engineering. *Inf. Softw. Technol.* **2015**, *67*, 79–107. [CrossRef]
19. Sharafi, Z.; Shaffer, T.; Sharif, B.; Guéhéneuc, Y.G. Eye-tracking metrics in software engineering. In Proceedings of the 015 Asia-Pacific Software Engineering Conference (APSEC), New Delhi, India, 1–4 December 2015; pp. 96–103. [CrossRef]
20. Maltz, M.; Shina, D. Eye movements of younger and older drivers. *Hum. Factors* **1999**, *41*, 15–25. [CrossRef]
21. Dalrymple, K.A.; Jiang, M.; Zhao, Q.; Elison, J.T. Machine learning accurately classifies age of toddlers based on eye tracking. *Sci. Rep.* **2019**, *9*, 6255. [CrossRef]
22. Shen, J.; Elahipanah, A.; Reingold, E.M. Effects of context and instruction on the guidance of eye movements during a conjunctive visual search task. *Eye Mov.* **2007**, 597–615. [CrossRef]
23. Alemdag, E.; Cagiltay, K. A systematic review of eye tracking research on multimedia learning. *Comput. Educ.* **2018**, *125*, 413–428. [CrossRef]
24. Scherer, R.; Siddiq, F.; Tondeur, J. The technology acceptance model (TAM): A meta-analytic structural equation modeling approach to explaining teachers' adoption of digital technology in education. *Comput. Educ.* **2019**, *128*, 13–35. [CrossRef]
25. Stull, A.T.; Fiorella, L.; Mayer, R.E. An eye-tracking analysis of instructor presence in video lectures. *Comput. Hum. Behav.* **2018**, *88*, 263–272. [CrossRef]
26. Burch, M.; Kull, A.; Weiskopf, D. AOI rivers for visualizing dynamic eye gaze frequencies. *Comput. Graph. Forum* **2013**, *32*, 281–290. [CrossRef]
27. Dzeng, R.-J.; Lin, C.-T.; Fang, Y.-C. Using eye-tracker to compare search patterns between experienced and novice workers for site hazard identification. *Saf. Sci.* **2016**, *82*, 56–67. [CrossRef]
28. Klaib, A.F.; Alsrehin, N.O.; Melhem, W.Y.; Bashtawi, H.O.; Magableh, A.A. Eye tracking algorithms, techniques, tools, and applications with an emphasis on machine learning and Internet of Things technologies. *Expert Syst. Appl.* **2021**, *166*, 114037. [CrossRef]
29. König, S.D.; Buffalo, E.A. A nonparametric method for detecting fixations and saccades using cluster analysis: Removing the need for arbitrary thresholds. *J. Neurosci. Methods* **2014**, *227*, 121–131. [CrossRef]
30. Romero, C.; Ventura, S. Educational data mining: A survey from 1995 to 2005. *Expert Syst. Appl.* **2007**, *33*, 135–146. [CrossRef]
31. Bogarín, A.; Cerezo, R.; Romero, C. A survey on educational process mining. *WIREs Data Min. Knowl. Discov.* **2018**, *8*, e1230. [CrossRef]
32. González, Á.; Díez-Pastor, J.F.; García-Osorio, C.I.; Rodríguez-Díez, J.J. Herramienta de apoyo a la docencia de algoritmos de selección de instancias. In Proceedings of the Jornadas Enseñanza la Informática, Ciudad Real, Spain, 10–13 July 2012; pp. 33–40.
33. Arnaiz-González, Á.; Díez-Pastor, J.F.; Rodríguez, J.J.; García-Osorio, C.I. Instance selection for regression by discretization. *Expert Syst. Appl.* **2016**, *54*, 340–350. [CrossRef]
34. Campbell, D.F. *Diseños Experimentales y Cuasiexperimentales en la Investigación Social [Experimental and Qusai-Experimental Designs for Research]*, 9th ed.; Amorrortu: Buenos Aires, Argentina, 2005.
35. Cerezo, R.; Fernández, E.; Gómez, C.; Sánchez-Santillán, M.; Taub, M.; Azevedo, R. Multimodal Protocol for Assessing Metacognition and Self-Regulation in Adults with Learning Difficulties. *JoVE* **2020**, *163*, e60331. [CrossRef]
36. Mudrick, N.V.; Azevedo, R.; Taub, M. Integrating metacognitive judgments and eye movements using sequential pattern mining to understand processes underlying multimedia learning. *Comput. Hum. Behav.* **2019**, *96*, 223–234. [CrossRef]
37. Munoz, D.P.; Armstrong, I.; Coe, B. Using eye movements to probe development and dysfunction. In *Eye Movements: A Window on Mind and Brain*; van Gompel, R.P.G.; Fischer, M.H.; Murray, W.S.; Hill, R.L., Eds.; Elsevier: Amsterdam, The Netherlands, 2007; pp. 99–124. [CrossRef]
38. Sulikowski, P.; Zdziebko, T. Deep Learning-Enhanced Framework for Performance Evaluation of a Recommending Interface with Varied Recommendation Position and Intensity Based on Eye-Tracking Equipment Data Processing. *Electronics* **2020**, *9*, 266. [CrossRef]
39. Moghaddasi, M.; Marín-Morales, J.; Khatri, J.; Guixeres, J.; Chicchi, G.I.A.; Alcañiz, M. Recognition of Customers' Impulsivity from Behavioral Patterns in Virtual Reality. *Appl. Sci.* **2021**, *11*, 4399. [CrossRef]
40. Qin, L.; Cao, Q.-L.; Leon, A.S.; Weng, Y.-N.; Shi, X.-H. Use of Pupil Area and Fixation Maps to Evaluate Visual Behavior of Drivers inside Tunnels at Different Luminance Levels—A Pilot Study. *Appl. Sci.* **2021**, *11*, 5014. [CrossRef]
41. Giraldo-Romero, Y.-I.; Pérez-de-los-Cobos-Agüero, C.; Muñoz-Leiva, F.; Higueras-Castillo, E.; Liébana-Cabanillas, F. Influence of Regulatory Fit Theory on Persuasion from Google Ads: An Eye Tracking Study. *J. Theor. Appl. Electron. Commer. Res.* **2021**, *16*, 1165–1185. [CrossRef]
42. Sulikowski, P. Evaluation of Varying Visual Intensity and Position of a Recommendation in a Recommending Interface Towards Reducing Habituation and Improving Sales. In *Advances in E-Business Engineering for Ubiquitous Computing, ICEBE 2019, Proceedings of the International Conference on e-Business Engineering Advances in E-Business Engineering for Ubiquitous Computing, Shanghai, China, 12–13 October 2019*; Chao, K.M., Jiang, L., Hussain, O., Ma, S.P., Fei, X., Eds.; Springer: Cham, Switzerland, 2019; pp. 208–218. [CrossRef]
43. Sulikowski, P.; Zdziebko, T.; Coussement, K.; Dyczkowski, K.; Kluza, K.; Sachpazidu-Wójcicka, K. Gaze and Event Tracking for Evaluation of Recommendation-Driven Purchase. *Sensors* **2021**, *21*, 1381. [CrossRef]

44. Bortko, K.; Piotr, B.; Jarosław, J.; Damian, K.; Piotr, S. Multi-Criteria Evaluation of Recommending Interfaces towards Habituation Reduction and Limited Negative Impact on User Experience. *Procedia Comput. Sci.* **2019**, *159*, 2240–2248. [CrossRef]
45. Lee, T.L.; Yeung, M.K. Computerized Eye-Tracking Training Improves the Saccadic Eye Movements of Children with Attention-Deficit/Hyperactivity Disorder. *Brain Sci.* **2020**, *10*, 1016. [CrossRef]
46. Peysakhovich, V.; Lefrançois, O.; Dehais, F.; Causse, M. The Neuroergonomics of Aircraft Cockpits: The Four Stages of Eye-Tracking Integration to Enhance Flight Safety. *Safety* **2018**, *4*, 8. [CrossRef]
47. Bissoli, A.; Lavino-Junior, D.; Sime, M.; Encarnação, L.; Bastos-Filho, T. A Human–Machine Interface Based on Eye Tracking for Controlling and Monitoring a Smart Home Using the Internet of Things. *Sensors* **2019**, *19*, 859. [CrossRef]
48. Brousseau, B.; Rose, J.; Eizenman, M. Hybrid Eye-Tracking on a Smartphone with CNN Feature Extraction and an Infrared 3D Model. *Sensors* **2020**, *20*, 543. [CrossRef]
49. Vortman, L.; Schwenke, L.; Putze, F. Using Brain Activity Patterns to Differentiate Real and Virtual Attended Targets during Augmented Reality Scenarios. *Information* **2021**, *12*, 226. [CrossRef]
50. Kapp, S.; Barz, M.; Mukhametov, S.; Sonntag, D.; Kuhn, J. ARETT: Augmented Reality Eye Tracking Toolkit for Head Mounted Displays. *Sensors* **2021**, *21*, 2234. [CrossRef]
51. Wirth, M.; Kohl, S.; Gradl, S.; Farlock, R.; Roth, D.; Eskofier, B.M. Assessing Visual Exploratory Activity of Athletes in Virtual Reality Using Head Motion Characteristics. *Sensors* **2021**, *21*, 3728. [CrossRef] [PubMed]
52. Scalera, L.; Seriani, S.; Gallina, P.; Lentini, M.; Gasparetto, A. Human–Robot Interaction through Eye Tracking for Artistic Drawing. *Robotics* **2021**, *10*, 54. [CrossRef]
53. Maimon-Dror, R.O.; Fernandez-Quesada, J.; Zito, G.A.; Konnaris, C.; Dziemian, S.; Faisal, A.A. Towards free 3D end-point control for robotic-assisted human reaching using binocular eye tracking. In Proceedings of the IEEE International Conference on Rehabilitation Robotics, London, UK, 17–20 July 2017; pp. 1049–1054. [CrossRef]
54. Antoniou, E.; Bozios, P.; Christou, V.; Tzimourta, K.D.; Kalafatakis, K.; Tsipouras, M.G.; Giannakeas, N.; Tzallas, A.T. EEG-Based Eye Movement Recognition Using Brain–Computer Interface and Random Forests. *Sensors* **2021**, *21*, 2339. [CrossRef] [PubMed]
55. IBM Corp. *SPSS Statistical Package for the Social Sciences (SPSS)*; Version 24; IBM: Madrid, Spain, 2016.
56. R Core Team. *R: A Language and Environment for Statistical*; Version 4.1.0; R Foundation for Statistical Computing: Vienna, Austria, 2021. Available online: http://www.R-project.org/ (accessed on 6 June 2021).
57. Hall, M.; Smith, L.A. Practical feature subset selection for machine learning. *Comput. Sci.* **1998**, *98*, 181–191.
58. Harris, E. Information Gain versus Gain Ratio: A Study of Split Method Biases. 2001, pp. 1–20. Available online: https://www.mitre.org/sites/default/files/pdf/harris_biases.pdf (accessed on 15 May 2021).
59. Cramér, H. *Mathematical Methods of Statistics (PMS-9)*; Princeton University Press: Princeton, NJ, USA, 2016. [CrossRef]
60. Arthur, D.; Vassilvitskii, S. K-means++: The advantages of careful seeding. In Proceedings of the SODA '07: Actas del Decimoctavo Simposio Anual ACM-SIAM Sobre Algoritmos Discretos, Philadelphia, PA, USA, 7–9 January 2007; pp. 1027–1035.
61. Bezdek, J.C.; Dunn, J.C. Optimal Fuzzy Partitions: A Heuristic for Estimating the Parameters in a Mixture of Normal Distributions. *IEEE Trans. Comput.* **1975**, *24*, 835–838. [CrossRef]
62. Zadeh, L.A. Fuzzy Sets and Information Granularity. *Fuzzy Sets Fuzzy Logic. Fuzzy* **1996**, 433–448. [CrossRef]
63. Daszykowski, M.; Walczak, B. Density-Based Clustering Methods. In *Comprehensive Chemometrics*, 2nd ed.; Brown, S., Tauler, R., Walczak, B., Eds.; Elsevier: Amsterdam, The Netherlands, 2020; Volume 2, pp. 635–654.
64. Hahsler, M.; Piekenbrock, M.; Doran, D. dbscan: Fast Density-Based Clustering with R. *J. Stat. Softw.* **2019**, *91*, 1–30. [CrossRef]
65. Hubert, L.; Arabie, P. Comparing partitions. *J. Classif.* **1985**, *2*, 193–218. [CrossRef]

Case Report

Remote Virtual Simulation for Incident Commanders—Cognitive Aspects

Cecilia Hammar Wijkmark [1], Maria Monika Metallinou [1,*] and Ilona Heldal [2]

1. Fire Disaster Research Group, Department of Safety, Chemistry and Biomedical Laboratory Sciences, Western Norway University of Applied Sciences, 5528 Haugesund, Norway; cecilia.hammar.wijkmark@hvl.no
2. Department of Computer Science, Electrical Engineering and Mathematical Sciences, Western Norway University of Applied Sciences, 5063 Bergen, Norway; ilona.heldal@hvl.no
* Correspondence: monika.metallinou@hvl.no; Tel.: +47-988-25-104

Abstract: Due to the COVID-19 restrictions, on-site Incident Commander (IC) practical training and examinations in Sweden were canceled as of March 2020. The graduation of one IC class was, however, conducted through Remote Virtual Simulation (RVS), the first such examination to our current knowledge. This paper presents the necessary enablers for setting up RVS and its influence on cognitive aspects of assessing practical competences. Data were gathered through observations, questionnaires, and interviews from students and instructors, using action-case research methodology. The results show the potential of RVS for supporting higher cognitive processes, such as recognition, comprehension, problem solving, decision making, and allowed students to demonstrate whether they had achieved the required learning objectives. Other reported benefits were the value of not gathering people (imposed by the pandemic), experiencing new, challenging incident scenarios, increased motivation for applying RVS based training both for students and instructors, and reduced traveling (corresponding to 15,400 km for a class). While further research is needed for defining how to integrate RVS in practical training and assessment for IC education and for increased generalizability, this research pinpoints current benefits and limitations, in relation to the cognitive aspects and in comparison, to previous examination formats.

Keywords: cognitive aspects; remote; virtual simulation; incident commander; user experiences; problem solving; decision making; assessment; learning

1. Introduction

Fire and Rescue Service (FRS) personnel respond to a wide range of emergencies affecting the civil society. The Incident Commander (IC) on the first (lowest) level in the command chain (IC-1) is often the first officer arriving at the incident scene, and thereby responsible for the initial assessment, decisions on the initial actions, and for providing accurate and informative reports to higher officers and/or the command center.

Incident commanders are devoted firefighters who have acquired additional competence for leading responses. An IC at the first level of command (IC-1) will usually lead four or five firefighters' actions with relevant equipment (a firetruck and a water truck) during handling routine incidents (for the emergency services) and the initial phase of more serious incidents, until an IC trained at a higher level arrives at the scene. There are several levels of command and related training courses, in many European Countries four or five levels [1–3] where the levels reflect the extent and severity of the incidents one may take the command over the response. The number of persons with higher qualification is lower for each level. The total force of the Swedish Fire Service consists of 12,500 responders (of which 2/3 are employed part-time, i.e., have other regular jobs as their main occupation). Sweden has about 2500 responders qualified as IC-1.

The education to become a firefighter includes practical training to acquire technical skills (handling equipment and performing operations according to procedures). The additional education to become IC-1 focuses on improving "non-technical skills" [4] as situational awareness, decision making, communication, and leadership. For firefighters, education is offered in classroom sessions and practical training at the training field. When practical training is scenario-based, it is called Live Simulation (LS). The scenarios unfold in a physical, controlled environment, using real fire (burning wood or gas), smoke, vehicles, and people acting according to a predefined setup for an arranged incident scenario. The physical objects and environment in LS are considered to allow naturalistic experiences, and thereby trigger cognitive processes in a similar way as real incidents do. Since the training of firefighter students is coordinated with the training of IC-students, the latter can command a student-firefighter team, thus practicing communication and leadership. However, LS has limitations as a method for training ICs, since the training facilities (involving a limited number of steel-and-concrete buildings, which have already sustained numerous fires) may not provide adequate variation and detailed cues to train situational awareness and decision making.

During the last decade, Virtual Simulation (VS) has become a mature method for practice-based training, implemented by several organizations.

There are contradictory opinions among stakeholders regarding the effectiveness of VS for training from the different educational fields, such as medicine, nursing, architecture, and management. However, also critical voices recognize the possible complementary value of VS training as a supplement to LS training [5].

Since COVID-19 hindered many training possibilities, especially for groups in laboratories or training grounds, the focus on allowing remote training (in the format of Remote Virtual Simulation, RVS), supporting the targeted cognitive processes has increased. Training and learning interventions are often discussed from the angle of cognitive science [6], due to the influence of cognitive load [7] for understanding and solving tasks. Additionally, higher cognitive processes such as recognition, decision making, and problem solving are essential during an emergency response [8].

In March 2020, the Swedish Civil Contingencies Agency (MSB), responsible for firefighter and IC-education in Sweden, stopped all on-site training to avoid gathering of people. By that time, one class (22 students) only lacked the final LS examination to qualify. At the same time, several of the FRSs needed the qualification, to increase their resilience during the pandemic. The MSB, having experience with VS on-site for basic IC training, decided to conduct the final exams using RVS for this class, based on the successful results of a pilot-test. This was the first IC-examination in remote virtual environments, to our current knowledge. Based on the action case research approach suggested by Braa and Vidgen [9], and the theoretical framework of cognitive science [10–12] this case report investigates the implementation of the practical part of the final examination for one class of IC-students, using RVS, at MSB, Sweden, through the following hypothesis:

Hypothesis 1 (H1). *RVS supports cognitive aspects of recognition, decision making, and problem solving adequately to allow students to demonstrate IC-skills.*

Hypothesis 2 (H2). *Through RVS examination, the instructors can assess the student's skills as ICs.*

Since RVS had not been used before, the first question was whether it could be used at all, if it would be accepted, and if so, how it related to earlier practices. Conducting an RVS pilot test was a necessary step prior to the final examination for one class in the RVS format. The results of the present action case may inform FRS professionals' educators how RVS can be used for training and assessment. The post-exam evaluation from instructors (performed through interviews) and students (performed through questionnaires) aims at answering the above hypotheses. The results also suggest that considering higher cognitive processes for evaluation of (R)VS tools may be a viable method for comparing and improving such tools and implementing them in future education.

The results are based on the implementation of RVS for the final examination of one (the first) IC-class (22 students). We acknowledge the low number of students as a limitation for the study. The scenarios used (described later) are according to the curriculum for IC-1, involving straightforward responses, without conflicting goals or high emotional pressure. Therefore, the results, may not be generalizable to training of higher levels of command. The conclusions are made in the Swedish context, with the specific resources, technical setting, educational structure, economic and organizational structure.

2. Theoretical Background

2.1. Cognitive Science

Cognitive Science represents a multidisciplinary approach to the human mind, often focusing on its problem-solving capabilities, as the differences between novices and experts [10,11,13]. Research on education often relies on contributions from Cognitive Science, especially for defining aspects improving learning [14] and tries to understand how the human brain functions. The Layered Reference Model of the Brain [12] decomposes cognitive processes into six layers (subconscious: Sensation, memory, perception, and action; and conscious: Meta cognitive processes and higher cognitive processes) encapsulating a total of 37 elemental cognitive processes. Among the elements of Layer 6 (higher cognitive processes), we find the cognitive elements of recognition, learning, decision making, and problem solving, which can be considered essential for training, contra not directly assessable subconscious elements [15].

VS for learning involves purposeful, computer generated graphical environments where the user can interact with the environment and representations of objects and humans, and based on specific rules, experience the effects of the interaction. This requires creating relevant scenarios unfolding in simulated real-life settings [16] with the potential to reveal cognitive learning processes through behavioral indicators [17]. RVS is conceptually not different from VS (such as to the pedagogical and cognitive aspects). However, it is different from VS regarding the technology and conduction, and this may influence the experience of both instructors and students in unknown ways.

To investigate H1 and H2, whether (R)VS supports the students' cognitive aspects to perform and demonstrate knowledge as IC, and the possibility of the instructors to assess it, we considered how virtual training and assessment interacts with the 16 higher cognitive processes of the brain (Layer 6) [12]. The exclusion process for some of the cognitive processes, is described below. Recognition (6.1) is crucial for performing as IC, also in the virtual environment. The used questionnaire includes questions about the perceived realism of the virtual space, as to buildings, vehicles, avatars, flame, and smoke, thus addressing the Cognitive Function, Recognition (6.1). Subtle cues can also be included in the scenarios by the instructors, to train and assess the student's recognition. Imagery (6.2) addresses the cognitive process of abstractly seeing visual images stored in the brain, without any sensory input. This cognitive process is not directly assessable and could only have been revealed by asking each student in interviews, which stored images they recalled, which was not done. Comprehension (6.3) is the action or capability of understanding, thus constructing a representation of the incident site. The cognitive aspect of comprehension draws parallels to the concept of situational awareness, as defined by Endsley [18–20] (involving recognition and interpretation of relevant cues as well as projection of the perceived situation in the time relevant for operation) and acknowledged by Flin et al. [4] as a very important non-technical skill. Comprehension can be assessed in VS by triggering events, effects or visualize cues, letting the instructors use firefighter avatars and asking the IC-students questions about various elements at the virtual incident site. This was actively used by the instructors during RVS examination. Learning (6.4) is about gaining knowledge or skill in some action or practice. Detailed learning objectives for IC-1 students involve procedural knowledge of the duties included in the role. These are thoroughly assessed. Reasoning, including Deduction and Induction (6.5, 6.6, and 6.7) are not stressed in the objectives of IC-1. Decision Making (6.8) is the process of choosing a course of action, among a set of alternatives. The human decision-making process is heavily affected by time constrains and level of risk. While analytical processes are used when time is ample, rule-based or intuitive processes are used when time is limited, as in incident command [4]. The Recognition-Primed Decision (RPD) model of rapid decision making [21,22] is often used when studying decisions at the incident site [23,24]. The IC-student must decide which actions shall be implemented (by the avatars) to resolve the incident and is thereby assessable. Problem Solving (6.9) is the way to the goal, mitigating consequences to the lowest possible level of damage using the available resources. If appropriate decisions were taken (and implemented through understandable orders to the avatars) the situation in the virtual environment will improve. Otherwise, adversity will increase, thereby making the cognitive process assessable. The cognitive processes of Explanation (6.10), Analysis (6.11), and Synthesis (6.12) may take place in the aftermaths of the active training or examination session and are thereby not assessable in the VS, but in the reflective feedback afterwards. Creation (6.13) is not expected to occur while training/assessing IC-1 students. Analogy (6.14) is a process in which a person understands a situation in terms of another situation. It may have links to the model of Recognition Primed Decision [21]. However, this model is associated with experience, which IC-1 students still do not have much of in the new role but may have from the role as a firefighter. This cognitive process has been considered not assessable. Planning (6.15) finds differences between the current and desired situation and governs decisions and actions. Planning involves also "instant pre-play" a cognitive process involving Imagery (6.2) to assess whether a choice of action is believed to give a favorable outcome for the affected people. Not directly assessable, but through 6.8 and 6.9 (Decision Making and Problem Solving). The last cognitive process, Quantification (6.16) has been considered less relevant for the job of the IC-1.

2.2. Simulation Training

Since cognitive aspects are extensively studied in healthcare, education or technology development, many influencing studies come from these areas. Virtual reality applications are highly domain specific, thus restricting generalizability, at least at the present state of maturity of the research area. Many influencing studies come from health care. Among others, research has been conducted on patient simulators [25–27], simulation for the operating room [28,29], prehospital care [30,31], pain [32], psychotherapy and cognitive support [33]. Other articles explore the qualities of simulators needed to engage health care practitioners [34,35]. Education expands in many different other domains using virtual reality technologies [36–38] or serious games in general for simulating the "work environment", e.g., Labster for Biotech subjects [39] or for a wide range of courses in virtual worlds such as Sloodle [40] or MaxWhere [41]. Other articles present technological advancements influencing cognitive aspects while using these new technologies, e.g., support of attention [42], navigation, and orientation (the different technologies require different support for navigation) [43–45] handling empathy [46] or emotion [47]. However, determining the added value of the different technologies for the various domains is demanding. It would be essential to know if more immersiveness contributes to more effective work [48] or learning [49] or how it is related to the design of the used environments [50]. There are questions about realism in virtual reality or serious games, and how the simulation fidelity is influenced by a buy-in effect of simulation technologies [51]. Often, simple analyses, e.g., a SWOT analysis [52] or ROI [53] may give a better insight into the added values. However, it is difficult to compare technologies in various domains and usage conditions.

2.3. Learning Approach

To take the role as an IC for the first level of command in a Swedish FRS, the person is required by law to have an "IC-1 course diploma" from the MSB College [54]. The pedagogical basis of IC education is based on reaching the third stage of Blooms taxonomy, i.e., remembering, understanding, and acting (through simulation training) [55]. Learning activities are organized accordingly in different learning spaces, as described in chapter

4. Procedures and actions during incident command are described in textbooks [56] and curriculums of fire academies [57]. The most important phases, and actions (the IC is expected to perform), visualised by Wijkmark and Heldal [58], are presented in Figure 1.

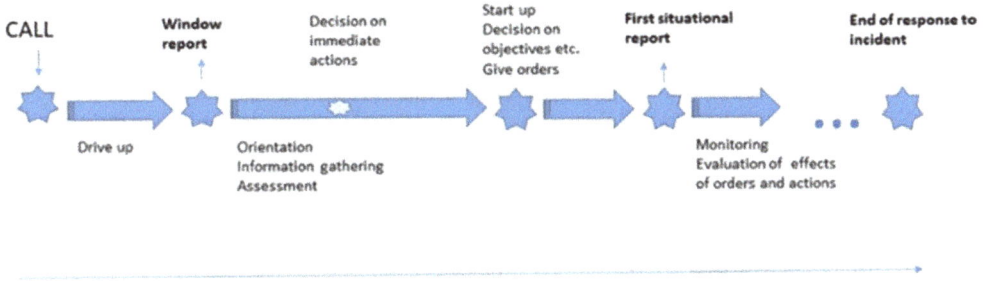

Figure 1. The most important phases (arrows) and actions (stars), where the IC-student should report to higher command or make decisions [58].

The student will go through all the phases of the response and thereby train and show the observable behaviors corresponding to the course sub-objectives, see Figure 1; confirm call; prepare team; initial orders; risk analysis; window report (by radio, describing what object is affected, what is the damage, and the current threat); stop the traffic; gather information (by talking to people on-site, perform reconnaissance); decision on actions (tactics, risk, make the optimal use of resources at hand); communication (team, higher command level in the FRS); collaborate with police and ambulance services; follow-up with a situation report (by radio); including the object, damage, threats, goal, actions that have been taken, and estimation of time; evaluate the effects of actions taken; end the incident operation. This is the IC-1 part of the seven-step model, the procedure, and command support tool applied in incidents in Sweden [56].

The focus of the IC-1 course is "routine incidents", i.e., house fires and car crashes, however, the content is not static. As the number of electric drive and hybrid electric vehicles has increased, this issue is now addressed in the education [59]. Moreover, overturn accidents with hazardous cargo are included, as Figure 2 shows. Increased consumption of different chemicals, generating increased transport [60], is a motivation for including these issues in the education.

Figure 2. An example of a CS setting facilitating a case-based discussion of a road-tanker incident in a model city on table-top "simulation".

3. Method

Cooperation between practitioners and scientists is often initiated by (at least) part of the organization who wish to explore a new technology for training purposes, thus making the collaboration process an *intervention to achieve a desirable change*. This points towards *action research* [61]. However, the aim of understanding the rich context of the domain, as well as the needs of the users (instructors and students) are also important for the scientists. This points towards *interpretation* and *case study* as a research method [62]. Braa and Vidgen [9] recognized the dilemmas often involved in "in-context" research and suggested an *action case* as a research methodology, often suitable when conducting information system research in the organizational context (such as the present study, conducted at MSB). The method recognizes the importance (and necessity) of balancing action (towards desirable change) with obtaining understanding, through interpretation. Studying the implementation of new techniques (without collaborating in the design of the technique), is suggested by Braa and Vidgen [9] to be a typical action case, if the participation of the organization in testing is adequate. The present work has the characteristics necessary to be typified as an *action case*. It is common that projects consist of several testing stages, i.e., in our case a pilot-test and a final examination conducted as RVS. These were compared to previous experiences in the training field through LS.

For the pilot test, a group of four instructors designed five scenarios, representing the challenges IC-1 students must resolve to qualify. These scenarios were prepared in the software XVR On-Scene, used by MSB, (XVR-sim, Delft, The Netherlands), and the remote technical setup was developed. Eight experienced ICs from different Swedish FRSs were invited to participate as "students". The instructors (i1–i4) and the pilot-test-participants were interviewed after the pilot-test. The results of the pilot-test were used as the foundation to the next step, the RVS examination. Of the 22 students participating in the RVS examination, 20 chose to answer the pre-exam and post-exam questionnaires.

The study was based on the battery of questions developed by Schroeder et al. [48] with added questions regarding the current incidents and to relate the experiences in the RVS to the LS environments. These added questions were inspired by the cognitive aspects presented in the "Layered Cognitive Model of the Brain" [12]. The pre-exam questionnaire covered background information of the participants (six questions) addressing the experience as firefighters, gaming experience, and familiarity to virtual simulation, followed by a post-exam questionnaire addressing the RVS examination experiences (25 questions). During the RVS, data were also collected in observations, and afterwards the instructors were interviewed. The instructor group was strengthened with one more instructor for the exam (i5), to provide redundancy in the case of illness. The five instructors/assessors were interviewed, after the exam was completed.

4. Learning Spaces

IC education often involves three different learning spaces: The Classroom Setting (CS), the Live Simulation (LS), and the Virtual Simulation (VS).

4.1. Classroom Setting (CS)

In the CS, natural-science-based lectures aim at creating a theoretical understanding of the potential hazards, combined with knowledge on command principles and legal aspects. This is often performed using cases and scenarios illustrated in PowerPoint slides [63], videos, pictures with added animation of fire and smoke, used in discussions or table-top training using models of cities, as shown in Figure 2. In CS, the focus is on discussion-based learning, reaching the first two steps of Bloom's taxonomy [55], remembering and understanding. CS can be at a fire academy campus or performed via a distance learning system, allowing the students to participate from their home or fire station.

4.2. Live Simulation (LS)

Live Simulation (LS) is included in practice-based training of firefighters and ICs worldwide. Fire academies have training facilities (buildings, fire trucks, and equipment), allowing for the simulation of several scenarios in a physical and the same geographical space. LS is used to allow firefighter students to train technical skills, and IC-students to practice decision-making competences, in a controlled environment. In simulation training, the IC-student will step into the role of the IC and lead a team of firefighter students in a simulated scenario. The IC-student must perform in the simulated incident, not just discuss or describe what she/he would have done in the situation. This is an important learning step, taking the student to the third step of Bloom's taxonomy, i.e., not simply remembering, and understanding, but also acting [55].

The steel and concrete buildings available in the fire colleges world-wide are built to withstand fire and water several times a day, for years. They must also represent different types of real-life buildings. As an example, the building shown in Figure 3 would represent an apartment building in one scenario and a mechanical workshop in another. Thus, they cannot look like any real-life buildings. Due to environmental and safety precautions, quantities and types of fuel are regulated, resulting in a controlled fire development. This limits the possible development of the fire, the cues, events, and consequences of the decision making and actions.

Figure 3. A building (**left**) used for LS, representing, e.g., a four-story apartment building, a mechanical workshop or a cruise ship, depending on the selected scenario. A family house with an attached garage (**right**), where the fire in the garage can only be represented by generated cold smoke, thus not behaving as a corresponding real fire.

The final examination of IC-students at MSB, have been performed in LS at the training ground, while the IC-candidate resolves the incident by following the steps described above. The instructors/assessors stand aside in the training field, observing and listening to the IC-1 student communicating with firefighters, bystanders on-site and via radio to the higher command level or dispatch center.

4.3. Virtual Simulation (VS)

On-site VS, using computer-simulated scenarios in 3D environments, has been used internationally and by MSB during recent years [64,65]. In VS, a student can act in the role of an IC in front of a large screen, move around in the virtual environment using a gamepad, talk to avatars (e.g., firefighters or bystanders) and make decisions on actions that are carried out in the simulated environment. In Figure 4 is an example of an apartment fire, that has spread to the roof (left) and a garage fire, while the affected family stands outside their home (to be compared to the LS settings in the previous chapter). The counterplay, i.e., the response by avatars and radio communication, is played by instructors, either in live role-play (by approaching the student) or through a speaker. Radios are used for communication as in a real incident. The setup in the room is schematically described in Figure 5a and an actual picture is shown in Figure 5b.

Figure 4. Example of how the IC-students' view may be at an incident involving an apartment fire where the fire has spread to the roof (**left**), and at an incident involving a fire in the garage attached to a family house (**right**).

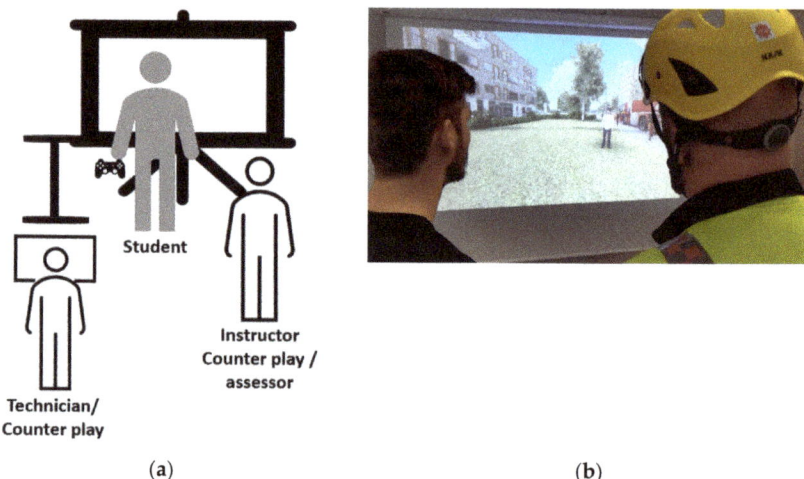

Figure 5. A schematic picture of the setup (**a**) and actual use (**b**) for VS on-site.

Based on the learning objectives, the instructor can build the scenario in a VS software tool, with prepared events and triggers, depending on the scenario and expected actions that the IC-student will take. During the training session, the scenarios are instructor controlled, giving the instructors the possibility to change the situation, and to "effectuate" the student's orders and act through various avatars. The IC-student acts in the incident, thus reaching the third step in Bloom's [55] taxonomy. At MSB, VS had been used for onsite basic IC-training, and not for examination, until March 2020.

5. Enablers for RVS Examination

5.1. Experiences of VS Training On-Site at MSB before the COVID-19 Pandemic

VS training had been used in the IC-1 ordinary training schedule, i.e., covering 2 days, for all students at one of the two MSB Colleges (Sandö) since 2018. During the VS training from January to September 2019, the experiences of 90 students (35% of all IC-1 students in MSB in 2019) were studied and analyzed [66]. An excerpt of the data providing evidence for the viability of VS as a training format (performed at MSB Sandö) is presented in Table 1.

Table 1. Students' response (n = 90) to VS training on-site, excerpt from [66]. Answers in Likert scale, 1 (low) to 5 (very high).

Questions	% Answers Likert 4 or 5	Average (Scale 1–5)	Standard Deviation
Experienced presence, compared to previous very high?	72%	3.90	0.83
Experienced presence in the simulated environment?	59%	3.63	0.86
Experienced being in the same env. as the "persons" you met?	68%	3.81	0.99
How easy was it to understand the training objectives?	80%	4.16	0.73
Would you like to perform similar training at your fire station?	100%	4.86	0.35
Would you like to perform similar training in your spare time?	80%	4.28	0.95

The results from this study provided instructors with extensive experience in developing and adjusting virtual scenarios, conducting VS training, and adjusting technological solutions. The data in Table 1 show the student information available to MSB when the COVID-19 pandemic struck. The number of students who answered the questionnaires were sufficiently large to provide internal validity. The acceptance and experience of VS by involved instructors were the foundation to support further action. Enabled by this experience, the instructors at MSB managed to adjust scenarios and develop the technical setup for the remote format of the examination and perform a pilot study, only days after the COVID-19 closure in March 2020.

5.2. The Pilot Test, before Deciding upon Remote Virtual Simulation Examination

In the few reported cases of using VS for assessment [1,67], the sessions have been held on-site, not remotely, which was a pressing need. However, the learning objectives, reflecting the necessary competencies for safe and effective incident command are the same. In LS, students and instructors are in the same physical space. The instructors/assessors can watch the student move, observe (see and hear) when he/she talks, and then observe the actions of the persons the student talked to and to observe the actions taken, e.g., if the leadership and the communication is satisfactory in relation to the assessment. Testing *how* the instructors/assessors would manage this, to reliably assess the students' performance in RVS, needed to be developed. This motivated the Pilot test, to explore the feasibility as well as test modes for transmitting to the instructors/assessors the necessary information to reliably assess the students.

Five scenarios were designed and built for RVS examination, see Table 2 for a brief overview. The technical setup, where the assessor could see the student's face at all time, hear everything said, and see what the student was looking at in the virtual environment, was developed, to provide the necessary assessment conditions as in LS (where the instructor can see the student at all time), see Figure 6. The audio and radio solution were setup using mobile phones and the standard digital communication tool in Sweden, RAKEL (RAdioKommunikation för Effektiv Ledning). The instructors could act as any of the persons involved in the incidents, e.g., another firefighter or a bystander, by choosing a corresponding avatar. The objectives of the Pilot test were to check the technology setup, the required bandwidth, the ease-of-use of technology mainly at the student site (which could be any fire station or the student's home), and to validate the scenarios and the assessment conditions.

Table 2. The scenarios used during the RSV examination.

Nr	Scenario Description	Learning Points Observed and Assessed
S1	Road traffic collision. A farmer has ended up in the ditch while attempting to avoid a collision with a deer. The farmer is not injured. On the pickup he has an IBC (Intermediate Bulk Container) with an unknown chemical. The tank faucet had been damaged and there was a leak. The chemical is Roundup, a herbicide that cannot be found in the decision support tool used by ICs in Sweden.	GENERAL LEARNING POINTS as presented in Figure 1. SPECIFIC FOR THIS SCNENARIO: - Gather information about the chemical and the tank. - If the chemical is unknown to the student, ask for support from the command center. - Decide on how to handle the chemical, and the leak. - Make sure the animal is handled.
S2	A garage attached to a Villa is on fire. The fire has started in a pile of junk in the garage attached to a villa. The family is safe outside.	GENERAL LEARNING POINTS as in Figure 1. SPECIFIC FOR THIS SCENARIO - Make sure no one is inside the villa - Gather information on what is in the garage and make correct decisions accordingly
S3	A fire in an apartment on the third floor. It is uncertain if anyone is inside the apartment initially. After a while, the friend of the owner of the apartment approaches the IC and explains that the owner is abroad, but her cat is in the apartment.	GENERAL LEARNING POINTS as in Figure 1. SPECIFIC FOR THIS SCENARIO - Make a suitable decision on tactics. - Gather information about the apartment and if someone is inside. - Inform the owner of the building about the end of the operation.
S4	Road traffic collision including three vehicles under an overpass. The collision is caused by timber on the road, that have come loose from a timber truck.	GENERAL LEARNING POINTS as in Figure 1. SPECIFIC FOR THIS SCENARIO - During reconnaissance, discover the timber and thereby the complexity of the incident. - Risk analysis and restrictions on where the firefighters can work. - Divide the incident into sectors and prepare orders for arriving firetrucks.
S5	Fire in a warehouse. Some youngsters have broken into the warehouse and started two fires before they left. There are caravans and vehicles, welding gas, etc. inside.	GENERAL LEARNING POINTS as in Figure 1. SPECIFIC FOR THIS SCENARIO - During reconnaissance, discover the other fire and thereby see the complexity of the incident. - Risk analysis and restrictions on where the firefighters can work. - Divide the incident into sectors and prepare orders for the arriving firetrucks. - Participate in a command meeting when the next level commander arrives, report on the actions taken and the plan.

Before the test, several of the eight expert ICs described their moderate expectations towards RVS, including concern of technical problems and difficulty in believing that RVS could be a satisfactory replacement for LS. The objectives of the pilot test were to check the technology setup, the required bandwidth, the ease-of-use of technology mainly at the student site (which could be any fire station or the student's home). They also had to give their comments on the scenarios, as well as the instructors' and assessors' role for running the scenarios. The instructors and assessors performed the counterplay and assessed these "expert-students" remotely. Valuable opinions regarding the setup as to what the assessor must see and hear to provide evidence-based assessment were expressed.

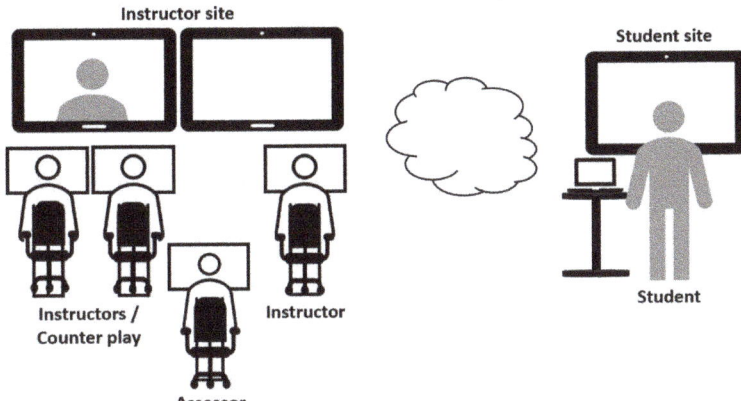

Figure 6. Schematic representation over the setup for RVS examination, where the student could participate from anywhere in the country.

The evaluation of the pilot test showed a positive turn in attitude towards RVS. All the participants agreed that the scenarios were designed to match the learning goals IC-1s need to achieve at a suitable level and corresponded to previous IC-1 LS examinations. They also agreed on the ease of use of technology at the student site. Using their computer keyboard, they could move in the environment on the incident scene, which was projected on a larger screen to allow the sense of higher presence [68]. No lagging was detected, and the communication via radio was working as within real incidents and previous LS examinations.

One pilot test participant, stated with a quite surprised tone, after the test: "This was really great. Why haven't you done this before? Everything you need [to perform in the role as an IC on the scene] is there". The instructors conducting the pilot test also expressed their experience as a positive surprise.

One of the researchers observed the instructors/assessors while "assessing the pilot-test students". It was noticed that the display showing the student's face was nearly not used, while the display showing what the student was looking at was in use most of the time. This means that the instructors extract useful information about what the student focuses attention on, and whether that is conscious "reading" of cues. This means that the assessors can follow the cognitive processes of the student, such as Recognition (6.1) and Comprehension (6.3) [12].

The pilot test compensated for the lack of experience in performing VS examination at MSB and was valuable for checking technical issues to perform the examination remotely. After the evaluation of the pilot test had been presented to the MSB management, it took only 15 days until the formal decision to perform an RVS examination was taken. The RVS examination was conducted during the period of 27 April–19 May 2020.

6. Results: RVS Examination

6.1. The RVS Examination—The Students' Experiences

The information gathered through pre-exam questionnaires revealed that all the students were men, with an average age of 40 (span between 32 and 56 years of age). The average number of years as a firefighter was 12, ranging from 3 to 31 years. Seven were part-time firefighters in rural areas, while 13 were full-time firefighters in cities. The experience of real fires among the IC-students varied from no real compartment fire (building fire) experience, to above 100 real fires.

Regarding familiarity with computer or mobile phone games, 70% never played computer games, and 60% never played mobile phone games. Only 15% stated that they played mobile phone games more than a few times per month, and no one played computer games more than a few times per month. Their previous knowledge or familiarity of VS for FRSs was limited. Three had prior knowledge or experience of the software used, gained from participation in VS projects where their FRS collaborated with MSB, and two had previous experience with the Response Simulator (https://www.vstepsimulation.com/response-simulator/rs-creator, accessed 17 February 2021).

Oral spontaneous comments from the students (documented by notes and video recordings) after completing the five scenarios, and/or free text comments written in the post-exam questionnaires are presented in Table 3.

Table 3. Students' experience of the RVS examination, oral comments (documented in notes and video recordings), and/or written free text comments on post-exam questionnaire.

Student	Comment
S1	I think it worked out well. Thanks to you [instructors], and it must be more of this in the course, especially remotely. It was gold [great], as close to real as it can get. And I did not have to drive 2000 km to the College [for the examination].
S3	This was great, it works great remotely
S4	I had a hard time interpreting a realistic picture of all impressions. It was hard to get the real feeling. Felt like I was talking all the time, and it was hard to feel the connection to the staff [firefighters].
S5	This is beyond my expectation. Interesting scenarios, the environment you built, giving orders works great [the firefighter avatars carry out the orders], and it feels like you are at the incident scene. This is the best substitute for being on-site.
S9	I was not comfortable in the situation. It is a good supplement, but I would have needed more real training [in LS before]. The scenarios were good, and I would have liked to train more times without the pressure of examination.
S14	This is more realistic than other methods for exercises.
S16	Overall, a great surprise. You do not have to pretend; all you see is what it is. Not like in the training ground.

One of the students, s4, comments on the cognitive difficulty of perceiving the represented situation. The comment "I had a hard time interpreting a realistic picture of all impressions" points towards experiencing a cognitive overload. However, this is the only negative comment on the cognitive part of the arranged examination. The second comment of the same student, "It was hard to feel the connection to the staff", addresses interpersonal aspects of simulation training, which was not in focus in the present article. (One of the instructors, has also commented that "Leadership", which is an interpersonal non-technical skill, is better taught in the training field, with real people).

After the final assessments of the five scenarios, a total of 4 h including breaks and feedback, the session was closed, and the students were asked to fill in the post-RVS questionnaire. The results are presented in Table 4. The results show that 90% of the students (18 of 20) would like (Likert 4 or 5) to perform a similar RVS training again, at their fire station, while 10% of the students (two persons) responded: "Neither or" (Likert 3). In addition, 75% perceived RVS as a suitable form (Likert 4 or 5) for IC-training.

In previous VS sessions, the students used a gamepad to move in the virtual environment, while in the RVS, the arrow keys were used to move around. No student stated any obstacles related to use of the keyboard for movements. For the question "How easy was it to move in the environment?", one student (5%) stated *hard* (Likert 2) and all the others stated no problems (Likert 3–5). In the previous VS study performed in 2019, 15% of the students expressed an unfamiliarity with the gamepad and considered it as an obstacle [66]. Consequently, using the keyboard was an improvement for some students.

Table 4. Students' response (*n* = 20) to the RVS examination. Answers in Likert scale, 1 (low) to 5 (very high).

Questions	% Answers Likert 4 or 5	Average (Scale 1–5)	Standard Deviation
Experienced presence, compared to previous very high?	70%	3.95	0.89
Experienced presence in the simulated environment?	75%	3.85	0.75
Experienced being in the same env. as the "persons" you met?	65%	3.60	0.99
How easy was it to understand the training objectives?	60%	3.80	0.89
How easy was it to communicate with others?	60%	3.80	1.01
Would you like to perform a similar training at your fire station?	90%	4.50	0.69
Would you like to perform a similar training in your spare time?	80%	4.28	0.95
To what extent do you consider RVS as a method for IC training?	75%	4.30	0.86

Regarding the students' experience of approaching and communicating with the firefighters (the instructors-controlled avatars), 60% stated that it was *easy* or *very easy* (Likert 4–5) and 10% stated *hard* (Likert 2). This suggests that future research should address the avatars' lay-out and the communication between the IC-student and the avatars to a greater extent.

On the question: *Please describe aspects that you found pleasant in the task*, 50% answered that they appreciated the good counterplay, the voice acting done by instructors, which enhanced the sense of realism of the situations.

6.2. The RVS Examination—The Instructors' Experiences

All five instructors conducting the RVS examination were convinced that the students were presented with similar challenges and performed similarly as in LS examinations. They also perceived the students' movement in the virtual environment and their communication with the avatars as easy and unproblematic.

The instructors stated that they could trustfully assess the students based on the learning objectives. One instructor explained the values of the virtual environment as "Everything that relates to the situation assessment, the development of the incident, like the spread of the fire and the extent of the damage, is possible to include in the virtual environment, which makes it extremely effective for assessment" (i1). Only one answer was given to the question regarding whether there are course objectives that cannot be assessed in RVS. This instructor (i2) commented: " . . . leadership might not be optimal here [in RVS], you need to train [to assess] this with real people, physically so to speak, to be able to train the basics".

A new possibility appreciated by the instructors was seeing the students' faces and reactions through their facial expressions and always seeing what the students were looking at. This cannot be achieved in LS, where the instructor cannot be sure about what the student is looking at. An instructor explains this in the following way: "I see and hear the student all the time. I can more easily assess communication and the orders given. I can see the exact picture of what he is looking at... It can sometimes be difficult to determine what the student is focusing on in a live exercise in the field" (i2).

The advantage of playing roles through avatars for instructors is explained by one instructor in the following way: "To have the opportunity [as instructor or assessor] to play the IC-student's personnel [firefighters] makes it possible to ask questions if orders are unclear. Also, later during the scenario, one can [with the firefighter avatar] walk up to the

IC and ask a related question to assess to what extent he or she understands the situation at hand" (i1).

6.3. Cognitive Aspects in Simulation Training

The different simulation formats, LS, and (R)VS are compared in Table 5, out of the model of higher cognitive processes. The compared formats support different higher cognitive processes to a different extent. This highlights the complementarity of the methods, and may enhance the knowledge about benefits and limitations of each. Table 5 summarizes the findings in the study with the focus on how the higher cognitive processes are supported in LS and (R)VS.

Table 5. Higher cognitive processes are supported in LS and (R)VS.

Higher Cognitive Processes	LS	(R)VS
6.1 Recognition Here, focused on perceived visual realism of the incident site, as to buildings, vehicles, involved participants, flames, and smoke.	Buildings, built to stand several fires per day and to represent different real-world objects. Real firetrucks and equipment are used. Old cars are used to represent cars in accidents. Involved participants are real people, often students or retired people hired as actors. Fire, smoke, evolvement, cues, and risks and cues are limited, due to safety and environmental regulations. Changes in the situation are not supported. Recognition is partly supported based on the above representations.	Buildings, vehicles, involved participants, flames, and smoke are chosen from a database. Events to trigger or change fire and smoke behavior illustrate cues and risks that are preprogrammed or changed during the training session. Changes in the situation are supported. Recognition is supported based on the above representations.
6.2 Imagery	The perceived realism of the incident scene is based a lot on imagery. This is very much dependent on the instructor's ability to describe the situation using the available method for training LS/(R)VS and individual experiences of the students. As we know the support for imagery is not included in training and assessment.	
6.3 Comprehension The action or capability of understanding. Involves constructing and internal representations based on existing knowledge. IC-students do not have experiences from the IC-perspective in an incident, although they have experienced from incident scenes as firefighters.	Existing knowledge related to the scenario may be affected by the fact that the LS objects are used for several scenarios and are familiar to all IC-students who were previously firefighter students. Therefore, it can be based on the previous experience of training at the LS training ground (i.e., where the fire can/cannot be placed, what are the possible scenarios), and by the additional information provided verbally by instructors. The team of firefighter-students are familiar with the training ground, and may "help" the IC-student by not asking when orders are unclear or safety measures do not meet the scenario. Few instructors live-play the police, ambulance, or bystanders, all looking the same. Comprehension partly supported.	The virtual environments and object, buildings, and avatars are all new to the IC-students. The instructors play the firefighter, police, and bystanders, all with different avatars. This makes it possible to use avatars to ask questions or react if the IC-student gives an unclear order. Comprehension can be supported over a wider specter.
6.4 Learning Learning acquisition of knowledge and skills resulting in a upgrade of the cognitive model. Confirmation of existing knowledge or deeper understanding are also recognized as learning [69].	Active experimentation [70] is not supported since the situation cannot evolve dynamically and one has very few tries in the training ground. Initial scenario design must be followed. Procedural learning is supported. Learning cannot be supported for all learning objectives.	Active experimentation [70] is supported, since the situation can evolve, the scenario can be changed, and more scenarios can be played. This will enhance learning. Procedural learning is supported. Learning can be supported for several learning objectives.
6.8 Decision making the process of choosing a course of action based on the current situation and the available resources. Especially for ICs the decision making is based on the above-mentioned aspects of cognition	Decision making is supported by the available stimuli of LS and the above-mentioned aspects.	Decision making is supported based on a wider specter of stimuli and the above-mentioned aspects.

The research hypothesis "RVS supports cognitive aspects of recognition, learning, decision making and problem solving for examining practical skills" has been confirmed. In the present study of the RVS examination, the students reported (Table 4) similar experiences as in Table 1 [66], showing results after introducing VS training on-site.

The second research hypothesis "Through RVS Examination, the instructors can assess student's practical skills as ICs" was confirmed, as well. The pilot study (with eight highly experienced ICs from different FRSs acting as IC-students) and the following real RVS examination of 22 students, were evaluated positively by all five instructors, the eight experienced ICs and 18, out of 20 students, who participated in the research. The instructors/accessors commented on the performance of the students to be "average", compared to the earlier LS examinations. Two of the students failed, which is "typical" for classes of this size. The instructors/assessors also assigned graded marks to the students (for the possible benefit of the research project, while the students received a Pass/No Pass result). However, this has not been done before in LS, so a detailed comparison of the students' performance was not possible. Both students and instructors seem to agree that the cognitive aspects for training and assessing ICs are supported well by (R)VS, while interpersonal skills are better supported by the live settings.

7. Discussion

Successfully performing an RVS examination (during the COVID-19 pandemic) may trigger more RVS training and examination also after the pandemic. The RVS examination proved to be technically feasible in Sweden, with the lowest bandwidth of 30/30 Mbps [71]. The impact of training in virtual environments, and transfer to real settings is a research objective. Since we always offer the best training available to every responder, it is impossible to conduct research involving a "non-treatment" control group. Research during the police-student education, with the possibility of offering training to the control group after the research was completed [72,73], shows that student-groups who trained on the communication procedure with a helicopter (one group LS, one group VS, control-group only read manual with procedures) performed similarly independent of the simulation format, and better than the control group. Similar results arose also from a study of procedural learning on tank-maintenance procedures [74]. The two simulation formats gave similar results upon assessing the students in the physical realm, and better than no-simulation training. Hall [75] studied the effect of VS training on fire ground ICs decision making, out of their self-evaluation and perceived confidence, and Gillespie [76] studied the transfer of virtual knowledge to the physical environment, connected to the acceptance of the virtual training. The LS physical examination format is very well established. Some virtual (on-site) assessments have been reported [1,67], while the present study is, to our knowledge, the first remote IC-1 examination mentioned in the research literature.

Psychological and social variables, which may have affected the students and enhanced their positive attitude towards RVS (for example, a wish to comply with the researchers) [77,78], are considered less prevalent in the remote setting, compared to VS training on-site (which was evaluated equally positively in 2019—see Table 1). Additionally, in an examination setting, the students are focused on their own performance, since "it counts" to pass the exam. The seriousness of the situation was likely to provoke honest reactions on behalf of the students, as to the perceived quality of the arrangement.

The study demonstrates the necessary, likely minimum, steps of familiarization and technology implementation in emergency response training for successful implementation of RVS examinations.

- The technology had previously been used for VS training on-site. Thereby, existing technical, scenario design, and conduction competence saved time and guaranteed usability.
- It was possible to perform a pilot test with experienced ICs. The positive evaluation motivated the final decision to use the RVS examination.
- A key component was the competence and interest of one champion and support from experienced VS instructors who were assigned time to participate.

The experience of successful RVS examinations has motivated a broader implementation of RVS training and examination at MSB, and this can inspire other fire academies in taking similar steps. This may provide further opportunities to study the implementation process as a future contribution to the (R)VS literature. However, it takes time to develop skills to design, build, and run RVS exercises with high quality. There is a risk that the organizations do not understand the competence needed and therefore do not allocate enough resources for the instructors to deliver scenarios of sufficient quality, which could result in less acceptance of RVS. While a stricter investigation of possibilities and challenges regarding the potential value for remote examinations may need further investigation, indeed the present study demonstrated added benefits for remote training as a solution to be included in future education. We believe that the instructors/assessors experienced "being closer to the student" (despite the physical distances) since they could continuously see what the student was looking at and hear what the student said.

Sweden is a long country (1572 km), with several sparsely populated regions. The Swedish fire and rescue services personnel therefore consist of 67% part-time firefighters and ICs, i.e., with other regular jobs. MSB has only two colleges offering an IC education, which makes the student travel costs high and the time away from the regular job (for part-time IC-students) and family (for all students) long. Enhancing and developing distance education by performing RVS training and examination may therefore also represent societal, human, and environmental benefits. Performing the exam in the RVS format on average saved each student a round trip of 9 h by car, based on 768 km distance on average, i.e., a total of 15,400 km or 38% of the Earth's circumference.

8. Conclusions

This study is an action case where the researchers participated in, and at the same time studied, the implementation of the VS training method and technology at the MSB, to the final step of conducting an RVS examination for IC-1 students. The cognitive aspects of recognition, learning, decision making, and problem solving were studied through questionnaires which the students filled post-training. The results indicate that the RVS, as implemented in the analyzed training and examination, adequately supported the above-mentioned cognitive aspects.

The existing VS implementation experiences at the MSB and corresponding studies of the students' cognitive benefits were the enablers, building competence in the organization, and thus making the COVID-19-forced RVS examination possible within a short preparation time.. This study demonstrated a proof of concept developed under time pressure, and with the precondition that students should be able to use standard PC equipment to perform their IC-1 final examination remotely. It demonstrates the possibilities and current challenges of RVS examination in the Swedish IC education. The RVS examination was performed satisfactorily and experienced positively by all involved parties. The main values of RVS for the students was that they could in fact graduate and they saved the travelling time and time away from home and regular jobs. The RVS was recognized by IC-1 students, instructors, assessors, and the MSB management, as high-quality training and examination methods, that have recently been implemented in the education of IC commanders at all levels.

Author Contributions: Conceptualization, C.H.W., M.M.M. and I.H.; methodology, C.H.W., I.H. and M.M.M.; scenarios selection, C.H.W.; field studies, C.H.W.; validation C.H.W., I.H. and M.M.M.; formal analysis, C.H.W. and M.M.M.; investigation, C.H.W. and I.H.; resources, C.H.W.; data curation, C.H.W. and M.M.M.; writing—original draft preparation, C.H.W., I.H. and M.M.M.; writing—review and editing, M.M.M., C.H.W. and I.H.; visualization, C.H.W.; supervision, I.H. and M.M.M.; project administration, I.H.; funding acquisition, I.H. and M.M.M. All authors have read and agreed to the published version of the manuscript.

Funding: The study was funded by the Western Norway University of Applied Sciences ICT PhD program (C.H.W.). This study was also partly funded by the Research Council of Norway, grant no. 298993 "Reducing fire disaster risk through dynamic risk assessment and management (DYNAMIC)" (I.H. and M.M.M.). The APC was funded by H.V.L.

Institutional Review Board Statement: The ethical review and approval were waived for this study since the involved humans (IC-students and instructors) performed activities normally included in their roles.

Informed Consent Statement: Informed consent was obtained from all subjects involved in the study.

Data Availability Statement: Anonymized background data available on request.

Acknowledgments: The authors appreciate the IC-1 students that participated in the study and the staff of the MSB College Sandø, for allowing the use of the premises for the study. A special thank to Sune Fankvist, for valuable input and support during this study.

Conflicts of Interest: The authors declare no conflict of interest. The funders had no role in the design of the study; in the collection, analyses, or interpretation of data; in the writing of the manuscript, or in the decision to publish the results.

References

1. Lamb, J.K.; Davies, J.; Bowley, R.; Williams, J.P. Incident command training: The introspect model. *Int. J. Emerg. Serv.* **2014**, *3*, 131–143. [CrossRef]
2. Bjølseth, T. *Grunnkurs for Brannkonstabel, Hefte 1*, 3rd ed.; Norsk Brannvernforening: Oslo, Norway, 2019; p. 18, ISBN 978-82-7485-099-4.
3. MSB. Available online: www.msb.se/sv/utbildning--ovning/alla-utbildningar/ (accessed on 10 April 2021).
4. Flin, R.; O'Connor, P.; Crichton, M. *Safety at the Sharp End; A Guide to Non-Technical Skills*, 4th ed.; Ashgate: Farnham, UK, 2015; p. 307, ISBN 978-0754646006.
5. Williams-Bell, F.M.; Murphy, B.M.; Kapralos, B.; Hogue, A.; Weckman, E.J. Using serious games and virtual simulation for training in the fire service: A review. *Fire Technol.* **2015**, *51*, 553–584. [CrossRef]
6. Fowler, C. Virtual reality and learning: Where is the pedagogy? *Br. J. Educ. Technol.* **2015**, *46*, 412–422. [CrossRef]
7. Koutromanos, G.; Sofos, A.; Avramidou, L. The use of augmented reality games in education: A review of literature. *Educ. Media Int.* **2015**, *52*, 235–271. [CrossRef]
8. Crandall, B.; Klein, G.; Klein, G.A.; Hoffman, R.R.; Hoffman, R.R. *Working Minds: A Practitioner's Guide to Cognitive Task Analysis*; MIT Press: Cambridge, MA, USA, 2006; ISBN 13: 978-0262532815.
9. Braa, K.; Vidgen, R.T. Interpretation, intervention, and reduction in the organizational laboratory: A framework for in-context information systems research. *Inf. Organ.* **1999**, *9*, 25–47. [CrossRef]
10. Friedenberg, J.; Silverman, G. *Cognitive Science: An Introduction to the Study of the Mind*, 3rd ed.; Sage Publications: Thousand Oaks, CA, USA, 2015; ISBN 978-1483347417.
11. Norman, D.A. Twelve issues for cognitive science. *Cogn. Sci.* **1980**, *4*, 1–32. [CrossRef]
12. Wang, Y.; Wang, Y.; Patel, S.; Patel, D. A layered reference model of the brain. *IEEE Trans. Syst. Man Cybern. Part C Appl. Rev.* **2006**, *36*, 124–133. [CrossRef]
13. de Groot, A.D. *Thought and Choice in Chess*; Mouton Publishers: The Hague, The Netherlands, 1965. [CrossRef]
14. Bransford, J.D.; Brown, A.L.; Cocking, R.R. (Eds.) *How People Learn: Brain, Mind, Experience, and School*; National Academy Press: Washington, DC, USA, 2000. [CrossRef]
15. Wilson, R.A.; Keil, F.C. (Eds.) *The MIT Encyclopedia of the Cognitive Sciences*; The MIT Press: Cambridge, MA, USA, 1999.
16. Bobrow, D.G.; Norman, D.A. Some principles of memory schemata. In *Representation and Understanding: Studies in Cognitive Science*; Bobrow, D.G., Collins, A.M., Eds.; Academic Press: New York, NY, USA, 1975.
17. Thornton, M.; Mosher, G. Quantifying cognitive processes in virtual learning simulations. In Proceedings of the 2014 ASEE North Midwest Section Conference, Iowa City, IA, USA, 16–17 October 2014.
18. Endsley, M.R. Design and evaluation for situation awareness enhancement. In Proceedings of the Human Factors and Ergonomics Society Annual Meeting, Anaheim, CA, USA, 1 October 1988; pp. 97–101. [CrossRef]
19. Endsley, M.R. Toward a theory of situation awareness in dynamic systems. *Hum. Factors* **1995**, *37*, 32–64. [CrossRef]
20. Endsley, M.R. Expertise and situation awareness. In *The Cambridge Handbook of Expertise and Expert Performance*; Ericsson, K.A., Charness, N., Feltovich, P.J., Hoffman, R.R., Eds.; Cambridge University Press: New York, NY, USA, 2006; pp. 633–651.
21. Klein, G. A recognition primed decision (RPD) model of rapid decision making. In *Decision Making in Action*; Klein, G., Orasanu, J., Calderwood, R., Zsambok, C., Eds.; Ablex: Norwood, NJ, USA, 1993; pp. 138–147.
22. Klein, G. *Streetlights and Shadows: Searching for the Keys to Adaptive Decision Making*; MIT Press: Cambridge, MA, USA, 2009.
23. Cohen-Hatton, S.R.; Butler, P.C.; Honey, R.C. An Investigation of operational decision making in situ: Incident command in the U.K. fire and rescue service. *Hum. Factors* **2015**, *57*, 793–804. [CrossRef] [PubMed]
24. Rake, E.L.; Njå, O. Perceptions and performances of experiences incident commanders. *J. Risk Res.* **2009**, *12*, 665–685. [CrossRef]

25. Good, M. Patient simulation for training basic and advanced clinical skills. *Med Educ.* **2003**, *37*, 14–21. [CrossRef] [PubMed]
26. Samei, E.; Kinahan, P.; Nishikawa, R.M.; Maidment, A. Virtual clinical trials: Why and what (special section guest editorial). *J. Med. Imaging* **2020**, 42801. [CrossRef]
27. Archer, F.; Wyatt, A.; Fallows, B. Use of simulators in teaching and learning: Paramedics' evaluation of a patient simulator. *Australas. J. Paramed.* **2007**, *5*. [CrossRef]
28. Gallagher, A.G.; Ritter, E.M.; Champion, H.; Higgins, G.; Fries, M.P.; Moses, G.; Smith, C.D.; Satava, R.M. Virtual reality simulation for the operating room. *Ann. Surg.* **2005**, *241*, 364–372. [CrossRef]
29. Robinson, S.; Radnor, Z.J.; Burgess, N.; Worthington, C. SimLean: Utilising simulation in the implementation of lean in healthcare. *Eur. J. Oper. Res.* **2012**, *219*, 188–197. [CrossRef]
30. Cohen, D.; Sevdalis, N.; Taylor, D.; Kerr, K.; Heys, M.; Willett, K.; Batrick, M.; Darzi, A. Emergency preparedness in the 21st century: Training and preparation modules in virtual environments. *Resuscitation* **2013**, *84*, 78–84. [CrossRef] [PubMed]
31. Alfes, C.M.; Reimer, A. Joint training simulation exercises: Missed elements in prehospital patient handoffs. *Clin. Simul. Nurs.* **2016**, *12*, 215–218. [CrossRef]
32. Ahern, M.M.; Dean, L.V.; Stoddard, C.C.; Agrawal, A.; Kim, K.; Cook, C.E.; Narciso Garcia, A. The effectiveness of virtual reality in patients with spinal pain: A systematic review and meta-analysis. *Pain Pract.* **2020**, *20*, 656–675. [CrossRef]
33. Riva, G. Virtual reality in psychotherapy. *Cyberpsychology Behav.* **2005**, *8*, 220–230. [CrossRef]
34. Rudolph, J.W.; Simon, R.; Raemer, D.B. Which reality matters? Questions on the path to high engagement in healthcare simulation. *Simul. Healthc.* **2007**, *2*. [CrossRef]
35. Engström, H.; Hagiwara, M.A.; Backlund, P.; Lebram, M.; Lundberg, L.; Johannesson, M.; Sterner, A.; Söderholm, H.M. The impact of contextualization on immersion in healthcare simulation. *Adv. Simul.* **2016**, *1*, 1–11. [CrossRef]
36. Aebersold, M. Simulation-based learning: No longer novelty in undergraduate education. *OJIN Online J. Issues Nurs.* **2018**, *23*, 1–13. [CrossRef]
37. Parong, J.; Mayer, R.E. Cognitive and affective processes for learning science in immersive virtual reality. *J. Comput. Assist. Learn.* **2021**, *37*, 226–241. [CrossRef]
38. Kara, N. A systematic review of the use of serious games in science education. *Contemp. Educ. Technol.* **2021**, *13*, ep295. [CrossRef]
39. Bonde, M.T.; Makransky, G.; Wandall, J.; Larsen, M.V.; Morsing, M.; Jarmer, H.; Sommer, M.O. Improving biotech education through gamified laboratory simulations. *Nat. Biotechnol.* **2014**, *32*, 694–697. [CrossRef] [PubMed]
40. Livingstone, D.; Kemp, J.; Edgar, E. From multi-user virtual environment to 3D virtual learning environment. *ALT-J* **2008**, *16*, 139–150. [CrossRef]
41. Horváth, I.; Sudár, A. Factors contributing to the enhanced performance of the maxwhere 3d vr platform in the distribution of digital information. *Acta Polytech. Hung.* **2018**, *15*, 149–173. [CrossRef]
42. Harada, Y.; Ohyama, J. The effect of task-irrelevant spatial contexts on 360-degree attention. *PLoS ONE* **2020**, *15*, e0237717. [CrossRef] [PubMed]
43. Howett, D.; Castegnaro, A.; Krzywicka, K.; Hagman, J.; Marchment, D.; Henson, R.; Rio, M.; King, J.A.; Burgess, N.; Chan, D. Differentiation of mild cognitive impairment using an entorhinal cortex-based test of virtual reality navigation. *Brain* **2019**, *142*, 1751–1766. [CrossRef] [PubMed]
44. Santos, B.S.; Dias, P.; Pimentel, A.; Baggerman, J.-W.; Ferreira, C.; Silva, S.; Madeira, J. Head-mounted display versus desktop for 3D navigation in virtual reality: A user study. *Multimed. Tools Appl.* **2009**, *41*, 161–181. [CrossRef]
45. Niehorster, D.C.; Li, L.; Lappe, M. The accuracy and precision of position and orientation tracking in the HTC vive virtual reality system for scientific research. *i-Perception* **2017**, *8*. [CrossRef]
46. Paiva, A.; Leite, I.; Boukricha, H.; Wachsmuth, I. Empathy in virtual agents and robots: A survey. *ACM Trans. Interact. Intell. Syst. TiiS* **2017**, *7*, 1–40. [CrossRef]
47. Costescu, C.; Rosan, A.; Hathazi, A.; Pădure, M.; Brigitta, N.; Kovari, A.; Katona, J.; Thill, S.; Heldal, I. Educational tool for testing emotion recognition abilities in adolescents. *Acta Polytech. Hung.* **2020**, *17*. [CrossRef]
48. Schroeder, R.; Steed, A.; Axelsson, A.-S.; Heldal, I.; Abelin, Å.; Wideström, J.; Nilsson, A.; Slater, M. Collaborating in networked immersive spaces: As good as being there together? *Comput. Graph.* **2001**, *25*, 781–788. [CrossRef]
49. Makransky, G.; Andreasen, N.K.; Baceviciute, S.; Mayer, R.E. Immersive virtual reality increases liking but not learning with a science simulation and generative learning strategies promote learning in immersive virtual reality. *J. Educ. Psychol.* **2020**, *113*, 719–735. [CrossRef]
50. Radianti, J.; Majchrzak, T.A.; Fromm, J.; Wohlgenannt, I. A systematic review of immersive virtual reality applications for higher education: Design elements, lessons learned and research agenda. *Comput. Educ.* **2020**, *147*, 103778. [CrossRef]
51. Alexander, A.L.; Brunyé, T.; Sidman, J.; Weil, S.A. *From Gaming to Training: A Review of Studies on Fidelity, Immersion, Presence, and Buy-In and Their Effects on Transfer in PC-Based Simulations and Games*; DARWARS Training Impact Group: Woburn, UK, 2005; Volume 5, pp. 1–14.
52. Engelbrecht, H.; Lindeman, R.W.; Hoermann, S. A SWOT analysis of the field of virtual reality for firefighter training. *Front. Robot. AI* **2019**, *6*, 101. [CrossRef]
53. Heldal, I.; Wijkmark, C.H. The RoI of simulation-based training vs live training of incident commanders. In Proceedings of the ITEC, Stockholm, Sweden, 14–16 May 2019.

54. Swedish Civil Protection Act and Regulation: Lag (2003:778) om Skydd mot Olyckor, Latest Update 2020:882, Förordning (2003:789). Available online: https://www.riksdagen.se/sv/dokument-lagar/dokument/svensk-forfattningssamling/lag-200 3778-om-skydd-mot-olyckor_sfs-2003-778 (accessed on 15 April 2021).
55. Bloom, B.S. *Taxonomy of Educational Objectives, Handbook I: The Cognitive Domain*; David McKay Co Inc.: New York, NY, USA, 1956.
56. Mattsson, M.; Eriksson, L. *Taktikkboken—En Håndbok i Systematisk Ledelse av Slokkeinnsatser mot Bygningsbranner (The Book of Tactics—A Handbook for Systematic Management of Firefighting Efforts Against Building Fires)*; Norsk Brannvernforening: Oslo, Norway, 2017; ISBN 978-82-7485-093-4.
57. Curriculum Incident Commander Level 1. The Swedish Civil Contingencies Agency, 2018. Available online: https://www.msb.se/siteassets/dokument/utbildning-och-ovning/alla-utbildningar/2018-00019-kursplan-raddningsledare-a.pdf (accessed on 16 April 2021).
58. Hammar-Wijkmark, C.; Heldal, I. Virtual and live simulation-based training for incident commanders. In Proceedings of the ISCRAM, Information Systems for Crisis Response and Management, Blacksburg, VA, USA, 24–27 May 2020; pp. 1154–1162.
59. Grant, C.C. *Firefighter Safety and Emergency Response for Electric Drive and Hybrid Electric Vehicles*; Fire Protection Research Foundation: Quincy, MA, USA, 2010; pp. 1–135.
60. Metallinou, M.M. Liquefied natural gas as a new hazard, learning processes in Norwegian fire brigades. *Safety* **2019**, *5*, 11. [CrossRef]
61. Baskerville, R.; Wood-Harper, A.T. A critical perspective on action research as a method of information systems research. *J. Inf. Technol.* **1996**, *11*, 235–246. [CrossRef]
62. Yin, R.K. *Case Study Research: Design and Methods*, 3rd ed.; Sage Publications: London, UK, 2003; ISBN 0-7619-2553-8.
63. Reis, V.; Neves, C. Application of virtual reality simulation in firefighter training for the development of decision-making competences. In Proceedings of the International Symposium on Computers in Education (SIIE), Tomar, Portugal, 21–23 November 2019; IEEE: Piscataway, NJ, USA, January 2020. [CrossRef]
64. Heldal, I. Contextual support for emergency management training: Challenges for simulation and serious games. In Proceedings of the 10th International and Interdisciplinary Conference, CONTEXT 2017, Paris, France, 20–23 June 2017; Brézillon, P., Turner, R., Pencopp, C., Eds.; Springer: Berlin/Heidelberg, Germany, 2017; pp. 484–497. [CrossRef]
65. Polykarpus, S.; Ley, T.; Poom-Valickis, K. Collaborative authoring of virtual simulation scenarios for assessing situational awareness. In Proceedings of the 18th ISCRAM Conference, Blacksburg, VA, USA, 23–26 May 2021.
66. Hammar-Wijkmark, C.; Metallinou, M.M.; Heldal, I.; Fankvist, S. The role of virtual simulation in incident commander education—A field study. *NIK Norsk Informatikkonferanse* **2020**, *1*, 1–12.
67. Lamb, K.; Boosman, M.; Davies, J. Introspect model: Competency assessment in the virtual world. In Proceedings of the ISCRAM, Kristiansand, Norway, 24–27 May 2015.
68. Heldal, I. The Usability of Collaborative Virtual Environments: Towards an Evaluation Framework. Ph.D. Thesis, Department of Technology and Society, Chalmers University of Technology, Göteborg, Sweden, 2004.
69. Sommer, M.; Braut, G.S.; Njå, O. A model for learning in emergency response work. *Int. J. Emerg. Manag.* **2013**, *9*, 151–169. [CrossRef]
70. Kolb, D.A. *Experiential Learning: Experience as the Source of Learning and Development*; Prentice Hall: Englewood Cliffs, NJ, USA, 1984.
71. Wijkmark, C.H.; Heldal, I.; Fankvist, S.; Metallinou, M.M. Remote virtual simulation for incident commanders: Opportunities and possibilities. In Proceedings of the 11th IEEE International Conference on Cognitive Infocommunications—CogInfoCom 2020, Online on MaxWhere 3D Web, 23–25 September 2020; 2020.
72. Moskaliuk, J.; Bertram, J.; Cress, U. Impact of virtual training environments on the acquisition and transfer of knowledge. *Cyberpsychol. Behav. Soc. Netw.* **2013**, *16*, 210–214. [CrossRef]
73. Bertram, J.; Moskaliuk, J.; Cress, U. Virtual training: Making reality work? *Comput. Hum. Behav.* **2015**, *34*, 284–292. [CrossRef]
74. Ganier, F.; Hoareau, C.; Tisseau, J. Evaluation of procedural learning transfer from a virtual environment to a real situation: A case study on tank maintenance training. *Ergonomics* **2014**, *57*, 828–843. [CrossRef] [PubMed]
75. Hall, K.A. The Effect of Computer-Based Simulation Training on Fire Ground Incident Commander Decision Making. Ph.D. Thesis, The University of Texas at Dallas, Dallas, TX, USA, May 2010.
76. Gillespie, S. Fire Ground Decision-Making: Transferring Virtual Knowledge to the Physical Environment. Ph.D. Thesis, Grand Canyon University, Phoenix, AZ, USA, 29 July 2013.
77. Wickström, G.; Bendix, T. The "Hawthorne effect"—What did the original Hawthorne studies actually show? *Scand. J. Work. Environ. Health* **2000**, *26*, 363–367. [CrossRef] [PubMed]
78. Merrett, F. Reflections on the Hawthorne effect. *Educ. Psychol.* **2007**, *26*, 143–146. [CrossRef]

Article

Framework for Preparation of Engaging Online Educational Materials—A Cognitive Approach

Žolt Namestovski [1] and Attila Kovari [2,3,4,*]

1 Hungarian Language Teacher Training Faculty, University of Novi Sad, 24000 Subotica, Serbia; zsolt.namesztovszki@magister.uns.ac.rs
2 Department of Natural Sciences, Institute of Engineering, University of Dunaújváros, 2400 Dunaujvaros, Hungary
3 The Institute of Engineering, Alba Regia Technical Faculty, Óbuda University, 8000 Szekesfehervar, Hungary
4 Department of Computer Science, GAMF Faculty of Engineering and Computer Science, John von Neumann University, 6000 Kecskemet, Hungary
* Correspondence: kovari@uniduna.hu

Abstract: This study examines the process of creating successful, engaging, interactive, and activity-based online educational materials, while taking the cognitive aspects of learners into account. The quality of online educational materials has become increasingly important in the recent period, and it is crucial that content is created that allows our students to learn effectively and enjoyably. In this paper, we present the milestones of curriculum creation and the resulting model, the criteria of selecting online learning environments, technical requirements, and the content of educational videos, interactive contents, and other methodological solutions. In addition, we also introduce some principles of instructional design, as well as a self-developed model that can be used to create effective online learning materials and online courses. There was a need for a self-developed, milestone-based, practice-oriented model because the models examined so far were too general and inadequate to meet the needs of a decentralized developer team, who work on different schedules, with significant geographical distances between them and do not place enough emphasis on taking cognitive factors into account. In these processes, special attention should be paid to having a clear and user-friendly interface, support for individual learning styles, effective multimedia, ongoing assistance and tracking of students' progress, as well as interactivity and responsive appearance.

Keywords: online educational material; cognitive aspects; instructional design; methodology; model

Citation: Namestovski, Ž.; Kovari, A. Framework for Preparation of Engaging Online Educational Materials—A Cognitive Approach. Appl. Sci. 2022, 12, 1745. https://doi.org/10.3390/app12031745

Academic Editor: Jing Jin

Received: 30 December 2021
Accepted: 1 February 2022
Published: 8 February 2022

Publisher's Note: MDPI stays neutral with regard to jurisdictional claims in published maps and institutional affiliations.

Copyright: © 2022 by the authors. Licensee MDPI, Basel, Switzerland. This article is an open access article distributed under the terms and conditions of the Creative Commons Attribution (CC BY) license (https://creativecommons.org/licenses/by/4.0/).

1. Introduction

Teachers have a big impact on students' learning outcomes, so more support for teachers' work will also lead to better student outcomes [1]. A lot of education research shows that the best learning experiences are those that make direct connections to students' existing knowledge and their life experiences [2]. In many cases, there is a lack of tools that help teachers apply curricula and teaching methods that consider students' individual learning abilities, and thus set curriculum goals that also effectively support individual learning opportunities [3]. Although many teacher tools are available through social networking sites and reusable lesson plans or activities, etc., effective assimilation of these materials still requires a reasonable degree of customization. It is often a daunting task for teachers to customize learning materials, since it takes more time to reuse existing resources than to implement self-developed materials [4,5]. Laboratory experiments play a very important role in the teaching of technical knowledge, as they establish a link between, and provide experience in, how theory and practice relate to each other. The possibilities provided by IT tools can even be used to perform studies similar to laboratory experiments in the framework of online education, and one of the best ways to do this is to take advantage of the possibilities provided by simulation environments [6]. The proportion of theoretical and practical knowledge in training varies from field to field, but practical and design-oriented education plays a role in developing analytical thinking skills [7]. Today, a number of up-to-date tools and methodologies are available to help students develop practical skills, such as taking advantage of active learning opportunities [8], the practical and team-oriented nature of project-based learning [9], the use of inverted classrooms [10], etc., which help in improving the effectiveness of the teaching–learning process [11]. Recently, online education has continued to spread rapidly, partly due to the mandatory introduction of online education in many countries as a result of the COVID-19 pandemic. With the growing popularity of online education, there is a great need for a pedagogically effective design model for online learning materials that facilitates the development and implementation of online learning environments [12]. Educational planning (ID), also known as educational systematic planning (ISD), is an exercise in creating educational experiences that should also be reflected in the developed online curriculum. A well-compiled online resource and the associated, appropriately selected teaching methodologies will make the acquisition of knowledge and skills more efficient, effective, and attractive [13].

Several models and methodologies can be adapted to designing online courses. One widely used model, the ADDIE model, offers five universal course design principles: analysis, design, development, implementation, and evaluation (ADDIE). Similar to any model, or design methodology, ADDIE has its advantages and disadvantages. The benefits of ADDIE are primarily that it provides structured guidance for planning and serves as a useful checklist for implementing the design process and course implementation, as well as placing a strong emphasis on the implementation process and evaluation. However, despite the advantages of ADDIE, there are some disadvantages as well. These include the fact that the analysis steps are not comprehensive enough in the design process and the model does not provide enough incentive for inspiration [14]. Huang's paper [15] summarizes guidelines for educational multimedia design from the Stanford University Virtual Labs Project and demonstrates the design of learning content that aims to develop interactive multimedia learning tools. In design, the emphasis depends on the visualization of learners on dynamic content and the control of what is learned. The paper points out that current methods for designing multimedia learning modules are not standardized and lack strong educational design [15]. Cognitive factors have been considered only to a limited extent during design and implementation.

Based on the above, we considered the development of a design methodology that takes cognitive skills into account during the design process and provides an opportunity to create inspiration in material development, while the steps of the implementation process can be broken down into elements that are easy to understand and implement. In order to develop successful online educational materials, the cognitive aspects of students need to be considered [16], for which the following theories and models are considered: Baddeley's model of working memory, cognitive load theory, Paivio's dual coding theory, individual learning styles, and situated learning. Baddeley's working memory model and Paivio's dual coding theory suggest that the information process of individuals is embedded in a dual channel: an auditory channel on the one hand, and a visual channel on the other [17].

2. Cognitive Theory in the Background

In order to produce effective online educational materials, it is necessary to know the factors influencing human cognitive abilities, such as working memory, cognitive load, dual coding theory, or individual learning styles, which are briefly summarized below.

2.1. Baddeley's Model of Working Memory

Working memory is a concept that emerged from the classical model of short-term memory [18], which was understood more as a set for the temporary storage of information before it was forwarded to long-term memory [19]. A more suitable model of short-term memory was finally projected, which was named working memory. The working memory model was a sub-mechanism schema that not only contained transient information but also handled it in a way that many fragments of verbal or visual information could be stored and combined. Under this model, Baddeley [20] proposed that there is an element in working memory that controls subcomponents or slave systems. This central unit, the central executive element, was responsible for overseeing the entire system, contributing to problem-solving tasks, and directing attention. Baddeley speculated that the central executive could delegate storage tasks to the two slave schemas in working memory, leaving the central executive with the capacity to execute heavier data processing commands. The two slaves are called the visuo-spatial sketch pad system and the phonological loop system. The first is supposed to preserve and deploy visual images, while the second stores and reviews verbal information and has a similarly significant developmental function in the sense that it simplifies language acquisition by retaining a new word in working memory until it can be learned [21].

Later, Baddeley [22] added that the introduction of a third subsystem into the model might be necessary, which became known as the episodic buffer. This emerged from a task previously classified as the central executive, acting firmly as a storage element that acts as a system with limited capacity to mix information sources from other slaves. Cognitive load theory helps in understanding the limitations of working memory and points out that there is a limit to the amount of information that can be managed simultaneously, which provides significant ideas for suggestions when creating multimedia educational materials [23].

2.2. Cognitive Load Theory (CLT)

Cognitive load theory [24] points out that working memory is limited in its capacity to selectively handle and process incoming sensory information. CLT is an important theory for understanding how learners focus and use their cognitive resources in learning and problem solving. This suggests that for education and learning to be successful, care must be taken to ensure that students' information processing skills are not overburdened. In the case of multimedia-supported education and learning, attention should be paid to the fact that students only have a limited amount of information processing capacity. As a result, great care must be taken in compiling educational materials when developing multimedia educational materials that provide relevant knowledge in relation to the knowledge to be acquired, given their limited information processing capacity [17].

Due to the above, it is important to know how the cognitive load caused by certain external stimuli and information can affect the information processing of students and the storage of relevant information in the long-term memory. If the cognitive load reaches too high, it prevents actual and productive learning, thus hindering the transfer of information itself. Inadequate educational environments, systems, and methodologies, such as gamification [25,26] or project-based learning [27], can significantly affect the effectiveness of learning. In teaching–learning, each learning process should be designed to minimize any cognitive processing that is irrelevant to learning and optimize the cognitive processing associated with the acquisition of information, and the construction of knowledge, taking into account the cognitive limits [28].

In terms of how the previously mentioned important factors should be taken into account in the planning of multimedia education materials, the material parts and layout that convey the relevant information need to be visually appealing [29] and intuitive in order to point out the essence [30]. However, student activities should continue to focus on the knowledge to be acquired, the purpose of the entertainment should be limited, primarily to maintain attention and interest, and this should in no way be burdensome for the student. In a poorly designed multimedia material, parts or activities that are primarily for fun can overload the working memory before the learner reaches the part of the material they want to learn, which can impair efficiency [17].

2.3. Dual Coding Theory (DCT)

According to the DCT theory, cognition involves two distinct subsystems. On the one hand, it is a verbal subsystem that deals directly with language, while on the other, it is a nonverbal subsystem designed to handle nonlinguistic objects and events. It is assumed that these systems consist of internal units of representation that are interconnected, thus realizing the transmission of nonverbal and verbal behavior independently or in collaboration with each other.

In some tasks, the verbal system is active (simple examples are crossword puzzles), while in others, the nonverbal picture system is in the foreground (e.g., puzzles). Cognition is the result of the interaction of these two systems, which also influences the way and manner in which the information is processed and exchanged [31]. In the case of multimedia content, both subsystems are relevant for improving development of content-specific parts of online curse materials.

2.4. Individual Learning Styles

James and Blank [32] defined learning style as the "complex manner in which, and conditions under which, learners most efficiently and most effectively perceive, process, store, and recall what they are attempting to learn". The learning style has three dimensions: the perceptual (physiological or sensory) mode, the cognitive (mental or information processing) mode, and the affective (emotional or personality traits) mode [33]. Considering learning styles in the design of teaching and learning materials will undoubtedly improve the effectiveness of learning different educational materials [34]. However, some basic aspects of adult education should not be forgotten either, as learning styles in adulthood follow different characteristics to those at a young age. Knowledge of the specificities of new technologies and individual learning styles will undoubtedly improve the effectiveness of learning, and these should be considered when compiling online e-learning materials [35]. The results of the study [36] suggest that simulation environments can be a good complement to teaching tools and an effective way to acquire state-of-the-art and sophisticated practical knowledge, which is a key requirement for engineers in the future [37].

Santo categorized "learning style" into three layers, thus providing some coherence to the wide range of models. The inner layer is about personality: "An underlying relatively stable dimension that controls learning behavior" [38]. The middle layer is about cognitive style and focuses on how learners process information, while the outer layer is about the environment in which students prefer to learn, including the nature of interaction with the instructor or other students [39].

3. Materials and Methods

This study examines the process of creating successful, engaging, interactive and activity-based online educational materials, while taking the cognitive aspects of learners into account. The study provides a multidimensional guide for the main concepts of the theoretical framework defined by the following key factors:

- Key principles of creating successful online educational materials.
- Team work.
- Aspects of e-learning, design, and creation.
- Platform.
- Multimedia.
- Methodological aspects.

The presented framework is primarily derived from the authors' own experiences with curriculum development, which have evolved through the participation in, and management of, a large number of projects implemented within the framework of the e-Region (www.e-regija.rs, accessed on 29 December 2021) online curriculum development organization. The guidelines and suggestions presented in this paper are based on the projects' implementation and feedback, as well as the whole process of online curriculum development.

4. Theoretical Framework for Creating Effective Online Educational Materials

The authors try, with tips, tricks, and practical concrete advice, to help colleagues working in public and higher education to create the most effective online curriculum possible. The next sections follow the steps defined in the methods section.

4.1. Key Principles of Creating Successful Online Educational Materials

One of the most important key aspects is the planning of students' online activities: continuous interactivity and sending feedback. One of the most effective ways to process the curriculum is to display it using multimedia, which is when we act on multiple senses in the form of images, video, text, and sound at the same time. The assessment can take the form of an online test (using different types of questions), but situated learning can also be used, where students need to make decisions in a given life situation, using the knowledge they have acquired [40].

4.2. Instructional Design Principles Depend on Cognitive Theories and Models

The cognitive theories listed above were developed by following the instructional design principles below:
- Clean and intuitive online learning environment, eye-catching, but without any unnecessary graphic or content elements (based on the principles of working memory and cognitive load theory).
- Presentation of the curriculum with subtitled multimedia, most often with a lecturer and in a virtual studio (based on the principles of supporting individual learning styles).
- Supporting individual learning styles (a variety of content is available to students: multimedia, e-book, summary videos).
- Continuous technical and content assistance.
- Monitoring and control of the learning processes (time spent in the system, login history).
- Interactivity and feedback (varied tasks, evaluation and situated learning).
- Responsive and device independence.
- Application of summary videos using infographics (based on the principles of dual coding theory).
- Multichannel communication (based on the principles of supporting individual learning styles).

The cognitive theories and models listed above all influence the production of effective online learning materials, so ignoring them would be to the detriment of efficiency. The cognitive aspects of learners have an influence on the process of creating successful, engaging, interactive, and activity-based online educational materials, which serves as the starting point for the design.

Support for individual learning styles can be implemented in several ways. For our developments, we support forms of learning that are independent of place and time and can be accessed from any smart device. In addition, during development, we pay special attention to the fact that the knowledge required to complete the training is available in at least three forms. These are usually subtitled educational videos, e-books, and summary videos.

If we take the three layers of Santo's learning style as a basis, instructional design should take the middle and outer layers into account. For the middle layer (information processing), different content should be provided, and individual learning pathways should be supported. In the case of the external layer (educational environment), however, the focus should be on the clarity of the platform, the user-friendliness, and the content and methodological elements of the educational materials appearing here.

In the case of synchronous or asynchronous education and adult education, it should be said that asynchronous education is more appropriate for adults, as synchronous education reduces access for working adults [39]. This research fully supports our developments, as a significant target group of our online educational material, which is implemented almost exclusively asynchronously, is the adult population.

4.3. Working in Teams

The creation of educational materials, which is effectively realized by an instructional development team, achieves high result rates and creates excellent teaching materials/courses with scientifically validated, unique methodological solutions. It is recommended that the online learning materials be created in a complex team, while proper planning and a precise definition of roles is also important.

If the online education material is created in a complex team, as mentioned above, either in-house or outsourced [41], then proper planning and a precise definition of roles will also be important. The development starts based on the customer's needs and ideas, and the development costs are usually paid by the customer. Project management coordinates both the implementation of the project, as well as the work and communication between teams. The instructional designer team develops and prepares the curriculum, uploads it to the platform, tests and evaluates it, and tracks user activity, as well as leads the process of writing the scenarios. Of course, they conduct their activities in coordination with all the other teams.

This team is successful because we look at a problem as individuals of different ages, perspectives, and statuses (teachers, college students, and PhD students). Within the team, everyone can specialize in a narrower area, such as scenario writing, multimedia editing, task creation, testing, and more. Each member of the team has a significant pedagogical background.

The cameraman and video post-production team are responsible for making the educational videos technically flawless.

This team of experts plays a major role mainly in the preparation phase. The experts are excellently acquainted with the given topic, and with their help, the scenario is prepared. The actor or lecturer uses the script to execute the curriculum, and from these pieces of content, instructional videos are made which serve as the main source of information and knowledge for the courses. The proposed structure of the team is shown in Figure 1.

Figure 1. The proposed structure of the team.

4.4. Aspects of E-Learning

The most common form of education in distance learning is e-learning. E-learning is an open form of exercise available on a computer network, independent of space and time constraints, which, by organizing the teaching–learning process, has effective, optimal knowledge transfer, learning methods, curriculum and student resources, tutor–student communication, and integrates computer-based, interactive, instructional software into a unified framework, making it accessible to the learner.

"E-learning can be viewed as an innovative approach for delivering well-designed, learner-centered, interactive, and facilitated learning environments to anyone, anyplace, anytime by utilizing the attributes and resources of various digital technologies along with other forms of learning materials suited for open, flexible, and distributed learning environments" [42].

According to DeGreeff, Burnett, and Cooley [43] who approach the need for e-learning from a societal perspective, it can be dated back to the overall phenomenon we now describe as a "rushing world". Central to this is the cumulative distribution of attention, the shortening and fragmentation of time spent on one thing, and the sequences of constant conversions into our next activity. Examined from another point of view, from a technical standpoint, all possibilities for the development of e-learning have gradually become more feasible: the continuous spread of computers and other mobile communication devices and their availability, together with the increasing mobility, and the rapid improvement in bandwidth and internet speed. This has greatly facilitated the spread of proprietary tools and collaborative approaches [44]. At the same time, software solutions coupled with these technologies have also made it possible for educators to create educational materials in a user-friendly way, which became accessible to learners through the Internet, without the need for advanced IT (e.g., programming) skills, and with the need for the involvement of special IT professionals not being required at all or only to a very limited extent. We are now in an age of big data and endless information. No matter where we are, we are inevitably faced with the situation of receiving news, either actively or passively [45].

The importance of online educational content has increased significantly in the context of the COVID-19 pandemic that began in 2020, as education could only be provided online in most cases.

"As a result of the COVID-19 pandemic, social conditions around the world have changed radically in a short time. In addition to social conditions, education has also been significantly affected by this global pandemic. A large number of educational institutions have been closed, and the only option for continuing education became (online) distance learning" [46]. Due to the circumstances, the transition to distance learning in this situation happened due to compulsion and not because of pedagogical development or innovation. However, in most cases, these emergency solutions were based on ill-considered strategies [46].

4.5. Design and Creation of Online Educational Materials

In order for e-learning to be used as a learning model, an appropriate training plan must be defined to enable the effective use of e-learning. In this process, it is very important to define the learning goals and to know under which criteria conditions they should be achieved. The criteria serve as a guideline for defining the expected outcomes of education, in which the educational design criteria for e-learning are inseparable from the commonly used development model of education [47].

The process of creating the educational material depends on appropriate instructional design and can be built on different models. The models define the steps of development and the relationship between them. One of the most common and most widely used instructional design models is the ADDIE model, which serves as an acronym for the following: analyze, design, develop, implement, and evaluate. In addition, often used is the Dick and Carey model, while the Nexius model is also noteworthy. The instructional design team of e-Region uses their own model, called the e-Region model, for planning, design preparation, scheduling, and tracking the course, and they built said model on the milestones shown in Figure 2.

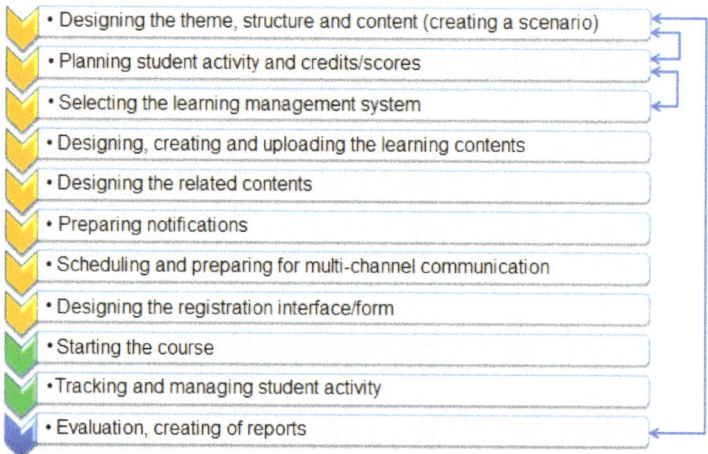

Figure 2. Theoretical model for design and creation of educational materials.

In the case of the presented model, the yellow phases indicate the development phase, the green ones the education phase, and the blue one the evaluation phase. The practical application of cognitive theories takes place in the milestones of the development phase.

In the phase of *designing the theme, structure, and content (creating a scenario)* the most important task is to write a scenario that includes the text of the tutorial videos with various comments (displayed content, tools used, etc.). The scenario also includes tasks and refers to various additional content, such as a summary video or contents related to online situational learning. It requires complex teamwork and is preceded by brainstorming.

Planning student activity and credits/scores is also one of the first steps in the planning process, as the possibilities are very limited only after the start of education. In our courses, the most points a student can earn are 100 points and they can complete the training with at least 50 percent.

Selecting the learning management system is also an important step, influenced by the project's financial plan, and the knowledge and experience of the development team. The learning management system determines the subsequent developments, the framework, but often also defines the limits of further planning.

The *designing, creating, and uploading the learning contents* phase is one of the most significant and time-consuming processes. This is where the tutorial videos are recorded, edited, and subtitled, and the tests and quizzes are created and uploaded to the platform.

Designing the related contents includes tasks such as making summary videos, developing e-books, and creating content for online situational learning.

The *preparing notifications phase* consists primarily of writing reminders and informational messages to be sent within the system, as well as preparing the text of the e-mails.

Scheduling and preparing for multichannel communication is a milestone in the creation of communication interfaces across different platforms. In many cases, in addition to platform-specific messages and emails, we use instant messaging applications and social networking sites to create groups around a course. In addition, we are constantly trying to help students solve problems and difficulties related to the technical and learning process.

In the step of *designing the registration interface/form*, the interface that will be the first one the students encounter is created. Here, they provide their contact details and practically register for the course.

Starting the course, as well as the opening and closing of the various modules and contents, using the results of our preliminary research, are scheduled for Sundays. In some cases, we used synchronous solutions such as video conferencing for these sections, but their popularity has steadily declined as the course has progressed.

Tracking and managing student activity is primarily concerned with the evaluation and administration of various tasks, as well as the ability to manage the communication in the forums. Scores are always administered in accordance with the current privacy regulations.

In the last phase, called *evaluation, creating of reports*, the collected and logged results are evaluated. In addition, we provide the customer with a detailed technical description that includes a list of all content produced (videos, tests, e-books), as well as their duration. The evaluation also includes an assessment of the student score, activity, and satisfaction questionnaire. These results will be examined later using statistical methods and will serve as empirical data for various scientific works.

With the examined models being too general, and not applicable to a decentralized development team, we found that there was a need to develop our own milestone model based on our own development experience. Our team members work on different schedules, from significant geographical distances, so it is important that their work be traceable to all members of the team, which is one of the basic conditions for effective cooperation. There was a need to develop our own model because our curriculum development team consists essentially of peer-to-peer developers, despite the fact that teachers and students are working on the same project.

The developed model is suitable for creating effective online learning materials and courses in any field of science in different learning management systems. This educational content focuses on multimedia learning tools and on interaction with dynamic contents. They support individual learning styles, provide effective/active learning, and build problem-solving and critical thinking skills. In the Supplementary Materials Section, we listed the online courses we created in 2021 based on the model and principles described above. The content of the courses ranges from language learning to self-knowledge and career choice, to the basic knowledge needed to start a business. It can also be used effectively in engineering education in an online educational environment.

4.6. Platform Used

Just as the location of traditional classroom education is not incidental, the online learning environment and the learning management system are also key. These platforms can be accessed via the Internet, and the digital content and teaching materials created by the instructors are placed here. As users access educational content through the framework, it is very important that it is clear and reliable. In addition, the cost of the learning management system, the available languages for the interface, and the need for installation and maintenance (an administrator must be employed, or the system is centrally maintained), as well as the availability and intensity of support, are all important. Just as in the case of traditional education, where location does not decisively determine the effectiveness of education, the same applies to online education. The framework can be in English or in another language, free or paid, with more or less possibilities, it can either be clear or difficult to review, while the effectiveness of online education is also determined by the personality, training, and professional knowledge of the lecturer themselves, as well as the applied professional methodology. This list should be supplemented (especially in the case of online education) with the implemented educational technology.

4.7. Educational Videos

In the current learning environments, multimedia materials are preferred by many teachers, which serves as both a major challenge and motivation [48]. Multimedia animation can contain words, pictures, sounds, images, and moving pictures [49]. Previous studies indicated that animation improves a learner's ability to remember facts and process information [50,51]. With sound, action, and images, animation can elucidate complex abstract concepts for learners' pictures [49]. Studies have also shown that multimedia animation as a teaching tool can help learners understand complex concepts, identify misconceptions, and positively impacts learners' motivation, satisfaction, learning achievement, digital tools, and information processing [52,53]. In addition, the effects of different multimedia materials on brain waves can be investigated [54], as the analysis of brain waves can also determine the level of attention and ultimately affect the effectiveness of learning [55,56], which anticipates one of the important applications of the future.

According to the results listed above, and our experience, online education is most effective when we teach with the help of educational videos, because multimedia (simultaneous display of text, sound, image and/or video, and interactive content) affects several of our senses at the same time. The educational videos are created by a dedicated cameraman and a post-production team. It is important that the finished content (audio and video) is perfect, as even a seemingly minor technical error can have a significant distracting effect. The videos begin with an eye-catching intro (5 s) that includes information about the course (course name and logo, possibly the name of the ordering company or sponsor). In the case of intros, care must be taken not to contain unnecessary information or overload the short-term memory of the viewers. The name of the lecturer and their affiliation should be displayed in a field at the beginning of the video. The videos should be subtitled, which can be optionally turned on/off so that hearing-impaired students can easily follow along as well. In addition, the captioning can be used to display any translations as well. When recording educational videos, it is recommended that instructors appear on the screen (give authenticity to the video) and that the keywords for that section appear next to them. The video can be recorded in a classroom or even outdoors, but care must be taken not to contain distractions or any distracting elements (focus on the presenter and blur the background).

Another good solution is the green box-recorded educational video, where you have the option to delete the entire background and display the presenter in a new virtual environment/studio [7].

On the other hand, in addition to the technical aspects of the videos, the preparation and presentation of the performers are also significant. The presentation should be enthusiastic, vibrant, friendly, and engaging, and the speakers need to be properly prepared for this. They can write notes, but it is also recommended that they practice their presentations before recording (even in front of a mirror). Care should be taken to seem as if having a personal conversation, and, if possible, it is recommended to establish a more informal relationship through the presentation. It is important for the beginning of the performance to be particularly good because the audience will make a decision about the quality of the performance after only a few seconds.

Thus, it is recommended to start the lecture with the following things:

1. A question.
2. With a personal story related to the topic of the lecture.
3. With interesting or surprising facts (which are also related to the topic of the lecture).

The appropriate choice of these can have an attention-grabbing effect, which helps the learner to show more interest in the curriculum. It is also important to plan the length of the videos. We often find that lecturers consider school hours, 45 min, to be authoritative. A study conducted on the scope of videos [57], covering 6.9 million views (Figure 3), has proved the following:

- Videos less than 6 min long provided a result of nearly 100%, meaning almost everyone watched the entire recording.
- Videos of 9 and 12 min were only nearly 50% watched.
- 12 and 40 min videos provided a watch time of only 20%.

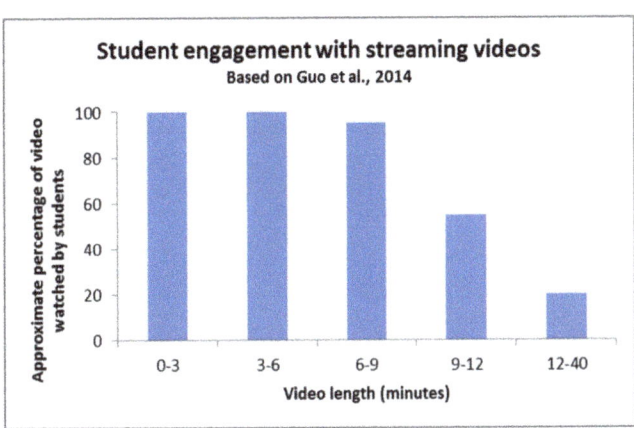

Figure 3. Educational video view rate (edited based on [57]).

Based on these findings, it can be said that the optimal length of tutorial videos is between 3 and 9 min. Of course, this does not mean that a 60 min video should be divided into ten 6 min sections, rather that instructional designer teams and experts need to define content units that can be properly presented in that amount of time.

On the other hand, students' attention can be increased or even be maintained with questions embedded in an educational video. The essence of this method is that while watching the tutorial video, questions pop up from time to time and the learner has to answer them. The questions are asked about the curriculum and the answers received are summarized by the system, so various statements can be created with their help.

4.8. Methodological Aspects

To publish educational content in the form of an online course, it is worth keeping in mind the following things, which also apply to online educational processes in general. Based on our experience and research results, it can be concluded that in addition to the regularities described above, many other factors influence the effectiveness of learning, the motivation of the participants, and their satisfaction.

4.8.1. Content Divided into Modules and Weeks

When planning the content, it is recommended to combine study materials on similar topics into one module at a time. When designing, it is worth following the one-module/one-week principle. Thus, the time allotted to learning will be divided into weeks, so the problem that users' activity increases radically before a deadline will not be as pronounced. Depending on the potential workload of our target group, it is recommended to set the weekly activity for an online course between 2 and 5 h. Based on our experience so far, deadlines and module changes should be planned for Sundays.

4.8.2. Diverse and Interactive Activities

When planning students' activities, all online tools and technical possibilities, which allow students to collaborate with others or to supplement static elements with interactive content, should be considered. We have achieved significant results with embedded questions in videos to help sustain audience attention. A mandatory element of the courses is the final online test, which can be used to assess the effectiveness of learning. The forums are the best embodiment of the community knowledge and experience that can be created on the surface of a course with hundreds or possibly thousands of students. In our experience, the intensity and direction of forum activity is largely determined by the topic of each sub-forum. It is recommended to use questions that are a bit provocative and affect emotions as well. If the content allows it, we can also perform exciting tasks by evaluating the inputs made on different interfaces as well as the students.

4.8.3. Awareness of Requirements and Expectations during the Online Learning Process

Before starting the course, it is recommended that participants or potential participants are informed of what exactly will be expected from them, exactly how much time they have to spend learning the management system, and what activities this process can be divided into (watching videos, forum activity, making submissions, completing tests, etc.) [12]. This time (according to our measurements so far) is three times the length of the tutorial videos, so there is time for taking notes as well as other activities. Of course, it is important for them to know what parts and modules the course is made up of and what dynamics are used to plan the teaching. It is also recommended to create a set of rules for the courses. If prior knowledge is required to complete the course, this should also be communicated to those interested.

4.8.4. Ongoing Assistance

An online course is usually entered by people with heterogeneous knowledge. This applies to the professional content of the training, but also to the IT competencies. That is why it is important to keep helping online learners. It is important to tell them which interfaces/systems the learning takes place on, and to create a registration and user guide for them. In addition, questions that arise during the course should be answered on an ongoing basis. The continuous tracking of students' scores also serves as a very strong motivating force, as they can receive feedback on the effectiveness of their activities after the activity is complete.

4.8.5. Multichannel Communication

According to our research, multichannel communication significantly motivates students. These can be the following channels: e-mail, the LMS's internal messaging system, forums on the course interface, a Facebook group, Instagram, etc. Of course, each environment has its own function and specificity. Deadlines and the tasks of the given week or module are sent out with the help of e-mails; on forums, there can be interactive professional discussions, and since the users spend a lot of time on social pages, these are suitable for both information transfer and communication. It is recommended to create groups on social media for this purpose. Of course, the various forms (application, satisfaction questionnaire) are an effective tool for communication and the further development of online content. We cannot prevent certain information from appearing duplicated, but the instructor may draw attention to this. In the case of online communication, however, it is important that if we already have a well-functioning communication interface with our target group, it should only be used in exceptional situations.

5. Discussion and Conclusions

Online education, as is clear from its definition, is a form of education in which no direct personal relationship is established between the teacher and the learner. This simple fact results in several methodological specialties, and the triangle of student, teacher, and educational material must be supplemented with the educational technology solutions that are necessary for the realization of online education. Educational technology solutions are various applications or software that allow the presentation to be recorded or transmitted, as well as complemented by various interactive and communication tools. Also included are technical devices, such as a computer, tablet, or smartphone, and the Internet itself, which are also needed for the implementation of online education. With all this in mind, it should be noted that the identity and role of the teacher remains key in the educational processes, and this fact is even more pronounced in the case of younger students [58]. The online form of education has also created a new educational situation by the fact that as opposed to the traditional form of education, enclosed between the four walls of the classroom, online education forms a multiple open system. Parents, the head of the institution and colleagues also gain insight into the educational content and the entire educational process itself. Therefore, it can be concluded that this situation definitely requires teachers to move away from their comfort zones and adapt the role of online educators. In addition, it is also a new situation to enter the private space, be that the living room, kitchen, or the child's room, even if only virtually, and it gains insight from strangers.

In online education, the curriculum should be divided into smaller sections, which are much shorter than a traditional 45 min school lesson. This can be a few minutes of a tutorial video or a video call, but in the case of microlearning, it is, in many cases, SMS message-length information (160 characters). These curricula and the whole educational process should be supplemented with constant feedback, evaluation, and motivation, and should be maintained and even increased where possible. Of course, the forms of work that were used in traditional education, such as frontal, individual, and pair work, are also excellent for online education when using the appropriate software tools.

In addition to the principles and benefits listed above, a further advantage of digital education, due to the new situation caused by the coronavirus, is the lack of need for meeting face-to-face, which, in the declared emergency, has resulted in distance learning remaining the only possible form of education. What is new, in the current situation, is that in the case of e-learning in the traditional sense, most of the online students are unknown to the lecturer, while in the current situation, the teacher continues to work online with an already well-known class or grade. The situation raises further questions, namely how well or differently the learners behave in an online space, what results does the lack of personal presence have, and how much does all this affect the activity and results of each student [59].

In our opinion, in the current situation, instead of the different platforms and strategies, the most important thing is for educators to master the basic methodological principles of online education and to apply them effectively in different educational situations. Of course, compared to the learning materials used in traditional teaching methods, the preparation of teaching materials compiled according to the methodology presented requires much more work and knowledge, but we believe that this will pay off later. It also arises as to what information from several sources may have an impact on the effectiveness of knowledge transfer, but we believe that the curricula prepared according to the described methodology are sufficiently attention-grabbing and highlighting, so that efficiency improvements are expected.

In this study, after defining the basic concepts, some of the key points in creating an effective online educational material are reviewed. We covered the general guidelines, the criteria for choosing the platform, the specifics of the educational videos, and the use of other tools. Based on our own experience, we have elaborated on the elements of online instructional design that can help us create more effective content. These are content divided into modules/weeks, diverse and interactive activities, multichannel communication, etc.

Supplementary Materials: Website S1: Tanulj velünk magyarul! (Learn Hungarian with us!) (https://tinyurl.com/tanuljvelunkmagyarul); Website S2: Tanulj velünk magyarul2! (Learn Hungarian with us2!) (https://tinyurl.com/tanuljvelunkmagyarul2); Website S3: Ki vagyok én? (Who am I?) (https://tinyurl.com/kivagyoken); Website S4: Vállalkozz, Vajdaság! (Do business, Vojvodina!) (https://vallalkozzvajdasag.prosperitati.rs/); Website S5: Út a siker felé (The road to success) (https://elearning.easygenerator.com/5589ab2a-ef7b-4f61-92eb-298cdd12ecbf/); Website S6: Web 2.0 Online eszközök használata a tanórán és azon kívül (Web 2.0 Use online tools in and out of class) (https://classroom.google.com/u/0/c/MjQ4NjU1MjQyMjU1) (accessed on 30 December 2021).

Author Contributions: Conceptualization, Ž.N. and A.K.; investigation, Ž.N.; methodology, Ž.N. and A.K.; writing—original draft, Ž.N. and A.K.; writing—review and editing, Ž.N. and A.K. All authors have read and agreed to the published version of the manuscript.

Funding: This research received no external funding.

Institutional Review Board Statement: Not applicable.

Informed Consent Statement: Not applicable.

Conflicts of Interest: The authors declare no conflict of interest.

References

1. Tomlinson, C.A.; McTighe, J. *Integrating Differentiated Instruction and Understanding by Design: Connecting Content and Kids*; Association for Supervision and Curriculum Development: Alexandria, VA, USA, 2006.
2. Bransford, J.D.; Brown, A.L.; Cocking, R.R. *How People Learn: Brain, Mind, Experience, and School*; National Academy Press: Washington, DC, USA, 1999.
3. Lengyelné Molnár, T. Changing reading habits and methodological options resulting from digital transformation. *J. Appl. Tech. Educ. Sci.* **2019**, *9*, 27–42.
4. Maull, K.E.; Saldivar, M.G.; Sumner, T. Observing online curriculum planning behavior of teachers. In Proceedings of the Educational Data Mining 2010, Pittsburgh, PA, USA, 11–13 June 2010; pp. 121–130.
5. Horváth, I. Evolution of teaching roles and tasks in VR/AR-based education. In Proceedings of the 9th IEEE International Conference on Cognitive Infocommunications, Budapest, Hungary, 22–24 August 2018; pp. 355–360.
6. Ibrahim, W.; Morsi, R. Online engineering education: A comprehensive review. In Proceedings of the American Society for Engineering Education Annual Conference & Exposition Annual Conference, Portland, OR, USA, 12–15 June 2005; pp. 10.973.1–10.973.10.
7. Bourne, J.; Harris, D.; Mayadas, F. Online engineering education: Learning anywhere, anytime. *J. Eng. Educ.* **2005**, *94*, 131–146. [CrossRef]
8. Lima, R.M.; Andersson, P.H.; Saalman, E. Active Learning in Engineering Education: A (re) introduction. *Eur. J. Eng. Educ.* **2017**, *42*, 1–4. [CrossRef]
9. Mills, J.E.; Treagust, D.F. Engineering education—Is problem-based or project-based learning the answer. *Australas. J. Eng. Educ.* **2003**, *3*, 2–16.
10. Bishop, J.; Verleger, M.A. The flipped classroom: A survey of the research. In Proceedings of the ASEE Annual Conference & Exposition, Atlanta, GA, USA, 23–26 June 2013; pp. 23.1200.1–23.1200.18.
11. Grodotzki, J.; Upadhya, S.; Tekkaya, A.E. Engineering education amid a global pandemic. *Adv. Ind. Manuf. Eng.* **2021**, *3*, 100058. [CrossRef]
12. Chen, L.L. A model for effective online instructional design. *Lit. Inf. Comput. Educ. J.* **2016**, *6*, 2302–2308. [CrossRef]
13. Roblyer, M.D. *Introduction to Systematic Instructional Design for Traditional, Online, and Blended Environments*; Pearson Higher Education: Hoboken, NJ, USA, 2014.
14. Quinn, C. The Great ADDIE Debate. Available online: http://blog.learnlets.com/?p=1489 (accessed on 30 December 2021).
15. Huang, C. Designing high-quality interactive multimedia learning modules. *Comput. Med. Imaging Graph.* **2005**, *29*, 223–233. [CrossRef]
16. Gőgh, E.; Racsko, R.; Kovari, A. Experience of Self-Efficacy Learning among Vocational Secondary School Students. *Acta Polytech. Hung.* **2021**, *18*, 101–119. [CrossRef]
17. Sorden, S.D. A cognitive approach to instructional design for multimedia learning. *Inf. Sci.* **2005**, *8*, 263–279. [CrossRef]
18. Atkinson, R.C.; Shiffrin, R.M. Human memory: A proposed system and its control processes. In *Psychology of Learning and Motivation*; Elsevier: Amsterdam, The Netherlands, 1968; Volume 2, pp. 89–195.
19. Baddeley, A.D.; Hitch, G. Working memory. In *Psychology of Learning and Motivation*; Elsevier: Amsterdam, The Netherlands, 1974; Volume 8, pp. 47–89.
20. Baddeley, A.D. *Essentials of Human Memory*; Psychology Press: Hove, UK, 1999.
21. Cowan, N. Working memory underpins cognitive development, learning, and education. *Educ. Psychol. Rev.* **2014**, *26*, 197–223. [CrossRef]
22. Baddeley, A.D. Is working memory still working? *Eur. Psychol.* **2002**, *7*, 85–97. [CrossRef]
23. De Jong, T. Cognitive load theory, educational research, and instructional design: Some food for thought. *Instr. Sci.* **2010**, *38*, 105–134. [CrossRef]
24. Sweller, J. Cognitive load theory. In *Psychology of Learning and Motivation*; Elsevier: Amsterdam, The Netherlands, 2011; Volume 55, pp. 37–76.
25. Pinter, R.; Cisar, S.M.; Balogh, Z.; Manojlovic, H. Enhancing Higher Education Student Class Attendance through Gamification. *Acta Polytech. Hung.* **2020**, *17*, 13–33. [CrossRef]
26. Kovács, G. Gamification vs Game Addiction. *Comput. Learn.* **2020**, *3*, 12–20.
27. Schrauf, G. Importance of project-based learning in software development. *Comput. Learn.* **2019**, *2*, 27–39.
28. Sweller, J.; van Merriënboer, J.J.; Paas, F. Cognitive architecture and instructional design: 20 years later. *Educ. Psychol. Rev.* **2019**, *31*, 261–292. [CrossRef]
29. Szűts, Z. An Iconic turn in art history—The quest for realistic and 3D visual representation on the World Wide Web. In *The Iconic Turn in Education*; Benedek, A., Nyíri, K., Eds.; Peter Lang: Frankfurt, Germany, 2012; pp. 59–66.

30. Molnár, G.; Nagy, K.; Balogh, Z. The role and impact of visualization during the processing of educational materials, presentation options in education and in the virtual space. In Proceedings of the 10th IEEE International Conference on Cognitive Infocommunications, Naples, Italy, 23–25 October 2019; pp. 533–538.
31. Paivio, A.; Clark, J.M. Dual coding theory and education. In *Pathways to Literacy Achievement for High Poverty Children*; The University of Michigan School of Education: Ann Arbor, MI, USA, 2006.
32. James, W.B.; Blank, W.E. Review and critique of available learning-style instruments for adults. In *New Directions for Adult and Continuing Education*; Wiley: Hoboken, NJ, USA, 1993; Volume 59, pp. 47–57.
33. Tyng, C.M.; Amin, H.U.; Saad, M.N.; Malik, A.S. The influences of emotion on learning and memory. *Front. Psychol.* **2017**, *8*, 1454. [CrossRef]
34. Rajcsányi-Molnár, M.; Bacsa-Bán, A. From the Initial Steps to the Concept of Online Education: Teacher Experiences and Development Directions Based on Feedback from Online Education Introduced During the Pandemic. *Cent. Eur. J. Educ. Res.* **2021**, *3*, 33–48. [CrossRef]
35. McIver, D.; Fitzsimmons, S.; Flanagan, D. Instructional design as knowledge management: A knowledge-in-practice approach to choosing instructional methods. *J. Manag. Educ.* **2016**, *40*, 47–75. [CrossRef]
36. Demeter, R.; Kovari, A. Importance of digital simulation in the competence development of engineers defining the society of the future. *Civ. Szle.* **2020**, *17*, 89–101.
37. Horváth, I. An Analysis of Personalized Learning Opportunities in 3D VR. *Front. Comput. Sci.* **2021**, *3*, 673826. [CrossRef]
38. Santo, S.A. Relationships between learning styles and online learning: Myth or reality? *Perform. Improv. Q.* **2006**, *19*, 73–88. [CrossRef]
39. Speece, M. Learning Style, Culture and Delivery Mode in Online Distance Education. *US-China Educ. Rev. A Educ. Pract.* **2012**, *2*, 1–12.
40. Benedek, A.; Molnár, G.; Szűts, Z. Practices of Crowdsourcing in relation to Big Data Analysis and Education Methods. In Proceedings of the IEEE 13th International Symposium on Intelligent Systems and Informatics, Subotica, Serbia, 17–19 September 2015; pp. 167–172.
41. Rajcsányi-Molnár, M.; András, I. Online strategies and international market development in higher education. In *Metamorphosis: Glocal Dilemmas in Three Acts*; Új Mandátum Kiadó: Budapest, Hungary, 2013; pp. 172–193.
42. Khan, B. Learning features in an open, flexible and distributed environment. *AACE J.* **2005**, *13*, 137–153.
43. DeGreeff, B.L.; Burnett, A.; Cooley, D. Communicating and philosophizing about authenticity or inauthenticity in a fast-paced world. *J. Happiness Stud.* **2010**, *11*, 395–408. [CrossRef]
44. Molnár, G. Collaborative technological applications with special focus on ICT based, networked and mobile solutions. *Trans. Inf. Sci. Appl.* **2012**, *9*, 271–281.
45. Krájnik, I.; Demeter, R. The specific characteristics, economic aspects and importance of banking risk management in accounting training. *J. Appl. Tech. Educ. Sci.* **2021**, *11*, 265.
46. Molnár, G.; Námesztovszki, Z.; Szűts, Z. Switching to online education, experiences from Hungary and Serbia. In Proceedings of the XI International IT and Education Development Conference, Zrenjanin, Serbia, 30 October 2020; pp. 55–59.
47. Triyono, M.B. The Indicators of instructional design for e-learning in Indonesian vocational high schools. *Procedia-Soc. Behav. Sci.* **2015**, *204*, 54–61. [CrossRef]
48. Molnár, G. Challenges and Opportunities in Virtual and Electronic Learning Environments. In Proceedings of the IEEE 11th International Symposium on Intelligent Systems and Informatics, Subotica, Serbia, 26–28 September 2013; pp. 397–401.
49. Chiou, C.C.; Tien, L.C.; Lee, L.T. Effects on learning of multimedia animation combined with multidimensional concept maps. *Comput. Educ.* **2015**, *80*, 211–223. [CrossRef]
50. Holzinger, A.; Kickmeier-Rust, M.; Albert, D. Dynamic media in computer science education; content complexity and learning performance: Is less more? *J. Educ. Technol. Soc.* **2008**, *11*, 279–290.
51. Kocsó, E.; Cserné Pekkel, M. Using of Dynamic Animations to Illustrate Mathematical Theorems. *Trans. IT Eng. Educ.* **2020**, *3*, 1–15.
52. Dalacosta, K.; Kamariotaki-Paparrigopoulou, M.; Palyvos, J.A.; Spyrellis, N. Multimedia application with animated cartoons for teaching science in elementary education. *Comput. Educ.* **2009**, *52*, 741–748. [CrossRef]
53. Orosz, B.; Kovács, C.; Karuovic, D.; Molnár, G.; Major, L.; Vass, V.; Szűts, Z.; Námesztovszki, Z. Digital education in digital cooperative environments. *J. Appl. Tech. Educ. Sci.* **2019**, *9*, 55–69.
54. Katona, J.; Ujbanyi, T.; Sziladi, G.; Kovari, A. Examine the effect of different web-based media on human brain waves. In Proceedings of the 8th IEEE International Conference on Cognitive Infocommunications, Debrecen, Hungary, 11–14 September 2017; pp. 407–412.
55. Katona, J. Examination and comparison of the EEG based Attention Test with CPT and TOVA. In Proceedings of the IEEE 15th International Symposium on Computational Intelligence and Informatics, Budapest, Hungary, 19–21 November 2014; pp. 117–120.
56. Katona, J.; Kovari, A. Examining the learning efficiency by a brain-computer interface system. *Acta Polytech. Hung.* **2018**, *15*, 251–280.
57. Guo, P.J.; Kim, J.; Rubin, R. How video production affects student engagement: An empirical study of MOOC videos. In Proceedings of the First ACM Conference on Learning Scale, Atlanta, GA, USA, 4–5 March 2014; pp. 41–50.

58. Bartal, O.; Rajcsányi-Molnár, M. Teachers of the 21st Century and the Mobile-tools. *J. Appl. Tech. Educ. Sci.* **2020**, *10*, 53–66.
59. Balogh, Z.; Kuchárik, M. Predicting Student Grades Based on Their Usage of LMS Moodle Using Petri Nets. *Appl. Sci.* **2019**, *9*, 4211. [CrossRef]

Article

Clustering of Road Traffic Accidents as a Gestalt Problem

Milan Gnjatović [1,*], Ivan Košanin [2], Nemanja Maček [3] and Dušan Joksimović [1]

1. Department of Information Technology, University of Criminal Investigation and Police Studies, Cara Dušana 196, 11080 Belgrade, Serbia; dusan.joksimovic@kpu.edu.rs
2. Ministry of the Interior of the Republic of Serbia, Kneza Miloša 101, 11000 Belgrade, Serbia; ivan.kosanin@mup.gov.rs
3. School of Electrical and Computer Engineering, Academy of Technical and Art Applied Studies, Vojvode Stepe 283, 11000 Belgrade, Serbia; nmacek@viser.edu.rs
* Correspondence: milan.gnjatovic@kpu.edu.rs

Abstract: This paper introduces and illustrates an approach to automatically detecting and selecting "critical" road segments, intended for application in circumstances of limited human or technical resources for traffic monitoring and management. The reported study makes novel contributions at three levels. At the specification level, it conceptualizes "critical segments" as road segments of spatially prolonged and high traffic accident risk. At the methodological level, it proposes a two-stage approach to traffic accident clustering and selection. The first stage is devoted to spatial clustering of traffic accidents. The second stage is devoted to selection of clusters that are dominant in terms of number of accidents. At the implementation level, the paper reports on a prototype system and illustrates its functionality using publicly available real-life data. The presented approach is psychologically inspired to the extent that it introduces a clustering criterion based on the Gestalt principle of proximity. Thus, the proposed algorithm is not density-based, as are most other state-of-the-art clustering algorithms applied in the context of traffic accident analysis, but still keeps their main advantages: it allows for clusters of arbitrary shapes, does not require an a priori given number of clusters, and excludes "noisy" observations.

Keywords: traffic accident; clustering; spatially prolonged risk; Gestalt; proximity; open data

1. Introduction

Road traffic accidents represent a global health and social problem. It is estimated that approximately 1.35 million people die each year in traffic accidents, up to 50 million are injured, and the costs for countries are approximately equal to three percent of their annual gross domestic product [1]. In the EU, 22,700 people die each year in traffic accidents and 120,000 are seriously injured, while the external cost of road traffic accidents represents approximately two percent of the EU's annual gross domestic product [2].

It comes as no surprise that significant research efforts have already been devoted to the question of automatic detection of traffic-accident-prone areas. In this paper, we consider a somewhat more specific question. One way to increase traffic safety is by traffic monitoring and managing. However, in circumstances of limited human or technical resources, it is necessary to select "critical" road segments to be the subject of monitoring or managing. For example, Figure 1 provides a map of traffic accidents with injuries or death that occurred in "inner" Belgrade, Serbia, over the one-year period from January 2021 to December 2021. It shows a relatively dense distribution with no clear cluster separation. The research question considered in this paper can be stated as follows: given data on traffic accidents, how we should conceptualize, cluster, and select "critical" road segments? Thus, the reported study makes novel contributions at three levels:

- At the specification level, we conceptualize "critical segments" as road segments of spatially prolonged and high traffic accident risk (cf. Section 2);

- At the methodological level, we propose a two-stage approach to traffic accident clustering and selection (cf. Section 3);
- At the implementation level, we report on a prototype system and illustrate its functionality using publicly available real-life data (cf. Section 4).

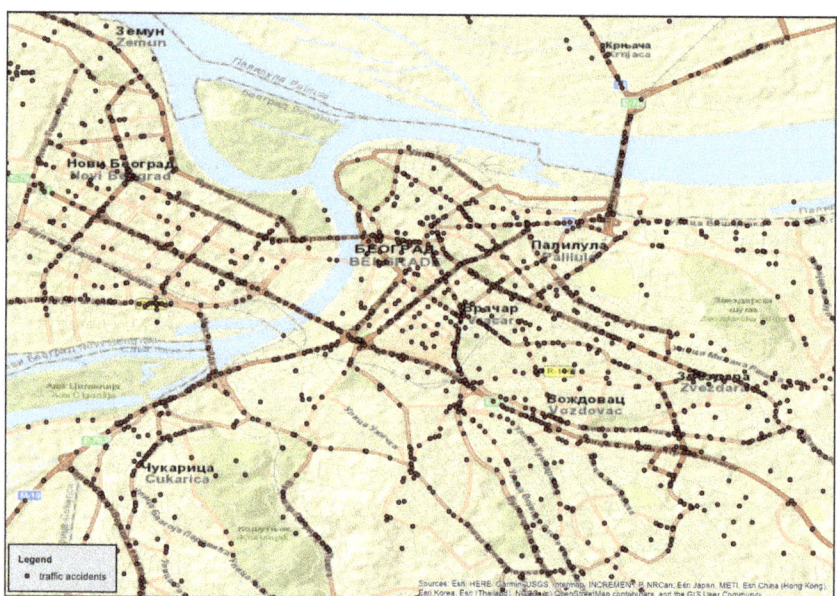

Figure 1. The map of road traffic accidents with injuries or death that occurred in "inner" Belgrade over the one-year period from January 2021 to December 2021. The map was generated using the ArcMap component of the Esri's ArcGIS suite (https://www.esri.com).

The point of departure for this study is that spatial clustering of traffic accidents is a Gestalt problem. One of the traditional problems considered by Gestalt psychologists is related to the question of how humans naturally group points on a two-dimensional plane. The approach presented in this paper is psychologically inspired to the extent that it introduces a clustering criterion based on the Gestalt principle of proximity. In line with this, the proposed algorithm is not density-based, as are most other state-of-the-art clustering algorithms applied in the context of traffic-accident analysis. On the other hand, it keeps their main advantages: it allows for clusters of arbitrary shapes, does not require an a priori given number of clusters, and excludes "noisy" observations.

The rest of this paper is organized as follows. Section 2 provides an overview of related work and describes the main idea underlying this study. Section 3 formally introduces a novel approach to spatial clustering and selection of road traffic accidents. Section 4 illustrates the functionality of a prototype system. Section 5 discusses the approach from the perspective of other relevant studies inspired by the Gestalt principles. Section 6 concludes the paper.

2. Related Work and Main Idea

The research question of traffic accident clustering has been devoted significant research attention [3–6]. For a more comprehensive overview, the reader may consult [3,7,8]. Here, we reflect on selected methodological aspects and emphasize the main idea of this particular study.

Some of the widely applied clustering algorithms (e.g., k-means type algorithms [9,10]) take the number of clusters as an input parameter (cf. also [11]). In practice, the observed data are clustered repetitively by varying the input number of clusters; then, the optimal number of clusters is selected with respect to some criterion. One of such criteria is based on the pooled within-cluster sum of squares around the cluster means [12]:

$$WCSS(t) = \sum_{i=1}^{t} \left(\frac{1}{2n_i} \sum_{j,k \in C_i} d_{jk} \right), \qquad (1)$$

where t is the number of clusters, C_i is the ith cluster, n_i is the number of observations assigned to cluster C_i, and d_{jk} is the pairwise distance between observations j and k. A plot of the within-cluster dispersion versus the applied number of clusters typically contains an elbow that indicates the optimal number of clusters [12]. An alternative method to determine the optimal number of clusters is described in [13].

In general, the requirement that the number of clusters should be given a priori represents a limitation. In addition, the k-means algorithm considers the entire dataset and generates spherical shape clusters that are not necessarily suitable to represent traffic-accident-prone areas [8]. To address these limitations, the density-based DBSCAN algorithm [14] is aimed at eliminating noise from data and allowing for clusters of arbitrary shapes. Instead of an a priori given number of clusters, this algorithm accepts two different parameters: the maximum neighborhood radius and the minimum number of points required to form a dense region. The OPTICS algorithm [15] is an extension of the DBSCAN algorithm that produces a density-based clustering structure of a dataset.

It is shown in [8] that the density-based clustering algorithms perform better than the k-means algorithm in the context of traffic-accident analysis. Similarly to them, the algorithm introduced in this paper allows for clusters of arbitrary shapes and does not require that the number of clusters is given in advance. The proposed clustering approach is not density-based, but inspired by the Gestalt principle of proximity [16]. According to this principle, when humans are confronted with a number of the same visual stimuli (e.g., points on a two-dimensional plane), the most natural form of grouping involves the smallest interval. For example, for the set of points given in Figure 2i, the most natural arrangement would be abc/def/ghi, while for the set in Figure 2ii the natural grouping would be adg/beh/cfi. It is important to note that the natural grouping is by no means impeded by increasing the number of points [16].

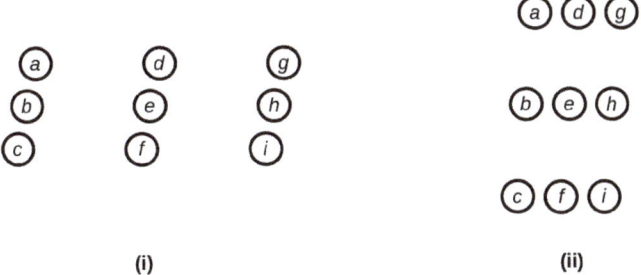

Figure 2. Illustration of the Gestalt principle of proximity. The most natural arrangement in (i) would be abc/def/ghi; the most natural arrangement in (ii) would be adg/beh/cfi (inspired by [16]).

We build on the Gestalt principle of proximity and introduce a novel approach to automatic spatial clustering of road traffic accidents. At the level of specification, our study aims at detecting road segments of spatially prolonged and high traffic accident risk. A road segment is considered to be of spatially prolonged risk if it is related to a nonempty set N of traffic accident locations, which can be considered close to each other by means of transitive closure. More precisely, let R be a relation defined on N as follows:

$$R = \{(n_i, n_j) \mid d(n_i, n_j) \leq \tau\}, \qquad (2)$$

where τ is a spatial threshold and $d(n_i, n_j)$ is spatial distance between traffic accidents n_i and n_j. A cluster is formed as a transitive closure of R, and detection of road segments of spatially prolonged risk is achieved by means of clustering, as explained in Section 3.2. Spatial threshold τ is an input parameter to the introduced clustering algorithm, and the selection of its particular value is discussed in Section 4.2.

In addition, a road segment is considered to be of high traffic accident risk if it can be considered dominant in terms of number of accidents. The adaptive selection of high-risk road segments are introduced in Section 3.3. Thus, our approach can be represented as a two-stage algorithm. The first stage is devoted to spatial clustering of road traffic accidents. The second stage is devoted to selection of dominant clusters.

3. Methods

In this section, we formally introduce our two-stage approach to road traffic accident clustering and selection. Section 3.1 introduces the basic notions. Section 3.2 describes a graph-based approach to spatial clustering of traffic accidents, and Section 3.3 introduces an approach to adaptive selection of clusters that are dominant with respect to the number of traffic accidents.

3.1. Basic Notions

A road traffic accident n_i is represented as follows:

$$n_i = (id_i, \varphi_i, \lambda_i), \tag{3}$$

where

- id_i is a unique identification number of n_i;
- φ_i and λ_i are positional coordinates of n_i, i.e., latitude and longitude expressed in radians, respectively.

Spatial distance between traffic accidents n_i and n_j is calculated based on the haversine formula [17]:

$$d(n_i, n_j) = 2 \cdot R \cdot \operatorname{atan2}(\sqrt{a(n_i, n_j)} \cdot \sqrt{1 - a(n_i, n_j)}), \tag{4}$$

where

$$a(n_i, n_j) = \sin^2 \frac{\varphi_2 - \varphi_1}{2} + \cos \varphi_1 \cos \varphi_2 \sin^2 \frac{\lambda_2 - \lambda_1}{2}, \tag{5}$$

function atan2 is an adoption of the arctangent function designed to calculate an unambiguous angle value, and $R = 6371 \cdot 10^3$ m (i.e., mean Earth radius). In addition, let τ be a spatial threshold value representing an input parameter to the clustering algorithm, and let $N = \{n_1, n_2, \ldots, n_k\}$ be a set of traffic accidents that occurred in a given period.

3.2. The Clustering Algorithm

The proposed clustering approach adapts the graph-based image segmentation algorithm introduced in [18] (cf. also [19]) and can be described as follows:

1. Throughout the algorithm execution, current clustering results are represented by integer array:

$$\mathcal{C} = (c(n_1), c(n_2), \ldots, c(n_k)), \tag{6}$$

where $(\forall\ 1 \leq i \leq k)(c(n_i) \in \{1, 2, \ldots, k\})$ and $c(n_i)$ represents the identification number of a cluster to which traffic accident n_i is currently assigned. In Step 1, each traffic accident is assigned to its own cluster, i.e.,

$$(\forall 1 \leq i \leq k)(c(n_i) = i). \tag{7}$$

2. Let $\mathcal{D}(N, \tau)$ be a set of all combinations of two traffic accidents (i.e., a set of all unordered pairs of traffic accidents) whose mutual distance is less than or equal to the threshold value τ. In other words, set $\mathcal{D}(N, \tau)$ contains pairs of traffic accidents that

are considered close to each other and are thus candidates to be in the same cluster. Without loss of generality, set $\mathcal{D}(N,\tau)$ can be defined as

$$\mathcal{D}(N,\tau) = \{(n_i, n_j) | \{n_i, n_j\} \subset N \wedge i < j \wedge d(n_i, n_j) \leq \tau\}. \tag{8}$$

3. We generate a sequence that contains all elements from $\mathcal{D}(N,\tau)$ ordered by nondecreasing distance between traffic accidents.

$$\hat{\mathcal{D}}(N,\tau) = (n_{i_1}, n_{j_1}), (n_{i_2}, n_{j_2}), \ldots, (n_{i_m}, n_{j_m}). \tag{9}$$

4. We iterate through sequence $\hat{\mathcal{D}}(N,\tau)$ from the first to the last position. For each ordered pair $\delta_p = (n_i, n_j)$ in $\hat{\mathcal{D}}(N,\tau)$, if traffic accidents n_i and n_j belong to different clusters $c(n_i)$ and $c(n_j)$, then those clusters are merged, i.e.,

$$\begin{aligned}
&\text{for } (1 \leq p \leq |\hat{\mathcal{D}}(N,\tau)|) \ \{ \\
&\quad \text{let } \delta_p = (n_i, n_j) \\
&\quad \text{if } (c(n_i) \neq c(n_j)) \text{ then} \\
&\quad\quad \text{for } (1 \leq q \leq |\mathcal{C}|) \\
&\quad\quad\quad \text{if } (c(n_q) = c(n_j)) \text{ then } (c(n_q) \leftarrow c(n_i)) \\
&\}
\end{aligned} \tag{10}$$

Thus, the clustering is performed by means of transitive closure of the undirected graph over set N defined in Step 3 (cf. sequence $\hat{\mathcal{D}}(N,\tau)$).

The clustering results are represented by array $\mathcal{C}$ after Step 4 is completed. In general, array $\mathcal{C}$ generated in this algorithm stage contains information on t clusters, where $1 \leq t \leq k$ (i.e., the number of cluster is equal to the number of distinct values in $\mathcal{C}$).

3.3. Cluster Selection

In the second algorithm stage, a subset of clusters that are dominant with respect to the number of traffic accidents is adaptively selected. Let $\chi(\mathcal{C})$ be the histogram of array $\mathcal{C}$, i.e.,

$$\chi(\mathcal{C}) = \{(c_1, p_1), (c_2, p_2), \ldots, (c_t, p_t))\}, \tag{11}$$

where

- c_i is the identification number of a cluster contained in array $\mathcal{C}$,
- p_i is the number of traffic accidents assigned to cluster c_i,
- and $1 \leq i \leq t$.

The adaptive cluster selection algorithm represents an adaptation of the method of threshold selection for image binarization introduced in [20] (pp. 120–121; cf. also [21]) and can be described as follows.

1. The starting threshold value μ_0 is set to the average number of traffic accidents per cluster:

$$\mu_0 = \frac{1}{|\chi(\mathcal{C})|} \sum_{i=1}^{|\chi(\mathcal{C})|} p_i. \tag{12}$$

2. Given a current threshold value μ_i, where $i \geq 0$, set $\chi(\mathcal{C})$ is divided into two disjoint subsets based on μ_i:

$$\begin{aligned}
\chi_1 &= \{(c,p) \mid (c,p) \in \chi(\mathcal{C}) \wedge p \leq \mu_i\}, \\
\chi_2 &= \{(c,p) \mid (c,p) \in \chi(\mathcal{C}) \wedge p > \mu_i\},
\end{aligned} \tag{13}$$

and the subsequent threshold value μ_{i+1} is calculated as

$$\mu_{i+1} = \frac{1}{2}\left(\frac{1}{|\chi_1|}\sum_{i=1}^{|\chi_1|}p_i + \frac{1}{|\chi_2|}\sum_{i=1}^{|\chi_2|}p_i\right). \tag{14}$$

3. If the change in threshold is not significant, i.e.,

$$|\mu_i - \mu_{i+1}| \leq \frac{1}{2}, \tag{15}$$

the calculation is completed and the final threshold μ is set to μ_{i+1}. Otherwise, the process returns to Step 2.

Finally, a subset of clusters that are dominant with respect to the number of traffic accidents is adaptively derived by applying the calculated threshold value μ:

$$\mathbb{C} = \{c \mid (c,p) \in \chi(\mathcal{C}) \land p > \mu\}. \tag{16}$$

4. Results

This section reports on the prototype system and describes the results obtained when it was applied to real-life data.

4.1. Tools

A prototype system based on the approach introduced in Section 3 is implemented in the Racket programming language. To graphically represent spatial data and estimate areas covered by clusters, we applied the ArcMap component of the Esri's ArcGIS suite.

4.2. Spatial Threshold Selection

Spatial threshold τ introduced in Section 3.2 represents an input parameter to the clustering algorithm. We set threshold τ to 200 m for the following reason. The national urban speed limit is set to 50 $\frac{km}{h}$ [22] (cf. article 43). However, to account for the relationship between the posted speed limit and actual speeds in urban areas, we consider the minimum speeding offense of exceeding the speed limit by up to 20 $\frac{km}{h}$ [22] (cf. article 333). Therefore, we assume a driver operating her or his vehicle at a speed of 70 $\frac{km}{h}$ and define the spatial threshold as the distance traveled by this vehicle in ten seconds (i.e., $\tau \approx 200$ m).

Although spatial threshold τ is assigned a particular value, we recall that it is introduced as an input parameter. In general, its value is intended to be set according to external criteria, which may vary with the application context. Thus, the spatial threshold is not learned as a hyperparameter in the sense typically found in the field of machine learning. Instead, it is intentionally left to the practitioner to decide on the spatial threshold value, i.e., on the maximum distance between two traffic accident locations that are considered close to each other.

4.3. Data

We resort to a publicly available dataset on traffic accidents provided by the Ministry of Interior of the Republic of Serbia. To illustrate the functionality of the prototype system (cf. Section 4.4), we use a part of this dataset containing details on 15,366 road traffic accidents that occurred in Belgrade, the capital of Serbia, over the one-year period from January 2021 to December 2021 [23]. Those accidents can be divided in three groups:

- 11,294 road traffic accident with material damage;
- 3996 road traffic accidents with injuries;
- 76 road traffic accidents with death.

We consider only severe road traffic accidents from the last two groups, i.e., 4072 (3996 + 76) accidents with injuries or death. For each accident, the prototype system considers only its unique identification number and positional coordinates (i.e., latitude and longitude). The map showing a subset of road traffic accidents with injuries or death that occurred in "inner" Belgrade during 2021 is given in Figure 1.

To estimate the stability of results through time (cf. Section 4.5), the algorithm is applied to data on traffic accidents with injuries or death that occurred in one of the "inner" Belgrade municipalities—i.e., the municipality of Zvezdara—over the three-year period from January 2019 to December 2021 [23–25].

4.4. Algorithm Execution

In the first algorithm stage, 4072 traffic accidents are divided into 1439 clusters. The average number of accidents per cluster is 2.796, with a standard deviation of 8.909. In the second algorithm stage, only ten clusters are selected as dominant with respect to the number of traffic accidents. The average number of accidents per cluster is 73.3, with standard deviation of 69.103 (cf. Table 1).

Table 1. A summary of the clustering and selection results obtained when the introduced algorithm was applied to publicly available data on traffic accidents with injuries or death that occurred in Belgrade during 2021.

	First Stage (Clustering)	Second Stage (Selection)
Number of traffic accidents:	4072	733
Number of clusters:	1439	10
Average num. of accidents per cluster:	2.796	73.3
Standard deviation:	8.909	69.103

The map representation of the selected clusters is given in Figure 3. Although the map shows only "inner" Belgrade, it contains all ten clusters selected when the prototype system was applied to data on traffic accidents in the entire city. The numbers of traffic accidents assigned to each cluster are provided in the second row of Table 2. The cluster identification numbers given in this table correspond to those given in the legends of Figures 3 and 4.

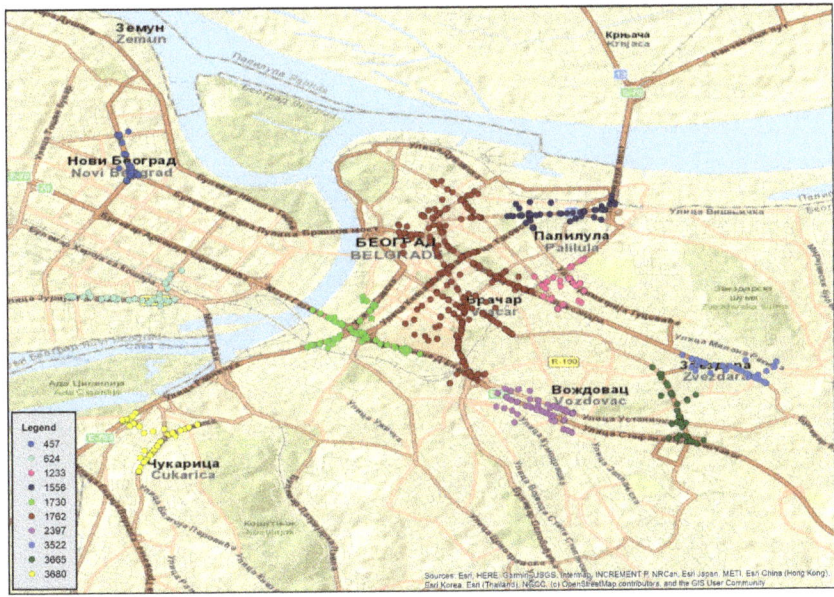

Figure 3. The map representation of the selected clusters, Belgrade, 2021. The map was generated using the ArcMap component of the Esri's ArcGIS suite (https://www.esri.com).

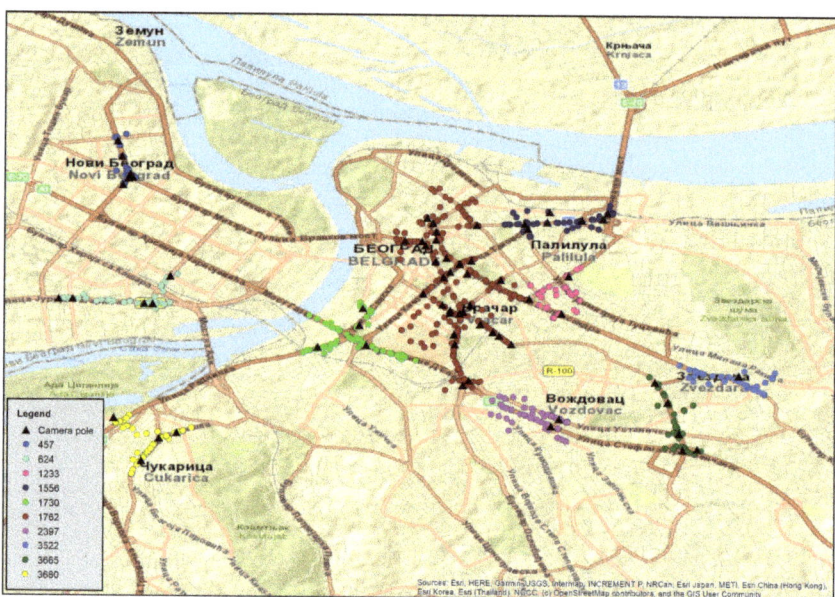

Figure 4. The map representation of the selected clusters (Belgrade, 2021) and camera poles (March 2022). The map was generated using the ArcMap component of the Esri's ArcGIS suite (https://www.esri.com).

Table 2. Description of the selected clusters obtained when the introduced algorithm was applied to publicly available data on traffic accidents with injuries or death that occurred in Belgrade during 2021. The cluster identification numbers given in this table correspond to those given in the legends of Figures 3 and 4.

Cluster ID:	1762	1730	2397	1556	3680	457	1233	3665	3522	624
Number of Traffic Accidents:	279	66	62	58	53	45	44	44	43	39
Area (km²):	1.75	0.30	0.33	0.22	0.20	0.06	0.23	0.15	0.13	0.29
Number of Camera Poles:	33	4	2	7	4	6	3	5	1	5

There is a set of well-established measures that are often applied to analyze results of traffic accident clustering by means of evaluating the tightness and separation of clusters: the silhouette coefficient [13], Calinski–Harabasz index [26], Davies–Bouldin index [27], etc. However, these measures are rather general (i.e., task-independent). Consequently, validation approaches based on these measures lack task-related criteria. In contrast to them, we apply a qualitative evaluation based on traffic-related criteria.

In line with this, the obtained results can be considered promising: ten selected clusters covering approximately 0.11 percent of the city area (i.e., 3.65 km² out of approximately 3233 km², cf. Table 2) capture 18 percent of all traffic accidents (i.e., 733 out of 4072, cf. Table 1).

For the purpose of further illustration, we compare the clustering results with the locations of traffic camera poles derived from the publicly available information provided by the Ministry of Interior of the Republic of Serbia [28]. To justify this decision, it is important to clarify the following:

- The locations of camera poles are determined by a third party, independent of this study.
- The introduced algorithm is agnostic of the camera pole locations, i.e., they are not considered in the clustering process.

- The traffic accident data used to generate clusters are collected during 2021. At the moment of conducting this study (i.e., March 2022), the considered traffic cameras still have not been put into use, i.e., they did not influence the traffic behavior in the observed period.

Thus, the particular camera pole locations can serve as an indirect "response" variable. Out of 464 camera poles installed in Belgrade, seventy are located within the selected clusters. The numbers of camera poles within each cluster are provided in Table 2. The map representation of the selected clusters and camera poles within them is given in Figure 4. It can be observed that the ten selected clusters, which cover 0.11 percent of the city area, capture 15 percent of the camera poles.

4.5. Stability of Results through Time

To estimate the stability of results through time, the introduced algorithm is applied to data collected in the same spatial area at different periods. The previous section considers the entire city of Belgrade, which has a surface area of approximately 3233 km^2. In this section, the same spatial threshold (i.e., τ = 200 m) is applied to just one of the "inner" Belgrade municipalities—the municipality of Zvezdara—which has a surface area of approximately 31.11 km^2 (i.e., 9.6 percent of the city surface area). In line with our goal to introduce an approach suitable for application in circumstances of limited human or technical resources for traffic monitoring and management, this municipality was selected as one of the "inner" municipalities with fewest camera poles. It contains only 16 out of 464 camera poles installed in Belgrade.

The algorithm is applied to publicly available data on traffic accidents with injuries or death that occurred in the municipality of Zvezdara over the three-year period from January 2019 to December 2021. The maps showing road traffic accidents that occurred in this municipality during 2021, 2020, and 2019 are given in Figure 5a,c,e, respectively. The corresponding map representations of the selected clusters are given in Figure 5b,d,f, respectively. The camera pole locations (March 2022) are represented for the purpose of completeness. A summary of the clustering and selection results is given in Table 3. The selected clusters are described in Table 4.

Table 3. A summary of the clustering and selection results obtained when the introduced algorithm was applied to publicly available data on traffic accidents with injuries or death that occurred in the municipality of Zvezdara during 2021, 2020, and 2019, respectively.

Year	Summary Data	First Stage (Clustering)	Second Stage (Selection)
2021	Number of traffic accidents:	317	116
	Number of clusters:	97	4
	Average num. of accidents per cluster:	3.186	29
	Standard deviation:	5.930	8.573
2020	Number of traffic accidents:	282	101
	Number of clusters:	95	5
	Average num. of accidents per cluster:	2.905	20.2
	Standard deviation:	5.020	11.214
2019	Number of traffic accidents:	349	136
	Number of clusters:	93	5
	Average num. of accidents per cluster:	3.699	27.2
	Standard deviation:	6.561	11.444

Table 4. Description of the selected clusters obtained when the introduced algorithm was applied to publicly available data on traffic accidents with injuries or death that occurred in the municipality of Zvezdara during 2021, 2020, and 2019, respectively.

2021	Cluster ID:	161	22	198	78	
	Number of Traffic Accidents:	43	28	25	20	
	Area (km^2):	0.132	0.193	0.171	0.189	
2020	Cluster ID:	28	5	6	140	182
	Number of Traffic Accidents:	42	19	16	12	12
	Area (km^2):	0.272	0.073	0.065	0.121	0.109
2019	Cluster ID:	194	72	192	116	80
	Number of Traffic Accidents:	50	23	22	21	20
	Area (km^2):	0.486	0.235	0.317	0.163	0.145

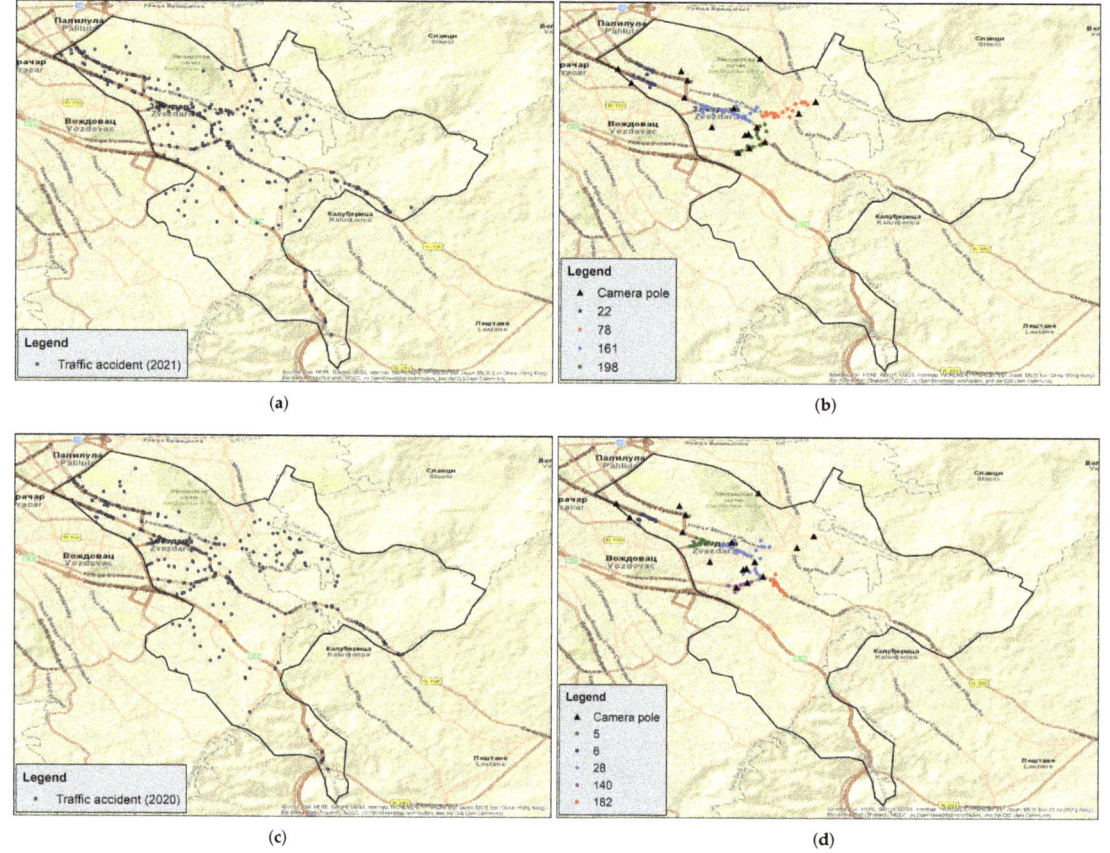

Figure 5. *Cont.*

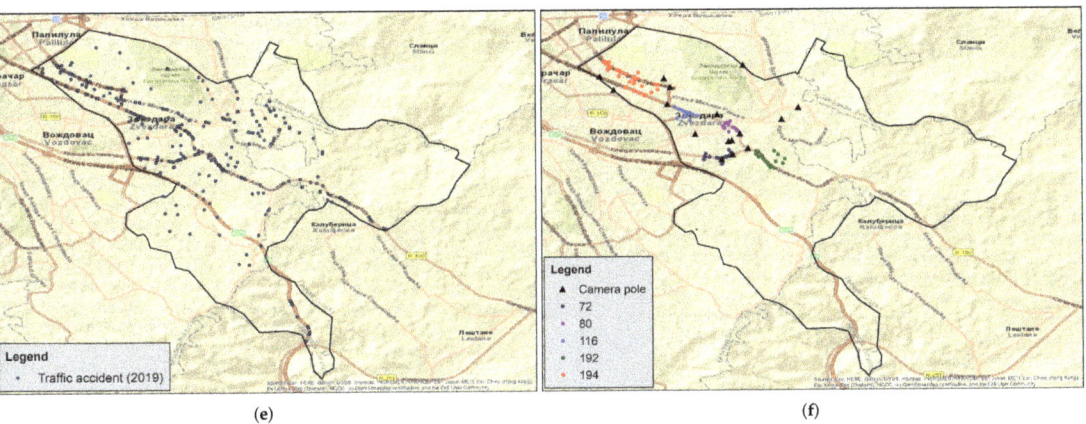

(e)   (f)

Figure 5. On the left: the maps showing road traffic accidents with injuries or death that occurred in the Municipality of Zvezdara during (**a**) 2021, (**c**) 2020, and (**e**) 2019. On the right: the corresponding map representations of the obtained clusters. The camera pole locations (March 2022) are represented for the purpose of completeness. (**a**) Traffic accidents, Zvezdara, 2021. (**b**) Selected clusters, Zvezdara, 2021. (**c**) Traffic accidents, Zvezdara, 2020. (**d**) Selected clusters, Zvezdara, 2020. (**e**) Traffic accidents, Zvezdara, 2019. (**f**) Selected clusters, Zvezdara, 2019. The maps were generated using the ArcMap component of the Esri's ArcGIS suite (https://www.esri.com).

The stability of results through time is considered in two aspects: the share of traffic accidents belonging to the selected clusters, and the overlapping surface area between the selected clusters in all three years. With regard to the first aspect, the following can be observed:

- In 2021, four selected clusters covering approximately 2.2 percent of the municipality surface area (i.e., 0.685 km^2 out of 31.11 km^2) capture 36.59 percent of all traffic accidents (i.e., 116 out of 317).
- In 2020, five selected clusters covering approximately 2.06 percent of the municipality surface area (i.e., 0.64 km^2 out of 31.11 km^2) capture 35.82 percent of all traffic accidents (i.e., 101 out of 282).
- In 2019, five selected clusters covering approximately 4.33 percent of the municipality surface area (i.e., 1.346 km^2 out of 31.11 km^2) capture 38.97 percent of all traffic accidents (i.e., 136 out of 349).

Thus, the share of traffic accidents belonging to the selected clusters is steady through the given three-year period (i.e, 36.59, 35.82, and 38.97 percent, respectively).

With regard to the second aspect, a significant overlapping between the selected clusters in all three years can be observed. The overlapping surface area is 0.353 km^2, which makes 51.53 percent of the selected surface area in 2021, 55.15 percent of the selected surface area in 2020, and 26.23 percent of the selected surface area in 2019.

5. Discussion

In addition to reporting the algorithm results, we discuss the introduced approach from the perspective of other relevant studies inspired by the Gestalt principles. The idea of applying human cognitive judgments reflecting the principles of visual Gestalt perception is not new. E.g., Ref. [29] introduces a clustering algorithm based on local k-dimensional neighbors of each point, allowing for an arbitrary number of clusters and arbitrary clusters shapes. However, their implicit conceptualization of proximity differs from the conceptualization adopted in our study. According to the conceptualization adopted in [29], the pattern of points given in Figure 6i contains three clusters: two large clusters and one "chain" cluster between them. In our approach, the proximity of points (i.e., locations) is defined by means of transitive closure, so the same pattern contains only one cluster (cf. Figure 6ii).

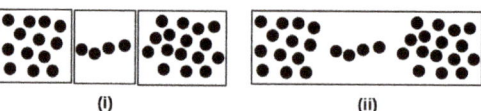

Figure 6. Emphasizing the difference in the conceptualization of the Gestalt principle of proximity in (i) the approach introduced in [29] and (ii) our approach.

More recently, two different approaches to saliency detection in digital images based on the Gestalt principles are proposed in [30,31]. Particularly, related to the Gestalt principle of proximity, these approaches consider color distance between image regions and implicitly include the transitive closure. However, both approaches restrict the selected image regions only to neighbors of a currently salient image region. Even the image segmentation algorithm introduced in [18], on which we build in this contribution, includes a pairwise region comparison predicate. In our approach, this restriction is not present and we comment briefly on this.

In [18], the difference between segments C_i and C_j is defined as the minimum weight edge connecting them, i.e.,

$$Dif(C_i, C_j) = \min_{\substack{v_i \in C_i, \\ v_j \in C_j, \\ (v_i, v_j) \in E}} w(v_i, v_j) , \qquad (17)$$

where v_i and v_j are two neighboring pixels (i.e., $(v_i, v_j) \in E$) belonging, respectively, to segments C_i and C_j, and $w(v_i, v_j)$ represents the color distance between v_i and v_j. In our approach, we consider spatial distance between traffic accident locations, but define the distance between two clusters in the same manner. On the other hand, to detect evidence of a boundary between segments C_i and C_j, the approach introduced in [18] assumes that their difference must be greater than their internal differences $Int(C_1)$ and $Int(C_2)$:

$$Dif(C_i, C_j) > \min\{Int(C_i) + \tau(C_i), Int(C_j) + \tau(C_j)\} , \qquad (18)$$

where threshold values are defined as inversely proportional to the size of a segment, i.e., $\tau(C) \sim \frac{1}{|C|}$. We relax this condition in our approach: in order to detect evidence of a boundary between two clusters, their distance must be greater than a constant threshold value (cf. Equation (8) in Section 3.2). The justification for this decision is related to the domain of this study. In line with our aim to detect road segments of spatially prolonged traffic accident risk, we do not require stronger evidence for boundary of relatively smaller clusters (and, therefore, do not consider internal cluster differences). The input parameter threshold allows for controlling the scale of observation: a larger threshold value causes a preference for larger clusters.

6. Conclusions

This paper introduced an approach to automatically detecting and selecting road segments of spatially prolonged and high traffic accident risk, intended for application in circumstances of limited human or technical resources for traffic monitoring and management. It also reported on a prototype system and illustrated its functionality using publicly available real-life data on road traffic accidents that occurred in Belgrade. The approach was positively evaluated in two aspects: (i) comparing the clustering results with the locations of traffic camera poles installed a posteriori; (ii) the stability of results through time.

To conclude, we first reflect on the comprehensiveness of the feature set that represents a traffic accident. Machine-learning-based approaches to traffic accident clustering typically deal with a number of features, including road features (e.g., road, surface, road type, vehicle type, etc.), environmental features (e.g., date, time, weather, etc.), and human features (e.g., participant's age and gender, violation of law, etc.) [3,7,32]. The application of those approaches assumes the existence of a dataset that is rather comprehensive in terms of features. However, the comprehensiveness of available datasets varies between different geographical areas and time periods. In contrast, a traffic accident in our approach is represented by two positional coordinates only, i.e., latitude and longitude, which increases the possibility of its application.

Related to the time complexity of the proposed approach, the clustering algorithm introduced in Section 3.2 represents the dominant component. Its running time can be factored as follows. Step 1 takes constant time. In Step 2, for a given set containing k traffic accidents, there are k^2 candidate elements for set $\mathcal{D}(N, \tau)$, i.e., this step takes $O(k^2)$ time. In the example given in Section 4, the number of traffic accidents was $k = 4072$, which means that approximately $k^2 \approx 16.6$ million candidate pairs were considered. However, the number of elements in set $\mathcal{D}(N, \tau)$, which corresponds to the memory footprint of Step 2, does not necessarily follow this pattern. E.g., set $\mathcal{D}(N, \tau)$ produced in the example contained only $m = 8407$ pairs. In general, the size of set $\mathcal{D}(N, \tau)$ depends on threshold value τ. Finally, it was shown in [18] that Steps 3 and 4 can be implemented in $O(m \log m)$ and $O(m\alpha(m))$ time, where α is the very slow-growing inverse Ackerman's function.

Author Contributions: Conceptualization, M.G.; methodology, M.G.; software, M.G. and N.M.; validation, N.M. and D.J.; formal analysis, N.M. and D.J.; investigation, M.G. and I.K.; resources, I.K.; data curation, I.K.; writing—original draft preparation, M.G.; writing—review and editing, I.K., N.M. and D.J.; visualization, I.K. All authors have read and agreed to the published version of the manuscript.

Funding: This research received no external funding.

Institutional Review Board Statement: Not applicable.

Informed Consent Statement: Not applicable.

Data Availability Statement: Publicly available datasets were analyzed in this study. These data can be found here: [23–25,28].

Conflicts of Interest: The authors declare no conflict of interest.

References

1. Arora, P. Final Report. Independent Evaluation of the United Nations Road Safety Trust Fund (UNRSF) Secretariat. 8 April 2021. DeftEdge Corporation. Available online: https://unece.org/sites/default/files/2021-04/TRANS_FinalReportUNRSF_Apri21_0.pdf (accessed on 25 March 2022).
2. European Parliament. EU Road Safety Policy Framework 2021–2030—Recommendations on Next Steps Towards "Vision Zero", P9_TA(2021)0407, Text Adopted. 6 October 2021. Available online: https://www.europarl.europa.eu/doceo/document/TA-9-2021-0407_EN.pdf (accessed on 25 March 2022).
3. Jeong, H.; Kim, I.; Han, K.; Kim, J. Comprehensive Analysis of Traffic Accidents in Seoul: Major Factors and Types Affecting Injury Severity. *Appl. Sci.* **2022**, *12*, 1790. [CrossRef]
4. Manap, N.; Borhan, M.N.; Yazid, M.R.M.; Hambali, M.K.A.; Rohan, A. Identification of Hotspot Segments with a Risk of Heavy-Vehicle Accidents Based on Spatial Analysis at Controlled-Access Highway. *Sustainability* **2021**, *13*, 1487. [CrossRef]
5. Santos, D.; Saias, J.; Quaresma, P.; Nogueira, V.B. Machine Learning Approaches to Traffic Accident Analysis and Hotspot Prediction. *Computers* **2021**, *10*, 157. [CrossRef]
6. Sun, Y.; Wang, Y.; Yuan, K.; Chan, T.O.; Huang, Y. Discovering Spatio-Temporal Clusters of Road Collisions Using the Method of Fast Bayesian Model-Based Cluster Detection. *Sustainability* **2020**, *12*, 8681. [CrossRef]
7. Bokaba, T.; Doorsamy, W.; Paul, B.S. Comparative Study of Machine Learning Classifiers for Modelling Road Traffic Accidents. *Appl. Sci.* **2022**, *12*, 828. [CrossRef]
8. Islam, M.R.; Jenny, I.J.; Nayon, M.; Islam, M.R.; Amiruzzaman, M.; Abdullah-Al-Wadud, M. Clustering algorithms to analyze the road traffic crashes. In Proceedings of the 2021 International Conference on Science & Contemporary Technologies (ICSCT), Dhaka, Bangladesh, 5–7 August 2021.

9. Lloyd, S. Least squares quantization in PCM. *IEEE Trans. Inf. Theory* **1982**, *28*, 129–137. [CrossRef]
10. Selim, S.Z.; Ismail, M.A. K-means-type algorithms: A generalized convergence theorem and characterization of local optimality. *IEEE Trans. Pattern Anal. Mach. Intell.* **1984**, *PAMI-6*, 81–87. [CrossRef] [PubMed]
11. Dukan, P.; Kovari, A. Cloud-based smart metering system. In Proceedings of the 2013 IEEE 14th International Symposium on Computational Intelligence and Informatics (CINTI), Budapest, Hungary, 19–21 November 2013; pp. 499–502.
12. Tibshirani, R.; Walther, G.; Hastie, T. Estimating the number of clusters in a data set via the gap statistic. *J. R. Stat. Soc. Ser. B* **2001**, *63*, 411–423. [CrossRef]
13. Rousseeuw, P.J. Silhouettes: A graphical aid to the interpretation and validation of cluster analysis. *J. Comput. Appl. Math.* **1987**, *20*, 53–65. [CrossRef]
14. Ester, M.; Kriegel, H.-P.; Sander, J.; Xu, X. A density-based algorithm for discovering clusters in large spatial databases with noise. In Proceedings of the Second International Conference on Knowledge Discovery and Data Mining (KDD'96), Portland, OR, USA, 2–4 August 1996; pp. 226–231.
15. Ankerst, M.; Breunig, M.M.; Kriegel, H.-P.; Sander, J. OPTICS: Ordering points to identify the clustering structure. In Proceedings of the ACM SIGMOD International Conference on Management of Data, Philadelphia, PA, USA, 31 May–3 June 1999; pp. 49–60.
16. Wertheimer, M. Laws of organization in perceptual forms. In *A Source Book of Gestalt Psychology*; Ellis, W.D., Kegan, P., Eds.; Trench, Trubner & Company: London, UK, 1938; pp. 71–88.
17. Sinnott, R.W. Virtues of the Haversine. *Sky Telesc.* **1984**, *68*, 158–159.
18. Felzenszwalb, P.F.; Huttenlocher, D.P. Efficient Graph-Based Image Segmentation. *Int. J. Comput. Vis.* **2004**, *59*, 167–181. [CrossRef]
19. Gnjatović, M.; Maček, N.; Adamović, S. A non-connectionist two-stage approach to digit recognition in the presence of noise. In Proceedings of the 10th IEEE International Conference on Cognitive Infocommunications (CogInfoCom), Naples, Italy, 23–25 October 2019; pp. 15–20.
20. Shih, F.Y. *Image Processing and Pattern Recognition: Fundamentals and Techniques*; John Wiley and Sons: Hoboken, NJ, USA, 2010.
21. Gnjatović, M.; Tasevski, J.; Borovac, B.; Maček, N. An entropy-based approach to automatic detection of critical changes in human-machine interaction. In Proceedings of the 9th IEEE International Conference on Cognitive Infocommunications (CogInfoCom), Budapest, Hungary, 22–24 August 2018; pp. 175–178.
22. National Assembly of the Republic of Serbia. Law on Road Traffic Safety. *Off. Gaz. Repub. Serb.* no. 41/2009-3, 53/2010-12, 101/2011-270, 32/2013-22 (decision of the Constitutional Court), 55/2014-61, 96/2015-106 (other law), 9/2016-178 (decision of the Constitutional Court), 24/2018-70, 41/2018-122, 41/2018-32 (other law), 87/2018-26, 23/2019-3, 128/2020-3 (other law). Available online: http://www.pravno-informacioni-sistem.rs/SlGlasnikPortal/eli/rep/sgrs/skupstina/zakon/2009/41/1/reg/20201026 (accessed on 1 March 2022).
23. Republic of Serbia. Data on Traffic Accidents for 2021 for the Territory of all Police Administrations and Municipalities. Available online: https://data.gov.rs/s/resources/podatsi-o-saobratshajnim-nezgodama-po-politsijskim-upravama-i-opshtinama/20220125-085458/nez-opendata-2021-20220125.xlsx (accessed on 1 March 2022).
24. Republic of Serbia. Data on Traffic Accidents for 2020 for the Territory of all Police Administrations and Municipalities. Available online: https://data.gov.rs/s/resources/podatsi-o-saobratshajnim-nezgodama-po-politsijskim-upravama-i-opshtinama/20210208-095135/nez-opendata-2020-20210125.xlsx (accessed on 1 March 2022).
25. Republic of Serbia. Data on Traffic Accidents for 2019 for the Territory of all Police Administrations and Municipalities. Available online: https://data.gov.rs/s/resources/podatsi-o-saobratshajnim-nezgodama-po-politsijskim-upravama-i-opshtinama/20200127-133136/nez-opendata-2019-20200125.xlsx (accessed on 1 March 2022).
26. Caliński, T.; Harabasz, J. A dendrite method for cluster analysis. In *Communications in Statistics*; Taylor & Francis: Oxfordshire, UK, 1974; Volume 3, pp. 1–27.
27. Davies, D.L.; Bouldin, D.W. A Cluster Separation Measure. *IEEE Trans. Pattern Anal. Mach. Intell.* **1979**, *PAMI-1*, 224–227. [CrossRef]
28. Ministry of Interior, Republic of Serbia. Camera Locations within Belgrade (In Serbian). Available online: http://www.mup.gov.rs/wps/wcm/connect/b152c15f-16eb-47b3-b9a4-c7f32c2cc1ba/Lokacij+Bg.pdf?MOD=AJPERES&CVID=n-sczZB (accessed on 25 March 2022).
29. Osbourn, G.C.; Martinez, R.F. Empirically defined regions of influence for clustering analyses. *Pattern Recognit.* **1995**, *28*, 1793–1806. [CrossRef]
30. Wang, Z.; Li, B. A two-stage approach to saliency detection in images. In Proceedings of the 2008 IEEE International Conference on Acoustics, Speech and Signal Processing, Las Vegas, NV, USA, 31 March–4 April 2008; pp. 965–968.
31. Wu, J.; Zhang, L. Gestalt saliency: Salient region detection based on Gestalt principles. In Proceedings of the the 2013 IEEE International Conference on Image Processing, Melbourne, VIC, Australia, 15–18 September 2013; pp. 181–185.
32. Kovari, A. Study of Algorithmic Problem-Solving and Executive Function. *Acta Polytech. Hung.* **2020**, *17*, 241–256. [CrossRef]

MDPI
St. Alban-Anlage 66
4052 Basel
Switzerland
Tel. +41 61 683 77 34
Fax +41 61 302 89 18
www.mdpi.com

Applied Sciences Editorial Office
E-mail: applsci@mdpi.com
www.mdpi.com/journal/applsci

www.ingramcontent.com/pod-product-compliance
Lightning Source LLC
LaVergne TN
LVHW070147100526
838202LV00015B/1910